OFFSHORE WIND

OFFSHORE WIND

A Comprehensive Guide to Successful Offshore Wind Farm Installation

KURT E. THOMSEN
Advanced Offshore Solutions
Tranbjerg, Denmark

ELSEVIER

AMSTERDAM • BOSTON • HEIDELBERG • LONDON
NEW YORK • OXFORD • PARIS • SAN DIEGO
SAN FRANCISCO • SINGAPORE • SYDNEY • TOKYO
Academic Press is an imprint of Elsevier

Academic Press is an imprint of Elsevier
225 Wyman Street, Waltham, MA 02451, USA
The Boulevard, Langford Lane, Kidlington, Oxford, OX5 1GB, UK

Notices

Knowledge and best practice in this field are constantly changing. As new research and
experience broaden our understanding, changes in research methods, professional practices,
or medical treatment may become necessary.

Practitioners and researchers must always rely on their own experience and knowledge in
evaluating and using any information, methods, compounds, or experiments described
herein. In using such information or methods they should be mindful of their own safety and
the safety of others, including parties for whom they have a professional responsibility.

To the fullest extent of the law, neither the Publisher nor the authors, contributors, or
editors, assume any liability for any injury and/or damage to persons or property as a matter of
products liability, negligence or otherwise, or from any use or operation of any methods,
products, instructions, or ideas contained in the material herein.

Library of Congress Cataloging-in-Publication Data
Application submitted.

British Library Cataloguing-in-Publication Data
A catalogue record for this book is available from the British Library.

ISBN: 978-0-12-385936-5

For information on all Academic Press publications
visit our Web site at *www.elsevierdirect.com*

Printed in the United States

11   12   13   14   15   10 9 8 7 6 5 4 3 2 1

To Jørgen W. Brorson, who made this adventure possible.

CONTENTS

PREFACE

Why Do You Need This Book?

This book is the result of 10 years of learning by doing. Ten years ago, no one had installed wind farms offshore commercially. Test sites that had been or were installed—in Denmark predominantly—consisted of turbines up to 600 kW and were limited in numbers. In 2001 the first semicommercial wind farm consisting of 20 Bonus (now Siemens) 2.0-MW turbines was installed at the port entrance of Copenhagen on a sandbank known as Middelgrunden. Those turbines were the result of a private initiative among the residents of Copenhagen called "Middelgrundens Vindmøllelaug." The initiative was backed by the utility "Københanvns Energi A/S," which signed the PPA for the 20 turbines to supply power. This made construction of the wind farm possible and installation was started in 2000.

When the project was planned and executed, there were no set standards for offshore work, no programming of activities, and no legal framework that could regulate the activities concerning HSE and the permitting of the project. All this was being developed as we moved forward. I wrote this book to record the findings, experiences, and methods that have proven solid enough to endure through the past ten years of offshore construction and have formed the basis of lawmaking, setting of best practice standards, and HSE regulations for working in the wind farm industry offshore.

What is here represents a work in progress. This means that the data, the statements, and the findings are the best information I can give the reader at this point in time. Therefore, the book will be updated over the coming years, and new information and more authors will contribute data that can be of use in the process of installing an offshore wind farm.

The intention is to create a robust basis for understanding the offshore wind farm industry. It is, however, also a book with some anecdotes from my years in the industry, and I hope provides the reader with some food for thought. It is my belief that this is a more interesting way to learn and to remember data that can sometimes be boring. By giving them life, they will stay in one's memory longer.

Who Should Read This Book?

This book is addressed to everyone involved in or soon to be involved in the offshore wind farm industry, whether a consultant, financier, engineer, or technician. It is designed to answer common questions that will be asked at the start, in the middle, and toward the end of a wind farm installation project.

The recorded experiences from the last ten years will help point the professional in the right direction. It will also give the financing society a chance to ask the technical questions that are necessary to determine whether project planning and execution sound sensible and have been adequately thought through. The engineer will be able to plan without walking into the biggest and most obvious obstacles that are common in the process of installing an offshore wind farm.

The health, safety, and environmental (HSE) professional will be able to figure out how to set up and execute the planning and monitoring of the processes from a safety point of view. The financier will be able to ask some more in-depth questions and certainly understand more of the process, the risks involved, and why the world looks different outside an office.

The intent is not to present "be all and end all" documentation but rather to provide a statement that the reader can use as a starting point for an offshore wind farm career. There is a huge need for more in-depth knowledge and information before the many statistics, hints, and fragments of advice can be of use. This is why one should start here and work through the knowledge base that is contained everywhere in the industry but, unfortunately, is scattered around the globe. The reader needs to locate and sort through the information in order to be able to work professionally. After reading this book, one should feel confident in one's knowledge about wind farms.

How Does the Author Feel about Wind Farms?

It is important for the reader to understand the author's context. Admittedly I am biased toward some ways of executing the offshore installation of foundations and turbines. Therefore, the recommendations and reservations in this book reflect my personal opinions. The recommendations and opinions are, of course, based on my years of experience in the industry.

It would be wrong, however, if I said that the methods and processes described here are the only valid ways of installing an offshore wind farm. So, whenever a method or process is described, I make an effort to document the shortcomings of all of the alternatives, as well as list other possible ways of carrying out the work. It is then up to the reader to decide which viewpoint to take toward the documentation and statements made in this book.

Let me emphasize though that the methods and processes described *do* work. Other solutions may work, too, and I make a sincere effort to state in unbiased terms the pros and cons of proposed alternatives. The litmus test that I will always apply to determine the viability of an alternative will be whether or not it is cost effective.

I believe we can engineer almost anything. Many alternative methods proposed in the offshore wind farm industry are technically possible and exist in other industries—mainly offshore oil and gas. But the cost of these technologies and methods may prohibit them from being used in the offshore wind farm industry. I make a point of demonstrating this whenever I list the alternatives.

The reader should always keep an open mind about what is stated in this book. It is essential to understand that the focus is on the object—installing an offshore wind farm—but the point of view may be different depending on who one talks to in real life. Therefore, the statements and recommendations should serve only as guidelines—things that have been done before, methods that may change—and all of this is intended to prepare the reader for further study, real work, and a fact-based opinion about this industry.

What Can You Get Out of This Book?

After finishing this book, it is my hope that the reader will have an in-depth understanding of the offshore wind farm industry. One should be able to make major decisions to map out the main planning and execution route for an offshore wind farm project that one is working to complete.

The reader should know the best choices to make and the consequences they will involve. This is important because every time a component in the offshore environment is changed, repercussions could well reach beyond the single component being altered.

As such, the offshore environment can pose challenges that are different from the onshore environment. Why is this so? A good example is if you change the size of the foundation due to poor ground conditions. Onshore you would ask the geotechnical engineer how much additional material to remove in order to get to a firm, stable ground. Once this is established, you can estimate how much extra concrete to pour in order to create the proper foundation for the turbine. This is fairly straightforward engineering and not really that complicated. It involves calculating amounts, an excavator, and additional time.

For the offshore wind farm, it is an entirely different beast. The first thing to ask for is an additional set of core drillings to establish how much the seabed varies and to what depth. This may very well change the entire foundation system to be used. As an example, the Baltic 2 project had such challenging ground conditions that it was necessary to install two entirely different types of foundations: monopiles and jackets. This was the result of the poor seabed conditions in the area. There is more to it however.

If the foundation or ground is different from the baseline set of characteristics, the entire iterative process of calculating the turbine–foundation interaction is different, and thereby the individual foundations need to be different. Furthermore, if the foundation is different—say, a jacket and a monopile—in the same wind farm offshore, it requires two sets of seafastening on the installation vessels, two types of hammers to drive piles, and so on. There are far-reaching consequences when greatly varying seabed conditions offshore are found compared to onshore.

This is why it is important to read and understand all of the factors of the various disciplines of an offshore wind farm project. The smallest component will change the larger system if it is important enough. After reading this book, the reader should be able to make detailed plans and to understand and account for the many variables that will impact the project from start to finish.

What is here is also intended to give the reader the opportunity to ask questions. The more that's read, the more thoroughly you will be able to understand the industry, giving one a continuously improving basis of forming one's own opinion. That is the goal of the book: to inform and to facilitate discussion.

This is not a recipe book in which the reader can look up any question and easily find an answer; it is meant to provide readers with opportunities to start their own thinking process and to develop methods and answers. This is the place to look for hints, advice, stories, and a possible road map to follow to establish the most direct route to the successful installation of an offshore wind farm.

I therefore suggest that you to read on and educate yourself. Have a few laughs over some of the things that have been done, and ponder some of the ideas that are freely floating around in the industry. All of this is here to give readers the best possible basis of educating themselves to become industry professionals—a few of whom I hope to meet offshore someday.

A final word of advice: When I started in the offshore wind farm industry, none of us knew what we were doing. We learned that in the most positive of environments we were lucky on the big and tough decisions, and we were blessed with two years of good weather. These conditions are not something to take for granted. As a good friend and colleague once said, "Good planning blessed with the maximum amount of luck available is not something you should frown upon."

There you have it. It is my firm belief that we can all do what we want to as long as we are open minded and ready to take in new knowledge and listen to sound advice. Don't be afraid to walk out in this industry as a novice. You will learn—from this book and from the fact that the *Titanic* was built by a team of the most skilled engineers available and Noah's ark was built by an amateur. So, the good news is that there are opportunities for all of us. We just have to be open to taking them—as I did.

ACKNOWLEDGMENTS

As with all things in life, no one can carry out an achievement on his own. This book is no different. It would not have been possible to deliver the manuscript without the support of my wife Åse and my family, who have been asking me with interest about my progress, encouraging me when things were going well, and pestering me when I didn't feel like writing and the laptop lay idle.

Furthermore, I would like to thank Rachel Pachter and Paul Quinlan for contributing parts concerning the permitting process. We have made an initial summary of the various processes, recognizing that this is a living thing, which will require revisiting on a regular basis. In the course of writing this book, many items in the permitting process in both Europe and the United States have changed and thus given rise to last-minute revisions.

Marilyn Rash, with the help of Deborah Prato and Samantha Graham, did a marvelous job of editing and proofreading the pages to make my "Denglish" into a perfectly understandable script. Even though various satellites refused to send the chapters back to me for proofing in a timely manner, we managed to get everything finished.

Finally, I would like to thank Tiffany Gasbarrini for undying support and pep talks during the entire process. If I am ever to experience divine patience, she must have been the one showing it. Tiffany gave me the opportunity to write this book, and I hope I did not disappoint her with the result.

ABOUT THE AUTHORS

Kurt Thomsen is the mastermind whose innovative ideas effectively invented functional offshore wind farms. An architect and crane operator by profession, he developed and patented the world's first "crane ship," before which there was no way to successfully transport and install wind turbines at sea. He has founded four companies, one of which, A2SEA A/S, has installed more than 800 wind turbines in the waters around Denmark, Sweden, Holland, and the United Kingdom.

The world's largest wind turbine manufacturers, including Vestas, Gamesa, GE, Siemens, Areva, and Nordex, have been working with Mr. Thomsen to develop new turbine designs that will facilitate better productivity in marine conditions. His current company, Advanced Offshore Solutions, has developed technology that enables the transport of 18 wind turbines simultaneously, and can complete installation of the lot within 24 hours.

A highly sought-after conference speaker and consultant, Mr. Thomsen has edited and translated technical books in English, German, Norwegian, and Danish. He is also a contributing columnist for *Cranes Today* magazine.

Rachel Pachter is the Permitting and Environmental Manager for Energy Management Inc., the developer of Cape Wind—the first proposed and permitted offshore wind project in the United States. Rachel has been working in U.S. offshore wind development since 2002.

Paul Quinlan is Managing Director of the North Carolina Sustainable Energy Association. He has extensive experience developing and promoting onshore and offshore wind energy policies. In addition, he regularly teaches and publishes reports on renewable energy topics.

CHAPTER ONE

What Is an Offshore Wind Farm?

Before we go any further in this book, it is probably a good idea to describe what exactly makes up an offshore wind farm. It will also be helpful to have a quick overview of many, if not all, of the technical terms used in this book so it will be easier to understand the descriptions.

Some people may think this superfluous; however, the year after installation of the Kentish Flats a letter to the editor of *The Times,* sent by an elderly couple, demonstrated to me very clearly that not everyone understands the concept. This couple was probably a bit extreme, but in essence they complained that they could not walk the beach anymore simply because the wind generated by the turbines made it impossible for them to keep their balance. So, just to make it clear—a turbine does not generate wind by using electricity; it generates electricity by using the wind as the energy source to turn the rotors. This in essence is what the editor wrote back, as kindly as possible, to the couple.

Anyway, now to get serious and down to the basics of wind energy. In this book, we are looking at the offshore wind farm installation and the offshore wind farm—or a wind farm in general; it can be described in the following manner.

A wind farm is made up of a number of wind turbines. Any wind farm will look more or less like another; Figure 1.1 shows a typical one. The wind farm is situated in an area with relatively shallow water not too far from the coastline and, of course, in an area where the mean wind speed is favorable.

A wind turbine consists of three main components—basically what can be seen from the outside:

- The tower, which is two or more steel tubes bolted together
- The nacelle, or the generator house, which is fitted on top of the tower
- The rotor, which consists of three blades connected to a central hub on the nacelle

Of course, the turbines cannot stand on the water (or the ground if onshore, for that matter). They need a foundation, as shown in Figure 1.2. It can just barely be seen sticking out of the water.

Figure 1.1 Aerial photograph of an offshore wind farm.

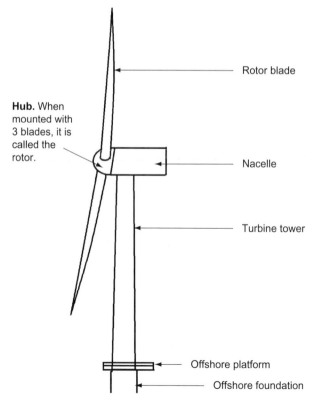

Rotor blade

Hub. When mounted with 3 blades, it is called the rotor.

Nacelle

Turbine tower

Offshore platform

Offshore foundation

Figure 1.2 Components of a wind turbine.

When a turbine is onshore, this is easy because the foundation is a concrete slab that is heavy enough to create sufficient moment and holding force to withstand the movements and bending moments of the wind acting on the turbines. When a turbine is offshore, the outcome is the same, but there are four additional factors to consider when designing the foundation:

- *Water depth:* The foundation must have an additional free-standing column.
- *Wave load:* The waves induce more loads and bending moments on the foundation than the turbine itself.
- *Ground conditions:* The foundation will not necessarily be fixed to the seabed immediately, but it may easily require additional depth before the ground has any bearing capacity due to the composition of the seabed.
- *Turbine-induced frequencies:* The turbine acts and counteracts the wave load, giving new and possibly higher loads to the foundation when the waves are added, which must be taken into consideration.

The design of offshore foundations for wind turbines is an entire science unto itself and is not covered in this book. However, the subdivision of the four most commonly used offshore wind turbine foundations are described in the following sections, along with the pros and cons of each type.

Monopile

A steel tube of a large diameter (4–8 m) that is driven into the seabed using a large hydraulic hammer is known as a monopile. It is able to stand upright because of the friction of the seabed on the sides and not having any vertical ground pressure on it. A monopile is commonly used in hard to semihard seabed conditions up to a water depth of around 25 m (Figure 1.3).

Gravity Base

A gravity-based foundation is a very heavy displacement structure usually made of concrete (Figure 1.4). The gravity base, which applies vertical pressure to the area below, stands on the seabed. The base is usually 15 to 25 m in diameter, and all of the forces and bending moments are transported through the base of the foundation. Typically, a gravity base is used on semihard, uniform seabed conditions and at shallower water depths, although a couple of projects (Thornton Bank, for example) have been installed in much deeper water using a gravity-based foundation.

The size and weight of the foundations (from 1500–4500 tons) make transport and installation cumbersome, and it is worth noting that the seabed

Figure 1.3 A monopile being lifted out of the transport vessel. *Courtesy of A2SEA.*

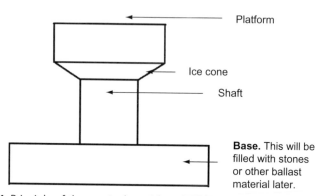

Figure 1.4 Principle of the gravity-based foundation.

must be prepared by dredging and backfilling material in order to install the foundation. So while concrete is cheap to build, it is extremely expensive and time consuming to install. Therefore, the gravity-based foundation is not the preferred solution.

Tripod

A tripod is a steel tube that protrudes out of the ocean surface (Figure 1.5). Under water there is a three-legged foundation; each "leg" ends in a pile sleeve, where an anchor pile is driven into the seabed to hold the foundation

Figure 1.5 Test tripod shown onshore in Bremerhaven. *Courtesy of BIS.*

in place. The advantage of the tripod is that while the area penetrating the wave zone is as small as a monopile, because it is a single tube, it is spread out like a camera tripod on the seabed. Thereby it provides enormous stability against bending moments.

In addition, the anchor piles, with their large distance to the center of the foundation, have the ability to resist very strong vertical forces as well as bending moments induced by the turbine and the waves. In particular, because there is only a single tube, it is possible to calculate wave-induced loads the same as a monopile. The tripod is commonly used by the oil and gas industry offshore at extreme water depths (25–50 m) and has proven itself to be very reliable.

The disadvantage of the tripod is that for an offshore wind turbine of reasonable size, it is very expensive to produce, difficult to handle in large numbers at a time, and takes much longer to install than a "regular" monopile.

Jacket

A jacket foundation is a lattice-type steel structure, usually square in footprint and constructed of thin tubes (Figure 1.6). The advantage of the jacket is that considering the size and water depth, its weight is fairly low. You can create massive resistance to force and bending moments simply by increasing the footprint of the foundation without significantly increasing the dimensions. It is used exclusively for large water depths, where it is the preferred solution. Like the tripod, it is fitted with corner pile sleeves where anchor piles are driven through to keep the jacket in place.

The disadvantages of the jacket, in general, are that the cost of manufacturing the many nodes in the structure is costly and must all be done

Figure 1.6 A jacket foundation being installed at the Beatrice test wind farm.

manually. In addition, the foundation is difficult to protect against ice loads as a result of the large footprint, even with the small tubes in the surface zone. This makes ice protection complicated and therefore expensive as well. Furthermore, the jacket is just as costly as the tripod, and currently turbines are not large enough to justify using jackets at great water depths.

Related Images

Photo: Hans Blomberg +46 70 550 0121

Lifting a full rotor from the installation vessel is delicate and specialized work. In port, it is reasonably easy, but the operation becomes very difficult offshore because it is not possible to put people outside the vessel parameter to hold the rotor steady. Lillgrund Wind Power Plant, 2007-08-01. Test of blade montage on M/S Sea Power in Nyborg, Denmark.

Crew transport between the turbines and from and to shore is an integral part of constructing and operating an offshore wind farm.

CHAPTER TWO

Obtaining Permits for Wind Farms

A permit is required before any offshore wind farm can be built anywhere in the world. The countries that engage in this industry handle the permitting process very differently. This can range from a meticulous and rigorous process with detailed specifications of the deliverables to no or very few regulations in place in some countries. This, of course, influences the scope of the wind farm owner's plans and the consistency of the regulatory process.

This book outlines the main parameters of three different regulatory processes—those of the United States, the United Kingdom, and Germany. These have been chosen because of the perception that they will be the main wind farm markets in the future and because they represent the mature, the young, and the emerging ones.

THE UNITED STATES

Permitting is the first hurdle in the development of offshore wind technology in the United States. The regulatory process for gaining approval remains an evolving one. Individual states, towns, electricity regions, and the federal government all maintain authority over offshore wind projects. Wind projects in the United States are currently being developed in three distinct offshore regions:

1. Outer continental shelf
2. Offshore in various states' waters
3. The Great Lakes

The outer continental shelf (OCS) is defined roughly as the area ranging from 3 nautical miles from a state's coastline out to 200 nautical miles. The offshore area under state jurisdiction is within 3 nautical miles of the coastline. However, Texas and the Gulf Coast of Florida have jurisdiction farther out into the coastal waters to approximately 10 miles.

The Great Lakes are located on the border between Canada and the United States. Both countries are considering installing offshore wind projects in the lakes. In the United States, adjacent states have jurisdiction up to the Canadian border, and certain federal regulatory statutes apply as well. The same is true of offshore wind projects being developed in state waters.

Offshore Wind Potential

The U.S. Department of Energy estimates that there are more than 4000 gigawatts (GW) of offshore wind potential in the United States.[1] The majority of development activity for offshore projects is taking place along the north and mid-Atlantic coasts and in the Great Lakes. More than 50 percent of the U.S. population is located along the coasts,[2] and America's electricity demand is highest along the coast.

In North America, onshore wind resources are generally greatest in the central portion of the country. Developing projects on the outer continental shelf reduces the need to run long transmission lines and incur electrical losses in order to provide clean and renewable power to the largest electric load centers.

Permits for the Outer Continental Shelf

The current path to obtaining permits was originally initiated in the federal Energy Policy Act of 2005 (EPAct 2005), which gave the U.S. Department of the Interior (DOI) regulatory authority over offshore renewable energy projects on the OCS. The Minerals Management Service (MMS) was designated as the agency within DOI responsible for crafting the regulations and leading the permitting process, which includes offering leases for offshore renewable energy projects.

The MMS was subsequently reorganized and renamed the Bureau of Ocean Energy Management Regulation and Enforcement (BOEMRE).[3] The regulations were finalized in June of 2009. The existing regulations allow for competitive bidding on areas of the OCS that BOEMRE opens up to lease.

In addition to a lease from BOEMRE, projects also need permission from additional federal agencies such as the U.S. Coast Guard, the Federal

[1] *www.nrel.gov/wind/pdfs/40745.pdf*

[2] *http://oceanservice.noaa.gov/facts/population.html*

[3] Following the British Petroleum oil spill in April 2010, the Secretary of the Interior renamed and reorganized the MMS.

Aviation Administration (FAA), the Environmental Protection Agency (EPA), and others. In addition, if a project is built in federal waters, state and local regulations apply to the cable needed to bring the power ashore once it is within state boundaries. Permission must also be obtained from the regional transmission authority.

The total permitting timeline currently is estimated to take more than 5 years. However, support for the rapid development of offshore wind is very strong among the highest-ranking politicians. Renewable energy and green job creation are high priorities in the United States, and many stakeholders are working closely to streamline the permitting process and encourage the timely development of projects in U.S. waters. On November 23, 2010, the Secretary of the Interior, Ken Salazar, announced a new initiative intended to do just that by reducing the permitting timeline and by identifying priority areas for offshore wind development.

Obtaining Permits for State Waters

Projects in state waters are not subject to the extensive regulations promulgated by BOEMRE for projects on the OCS, making obtaining permits less complicated. However, each state has its own regulations, and the review process differs from state to state. In addition, several federal agencies have regulatory authority offshore even in state waters. Project proponents must work with the Army Corps of Engineers (ACOE), the Federal Aviation Administration, the U.S. Coast Guard (USCG), and other federal agencies before building a project.

Obtaining Permits for the Great Lakes

Planned projects in the Great Lakes are still working through the permitting process, since the jurisdiction for the area is both state and federal. Multiple committees and groups have been formed that include wind farm owners, environmental nongovernmental organizations (NGOs), regulators, potential power purchasers, and others to work out the process.

Offshore Planning

Both state and federal governments have taken steps to comprehensively review the offshore environment for the development of offshore wind. In some cases the goal has been to provide wind farm owners with in put and opportunities to choose the least environmentally and socially conflicted areas. In other cases it has been done to speed up the environmental review process for proposed projects by initiating the review as early as possible.

Federal Planning

The office of the President of the United States issued an Executive Order in June 2009 establishing an interagency Ocean Policy Task Force. The task force is providing recommendations on Marine Spatial Planning (MSP) for the U.S. offshore waters. According to the website for the MSP process:

> The Task Force defines coastal and marine spatial planning as a comprehensive, adaptive, integrated, ecosystem-based, and transparent spatial planning process, based on sound science, for analyzing current and anticipated uses of ocean, coastal, and Great Lakes areas. Coastal and marine spatial planning identifies areas most suitable for various types or classes of activities in order to reduce conflicts among uses, reduce environmental impacts, facilitate compatible uses, and preserve critical ecosystem services to meet economic, environmental, security, and social objectives.[4]

This process is going on concurrently with the development of offshore projects.

Wind farm owners of projects are concerned, however, that the MSP process will impede and/or delay the development of offshore wind in the United States. As the cooperative efforts of the task force have been proceeding, they are not expected to delay offshore wind development.

State Planning

States can play several roles in the development of an offshore wind project. They can be involved in providing financial incentives, they can be a potential customer for offtake or facilitate the offtake (discussed later) of a project, and they certainly have permitting authority over projects and/or transmission cables. Several states with the potential for projects in state and nearby federal waters have taken the initiative to comprehensively review the offshore region for potential environmental, socioeconomic, cultural, and navigation conflicts and to search for areas that would be conducive to the development of offshore wind.

---

[4] *www.msp.noaa.gov/*

For example, Massachusetts developed the Massachusetts Ocean Management Plan that addresses the potential for offshore wind development within the Commonwealth's waters. The final plan was issued in December 2009 and states the following:

> In response to public comments on the draft plan, as well as additional information brought to bear during the public review period, the final plan adds strong new protections for critical marine life and habitats, identifies areas suitable for renewable energy development, and initiates a five-year program of high-priority research. The final plan includes stronger and more detailed siting and performance standards associated with important environmental resources and revised management provisions for regional planning authorities regarding wind energy development.[5]

The final plan identifies areas that have limited conflicts for developing offshore wind and plans for moving forward to develop potential offshore projects in state waters.

Another example is the Rhode Island Special Area Management Plan (SAMP). The SAMP is working toward designating offshore areas for specific purposes. The plan states the following:

> The major driver for the development of the Ocean SAMP was the determination by the Rhode Island Office of Energy Resources in 2007 that investment in offshore wind farms would be necessary to achieve Governor Donald Carcieri's mandate that offshore wind resources provide 15 percent of the state's electrical power by 2020. In response, the CRMC proposed the creation of a SAMP as a mechanism to develop a comprehensive management and regulatory tool that would proactively engage the public and provide policies and recommendations for appropriate siting of offshore renewable energy.[6]

The state of New Jersey conducted a broad environmental review process called an Ecological Baseline Study that provided environmental data for a large portion of its coast. The final study states:

> The objective of this study was to conduct baseline studies in waters off New Jersey's coast to determine the current distribution and usage of this area by ecological resources. The goal was to provide GIS and digital, spatial, and temporal data on various species utilizing these offshore waters to assist in determining potential areas for offshore wind power development.[7]

The dataset will be used by both wind farm owners and regulators to help site projects, as well as to assess the potential impacts of a project.

---

[5] *www.env.state.ma.us/eea/mop/final-v1/v1-text.pdf*

[6] *http://seagrant.gso.uri.edu/oceansamp/documents.html*

[7] *www.state.nj.us/dep/dsr/ocean-wind/vol1-cover-intro.pdf*

Requests for Proposals

One way in which states are involved in authorizing offshore wind projects is through a Request for Proposals (RFPs) process. States may issue RFPs for offshore wind projects, renewable energy projects, and/or energy-generation projects. An offshore wind project can obviously bid into any one of these. Several states have initiated an RFPs process that incorporates offshore wind, including Massachusetts, New York (both in the Atlantic Ocean and the Great Lakes), Rhode Island, Delaware, and New Jersey.

If a project wins the RFPs, it may be granted permitting priority (although if the project is located on the OCS, it still needs to undergo the BOEMRE process to secure a lease); development rights if it is in state waters; or, most important, a long-term Power Purchase Agreement (PPA).

Federal Permitting

The permitting process itself can be lengthy. As just discussed, the current process for projects on the OCS may take more than 5 years. Projects in state waters may take somewhat less time, and projects in the Great Lakes may vary. With respect to the procedure itself, the permitting process involves distinct actions. Certain processes must be followed and permits must be obtained. For example, a comprehensive environmental review process will likely take place, and that process will need to be completed before an agency can issue a permit or authorization.

The most detailed of the processes that must be completed is the one in the National Environmental Policy Act (NEPA). The NEPA process provides for a comprehensive review of all potential environmental and socioeconomic impacts. According to the EPA's website:

NEPA requires federal agencies to integrate environmental values into their decision-making processes by considering the environmental impacts of their proposed actions and reasonable alternatives to those actions.[8]

Any federal agency making a decision can lead the NEPA process, and a NEPA process done for a single project can be used to inform multiple permit decisions by a number of agencies for that same project.

All projects that need a federal permit (projects in state waters and the Great Lakes still need federal permits) are subject to NEPA. The scope of review for a NEPA process is broadly informed by comments received from

---

[8] *www.epa.gov/compliance/nepa/*

the public, regional stakeholders, and other agencies (i.e., federal, state, and local), as well as the lead agency. The result of this review is an Environmental Impact Statement (EIS) produced by the lead agency. The EIS is first released in draft form, and then once public comments are received and responded to, it is released in final form.

Even though the NEPA process is the overriding one that drives the public interest review, other regulations are activated when pursuing a permit. Some of the more pertinent regulatory processes that need to be satisfied include:

- Endangered Species Act
- Rivers and Harbors Act
- Clean Water Act
- Clean Air Act
- Marine Mammal Protection Act
- National Historic Preservation Act
- Coastal Zone Management Act

The Endangered Species Act (ESA) requires review by species experts if an endangered species resides in the vicinity of the project. In the case of offshore wind projects, marine mammals and avian species are likely candidates for this additional review. The ESA review is administered by the agency responsible for the species (e.g., National Marine Fisheries Service [NMFS] in the case of endangered whales).

The Rivers and Harbors Act is administered by the ACOE. The Army Corps of Engineers issues Section 10 permits for projects in the navigable waters. The ACOE also issues Section 404 permits under the Clean Water Act for any offshore dredging work. This may be necessary depending on the project's installation methodology. The EPA must provide an air permit under the Clean Air Act for the operation of construction vessels. This is applicable to projects located on the OCS. If the project has the potential to harm marine mammals in any way, the project may need authorization under the Marine Mammal Protection Act.

The National Historic Preservation Act (NHPA) requires review of the visual impacts to historical locations onshore, potential historic locations offshore such as shipwrecks, and any potential impacts to Native American concerns. Section 106 of the NHPA provides for a process by which stakeholders with an interest in historic preservation can work with the agency and project proponents to address concerns.

The Coastal Zone Management Act (CZMA) allows states located adjacent to projects on the OCS to review projects for consistency with an approved state ocean management plan.

The Permitting Process

Applicants who want to obtain a permit for a project on the OCS begin with the Bureau of Ocean Energy Management Regulation and Enforcement. BOEMRE has designed the process such that it engages a stakeholder group with nearby states before opening up areas for bid by wind farm owners. Once the stakeholder process is underway, BOEMRE issues a request for expression of interest (RFI) for a designated area offshore.

Wind farm owners then submit a response to the issued RFI, and the agency will decide based on the content of the information submitted whether it will proceed with a competitive or noncompetitive leasing process. If it is a noncompetitive process, wind farm owners can proceed along the permitting track. If it is a competitive process, a wind farm owner will be chosen based on a bidding process.

BOEMRE will take the lead on a NEPA review and will provide multiple approvals along the way, including approval of the Site Assessment Plan (SAP) and the Construction and Operations Plan (COP), and will furnish the applicant a lease allowing for the exclusive rights to the proposed project area.

When a lease is issued, the lessee is required to start paying rental fees, and once the project is up and running, the owner will pay lease fees (royalties) to the federal government. A percentage of the lease fees is shared with the adjacent state.

In addition to the preceding federal approvals, a project must receive permission from the Federal Aviation Administration, which will look into the impact of air flight navigation paths and provide a "No Hazard" determination, as well as approve or modify any existing lighting plans. The U.S. Coast Guard also must approve sea navigation safety equipment and markings. The USCG may be more involved in the review process if there are siting and navigation concerns.

The agencies permitting offshore projects will prescribe mitigation as part of final permitting decisions or as input to the lead agency. Mitigation for avian species, marine mammals, fisheries, and other resource areas are likely to be imposed.

State, Regional, and Local Permitting

Every project will encounter a variety of regulations as each attempts to connect the project through to the shore and into the upland area to ultimately connect to the electric grid. For projects on the OCS, once the cabling enters state waters, the required permits are expected to address permission to install and disturb subsea sediments per the Clean Water Act, permission to access tidelands, potential impacts to commercial or recreational shellfish operations, road access, upland and offshore wetland impacts, and a process to site energy transmission lines. These issues will be addressed by state permitting authorities as well as local ones. The processes vary from state to state. Project proponents likely need to go before state and local permitting boards for project approval.

In addition, the regional transmission authority needs to approve the interconnection. This review ensures that the power supply is able to be accepted by the electric grid. This process may result in an Interconnection Agreement.

Stakeholder Outreach

Project proponents, as well as states that support offshore wind, are engaging heavily in community and stakeholder outreach. Because the offshore wind industry is relatively new to coastal U.S. communities, education is a key part of the development process. Almost every step of the permitting process allows for public input, and stakeholders largely drive the content and scope of the NEPA process. Therefore, it is imperative that the public understand the general parameters of the proposed project and are provided with opportunities to learn about it and to ask questions.

THE UNITED KINGDOM

The United Kingdom maintains the largest market for offshore wind energy in the world. In July 2011, there was more than 1.3 GW of installed capacity in 15 operational wind farms, with nearly an additional 6 GW under construction, awaiting construction, or in planning.[9] The existing wind farms are located in the North Sea, the Irish Sea, and Scottish territorial waters.

Offshore Wind Potential

The United Kingdom has indicated offshore wind development will be an important resource to meet the country's target to have 15 percent renewable energy by 2020. The target originates from the 2009 Renewable Energy Directive of the European Union.[10] That directive established

---

[9] United Kingdom Department of Energy and Climate Change. *UK Renewable Energy Roadmap*. July 2011.

[10] Directive 2009/28/EC of the European Parliament and of the Council of 23 April, 2009.

renewable energy requirements for each member country based on a number of factors, including existing renewable generation and gross domestic production. The overall objective is for the European Union to secure at least 20 percent of its total energy from renewable energy sources by 2020.

Growth estimates in the United Kingdom anticipate adding another 10 to 26 GW of installed offshore wind capacity by 2020.[11] Over a longer time horizon, the United Kingdom could deploy more than 40 GW of offshore wind capacity by 2030—providing enough capacity to power the equivalent of all the homes in the country.[12]

Offshore Planning

In 2001, European Union member countries were required to establish laws, regulations, and administrative provisions to ensure that strategic environmental assessments (SEAs) are conducted to be sure that environmental protection is considered as a part of government plans or programs, such as offshore wind energy development.[13] SEAs require governments to include the following details:

- An outline of the contents or main objectives of the plan or program and its relationship with other relevant plans and programs.
- The relevant aspects of the current state of the environment and the likely evolution thereof without implementation of the plan or program.
- The environmental characteristics of areas likely to be significantly affected.
- Any existing environmental problems that are relevant to the plan or program including, in particular, those relating to any areas of a specific environmental importance, such as areas designated pursuant to Directives 79/409/EEC and 92/43/EEC (the Birds and Habitats Directives).
- The environmental protection objectives established at international, community, or member state level that are relevant to the plan or program and the way those objectives and any environmental considerations have been taken into account during its preparation.
- The likely significant effects on the environment, including issues such as biodiversity, population, human health, fauna, flora, soil, water, air, climatic factors, material assets, cultural heritage including architectural and

---

[11] *UK Renewable Energy Roadmap.*

[12] *UK Renewable Energy Roadmap.*

[13] Directive 2001/42/EC of the European Parliament and of the Council of 27 June 2001.

archaeological heritage, landscape, and the interrelationship between the preceding factors.

- The measures envisaged to prevent, reduce, and, as fully as possible, offset any significant adverse effects on the environment of implementing the plan or program.
- An outline of the reasons for selecting the alternatives dealt with and a description of how the assessment was undertaken, including any difficulties (e.g., technical deficiencies or lack of know-how) encountered in compiling the required information.
- A description of the measures envisaged concerning monitoring.
- A nontechnical summary of the information provided under the other items in this list.[14]

Early offshore wind projects in the United Kingdom included a series of demonstration projects, known as Round 1, and limited development, known as Round 2, which required an SEA and was limited to three strategic areas: the Greater Wash, the Thames Estuary, and Liverpool Bay. In 2007, the then Department for Business, Enterprise and Regulatory Reform initiated an SEA of U.K. waters to open up the seas to up to 33 GW of offshore wind energy. This SEA ultimately authorized The Crown Estate to initiate Round 3 of offshore wind leases.

Offshore Leasing

The Crown Estate is the corporate body, reporting to Parliament, that manages the property portfolio owned by the monarch as part of the hereditary possessions of the Crown. The Crown Estate is one of the largest property owners in the United Kingdom. The marine portion of the portfolio includes virtually the entire U.K. seabed out to the 12 NM territorial limit and rights to generate renewable energy on the U.K. Continental Shelf, which extends up to 200 miles from the coast.

Consequently, The Crown Estate solicits and manages offshore wind leases in the United Kingdom. It identified nine development zones for Round 3 of offshore wind leases. Each zone will be managed by a single development partner—either a company or consortium—that will oversee development of the zone. The nine partners were announced in January 2010.

---

[14] *www.offshore-sea.org.uk/site/scripts/documents_info.php?documentID=5&pageNumber=2*

Permitting and Consenting

Once leases have been established with The Crown Estate, an offshore wind project must proceed with permitting and consenting. The primary agency responsible for that will be the Marine Management Organisation (MMO) or the Infrastructure Planning Commission (IPC). The appropriate agency will be determined based on the total capacity of the offshore wind farm.

The MMO is responsible for the permitting and consenting of offshore wind facilities that are of 1- to 100-MW capacity. The agency consents these, among other marine construction activities, under a new marine licensing system that is part of the Marine Coastal Act of 2009, which started in April 2011. Marine license applicants will be subject to review under the Habitats Regulations[15] and Water Framework Directive[16] and may be required to conduct an environmental impact assessment.[17] The MMO encourages applicants to consult with the agency during the preapplication process if additional assessments are anticipated.

The IPC is the independent body that provides a streamlined examination of applications for nationally significant infrastructure projects, including offshore wind projects with a capacity of more than 100 MW. The application process begins with preapplication, which involves a developer signaling an intent to file an application with the IPC. During the preapplication stage, an offshore wind project is added to the Programme of Projects and the developer must consult with relevant local and regulatory authorities. In addition, the developer may request a screening opinion, which would identify information required in an environmental impact statement.

Once a full application is submitted, the IPC has 28 days from the day after receipt to decide whether or not to accept it for examination. If accepted, the application enters the pre-examination phase, a three-month process during which application acceptance must be publically announced and open for review. The IPC then conducts the six-month examination phase. This stage engages stakeholders who requested to register their view during the pre-examination phase. The final approval of a project is decided

---

[15] Council Directive 92/43/EEC of 21 May 1992 on the conservation of natural habitats and of wild fauna and flora.

[16] Directive 2000/60/EC of the European Parliament and of the Council of 23 October 2000 establishing a framework for Community action in the field of water policy.

[17] Council Directive 85/337/EEC of 27 June 1985 on the assessment of the effects of certain public and private projects on the environment.

by the IPC or relevant Secretary of State in the absence of National Policy Statement.

The IPC was originally established in October 2009 under the Planning Act of 2008. Following the General Election of May 2010, the Coalition Government's intention is to alter the infrastructure consenting process through a Decentralisation and Localism Bill. It is expected that the changes will retain the main elements of the streamlined process while improving democratic accountability.

Additional Industry Support

The U.K. government intends to provide additional resources to support the development of offshore wind industry projects. In the *UK Renewable Energy Roadmap*, published by the Department of Energy and Climate Change in July 2011, the government noted that there is a priority to establish

> ... an Industry Task Force to set out a path action plan to reduce the costs of offshore wind, from development, construction, and operations to £100/MWh by 2020. This will be supported by up to £30m, subject to value-for-money assessment, to foster collaboration between technology developers and support innovation in the production of components over the next 4 years. This builds on existing support to increase the rate of innovation and develop the supply chain.[18]

GERMANY
Offshore Wind Potential

Germany has a growing offshore wind industry, claiming 108 MW of installed capacity at the end of 2010. With dozens of projects permitted or planned, this figure is expected to grow to 3000 MW by 2015.[19] Offshore wind projects are being constructed in the territorial sea, extending 12 NM from shore, and the exclusive economic zone (EEZ) in the North Sea and the Baltic Sea.

Offshore wind development in these regions is attractive because the seas are shallow and close to urban populations. In addition, the electricity costs in northwestern Europe are high in comparison to other international markets, allowing offshore wind to become competitive more quickly than in other European markets. The region also has extensive onshore wind

---

[18] *UK Renewable Energy Roadmap.*

[19] Global Wind Energy Council (*www.gwec.net/index.php?id=129*).

development, thereby limiting the number of remaining onshore wind opportunities.

Similar to the United Kingdom, Germany has renewable energy targets originating from the 2009 Renewable Energy Directive of the European Union. The country's target is to reach 18 percent of consumption from renewable resources by 2020.[20] In 2011, Germany established higher national renewable energy targets following the nuclear accident at the Fukushima nuclear power plant in Japan. Regularly scheduled revisions to the Renewable Energy Sources Act, known as the *Erneuerbare-Energien-Gesetz* (EEG) in German, call for a full departure from nuclear power by 2022 and the establishment of aggressive renewable energy requirements. Under the revisions, the country must produce at least 35 percent from renewable resources by 2020, 50 percent from renewables by 2030, 65 percent from renewables by 2040, and 80 percent from renewables by 2050.

Governmental Permitting

The German government provides permitting oversight for offshore wind projects located in the exclusive economic zone with authority assigned to the Federal Maritime and Hydrographic Agency, known as *Bundesamt für Seeschifffahrt und Hydrographie* (BSH) in German. Permitting is conducted in accordance with the Federal Maritime Responsibilities Act and the Marine Facilities Ordinance.

An offshore wind facility can be approved provided the project does not impair the safety and efficiency of navigation and that it is not detrimental to the marine environment. In accordance with the Marine Facilities Ordinance, the regional Waterways and Shipping Directorate must grant consent and confirm that an offshore wind facility does not impair navigation safety and efficiency. Environmental impacts are considered by the BSH, with projects consisting of more than 20 turbines requiring an environmental impact assessment based on the Environmental Impact Assessment Act.

Throughout the application process, the BSH holds conferences and hearings that provide interested stakeholders and the public with an opportunity to become informed about the project and provide comment. In addition, the BSH collaborates early in the process with those German coastal states that are responsible for approving any electric cables laid in the territorial sea.

---

[20] Directive 2009/28/EC of the European Parliament and of the Council of 23 April 2009.

In the event applications are received for the same location, the Marine Facilities Ordinance directs the BSH to consider approval of the application that first meets all the application requirements. An application is considered to meet all application requirements when all documents needed for the decision are available to the agency.

State Permitting

Applications for the wind farms permits within the territorial sea extending into the 12-NM zone from the shore are submitted to the German coastal states of Lower Saxony, Schleswig-Holstein, and Mecklenburg-Vorpommern. These states also give permits for underground electrical cables traveling from offshore wind farms in the EEZ to mainland grid access points.

Offshore Wind Standing Committee

The Offshore Wind Standing Committee of the federal government and the coastal Länder (StAOWind) was set up to coordinate the offshore wind approvals procedures for those projects that can extend from the EEZ to the coastal states. StAOWind brings together the states of Schleswig-Holstein, Lower Saxony, Mecklenburg-Vorpommern, and Bremen; the Federal Ministry for the Environment, Nature Conservation and Nuclear Safety, the Federal Ministry of Economics and Technology, and the Federal Maritime and Hydrographic Agency (BSH). The German Energy Agency is responsible for the management of this committee, which has been meeting regularly since 2002.

OTHER SIGNIFICANT OFFSHORE WIND MARKETS

Denmark, the Netherlands, and Sweden also have significant offshore wind installations already in place, with increasing scale-up planned for the next several years. These efforts' intents are to reach established 2020 EU energy efficiency and carbon reduction goals. Denmark's efforts, in particular, to streamline the permitting process by minimizing the trials needed to obtain sign-off have been a great boon to the development of offshore wind in that region. The average time needed to register and apply for permits there is significantly shorter than it is in other EU member states, where it can take as long as five or seven years to navigate the process.

Additionally, France, Spain, Norway, and Belgium all have plans to enter the global offshore wind market in a meaningful way over the next decade.

Efforts are underway to establish an overarching EU policy governing offshore wind, but as might be expected, this will likely take some years to
achieve. Therefore, permitting and governance guidelines, which currently
vary widely from country to country, are likely to continue to do so for the
immediate future.

While offshore wind is being explored in emerging markets, such as
China and Japan, permitting has not been as thorny an issue in that area
of the world as it has proved to be in Europe and the United States.

It should be noted that entire volumes could be dedicated to the finer
points of offshore wind permitting across the globe. This book does not
aim to undertake that work, which would be impossible to achieve within
the context of this much broader discussion of offshore wind as a holistic
endeavor. Readers interested in granular detail regarding offshore wind permitting for a particular region are encouraged to consult resources specific to
the area under consideration; often such information can be obtained by
contacting a local or regional wind power organization.

● ● ●

Related Images

Placing a monopile in the pile guide prior to driving it into the seabed. The components are all big, heavy, and difficult to handle. This makes the job of installing both interesting, challenging, and—if inexperienced—dangerous.

Placing the transition piece after driving the monopile foundation offshore.

CHAPTER THREE

Project Planning

It is clear that in order to carry out the task of installing an offshore wind farm with all of the different and complex parts involved, it is necessary to plan the entire project from start to finish. This is a very comprehensive and complex process to engage in. The planning of the project starts with an outline of what is to be done, when, how, and by whom. This will take place in a simple top-down manner so as to divide the project into manageable pieces for the various departments and stakeholders in a company wishing to install an offshore wind farm.

Items such as site locating, permitting of the wind farm, due diligence of the project in terms of yield and earnings, budgeting, pricing, and tendering, the individual packages as well as the actual project management (PM) of the work packages if and when the farm is built, are all important issues that have to be dealt with to make a project happen. This logically incorporates a number of tasks to be delegated, and the allocation of resources must be considered and carried out. This is crucial to the successful planning and execution of any offshore wind farm.

PROJECT STRATEGY OUTLINE

When starting a project from scratch, it is necessary to decide which strategy should be pursued to deliver the wind farm on time, at cost, and with zero incidents. The idea is for the owner to outline a number of basic principles that must be followed and that should fit the organization and the "usual" strategy for project development and execution of the wind farm.

The importance of outlining the basic strategy cannot be underestimated. If a decision that was made in the early stages turns out to be impractical, or just plain wrong, the entire project from start to finish will suffer. The basic rule of 20/80 applies: 20 percent of the decisions made in the beginning will affect 80 percent of the outcome in the end, and vice versa. Therefore, an extensive effort should be put into outlining the project strategy. The following sections discuss some critical issues and how to handle them.

Organization

Is an internal organization already in place? Or must the entire staff be located and bought in? Will the initial organization be a mix of both? The differences are large, and the consequences of all three alternatives can have a huge impact on the outcome.

Although we discuss the organization a little later in the book, a few points have to be made at this stage concerning the maturity of the organization and whether it is internal or external. If there is an internal organization, what are the competencies that this organization possesses? What is missing in order to carry out the project? Will the personnel and other resources be hired or bought in as a subcontractor supply?

If it is the first project the organization is undertaking, these questions are very relevant. The offshore installation of a wind farm is different from any other offshore projects that take place today. The planning must be developed for a production-line type of work. You install the foundations, the scour protection, the cables, the substation, and finally the turbines. This is a project that can last up to two years from the time you start until the time the project is handed over to the owner.

A large number of skills are required at specific points, and therefore your organization must be geared to solving the various tasks connected to the single item that is to be performed. As an example, it is necessary to employ experts in foundation design and calculations to develop a cost-effective foundation solution. However, foundation design is influenced by issues such as the following components.

Metocean Conditions

How big are the waves? What is the predominant direction of the maximum return wave? This requires a specialist in hydrology, which is a position that most companies do not have on their list of employees. Therefore, you may have to hire someone with this specific skill or indeed outsource the entire design process.

Seabed Conditions

Whether you are dealing with a soft or hard subsurface determines the penetration depth of, for example, a monopile in order for it to be stable. However, the hydrologist must be consulted again about this issue because the seabed current will determine whether scour protection is necessary, and if it is, it must be decided how far-reaching the scour protection must be. What influence does this have on the behavior of the monopile?

The Turbine

What are the size and the behavior of the turbine on the foundation? What are the eigen frequencies and the forces and bending moments induced on the foundation? The size and impact of the turbine will influence the design of the foundation dramatically. And this again influences the type of foundation required for a specific set of seabed conditions.

On top of this, the forces, bending moments, and frequencies of the turbine are either enhanced or, in some areas, dampened by the metocean conditions where the turbine is to be installed. Again, this will affect the design of the foundation. Therefore, the process is an iterative research one in order to find the optimum solution, and revisiting of many single components in the entire scope of work is necessary before the best fit for a foundation will be found.

So, as you can see in the preceding, a number of other components influence the design of a turbine foundation. Furthermore, it is possible for several different types of foundations to be used on the same site, and this again is a factor that must be considered. It is also important to note that different types of foundations are developed for a variety of sites, but that there is no best-fit foundation for all sites. This makes the process more time and cost consuming, even though the various believers in the different types of foundations will try to persuade you to use their particular type. We will return to this topic later.

Health, Safety, and Environmental Compliance with Permitting

The permit to install an offshore wind farm is always accompanied with a set of health, safety, and environmental (HSE) conditions, which can be substantial and very strict. Therefore, it is important at an early stage to determine personnel requirements in order to develop the entire solution that must be compliant with the permitting text.

The main focus, of course, is to ensure that everyone involved in the project can do his or her job safely and without any impact on the environment. Therefore, the HSE system development is important. The HSE work starts with outlining the rules and regulations of the country where the project is to be executed. It is important to notice that the permit will contain a set of conditions that are additional requirements to be met outside of the rules and regulations of the country in which the project is located.

Normally, the HSE planning starts at the beginning phase. The client will request documentation of the compliance with the permitting text and will require proof of the presence of an HSE system and its general layout. The documentation required for the beginning phase is as follows:

- Actual status of the system—meaning documents showing an updated version of the HSE plan. Verification by relevant authorities of the system and the latest certificates, if applicable, are required.
- A record of lost time incidents (LTI) over the last three years per million work hours is the normal standard for documentation of the actual quality of the HSE operation in a company. A high number of LTIs indicates at first glance that the work procedures and the running of operations are handled poorly. A steady decline—even from a high number—demonstrates an increasing awareness and a well-functioning HSE operation. Finally, a steady and slowly declining number from a very small offset will indicate that the organization is taking security very seriously.

The client must make sure all contractors and suppliers have the newly established ISO 18000 for health and safety and the ISO 14000 for the environment. This is the minimum requirement for contractors should they want to bid on any project, service, or supply of goods. Being aware of these regulations in the wind industry in general, and in the offshore wind industry in particular, is critical; the importance of both standards should not be underestimated. They supply the basics for all operations when going offshore.

Often, the contractor and the client/owner for an onshore project establish that if the most inexpensive component—whether goods or services—is acceptable, then the entire project will use the lowest-priced materials and will still be able to deliver an acceptable standard. This practice is at best a big mistake in the onshore construction business and extremely dangerous in the offshore industry.

We will return to this subject in Chapter 7, Interface Management. For now, we should think about whether the least expensive acceptable component for, say, firefighting onboard a jack-up is "good enough" in a dangerous situation. Perhaps you would rather have a high-quality component that never breaks down so that you can worry about other issues instead of whether the equipment is going to work.

As just mentioned, the HSE systems should be considered very carefully. It should also be noted that they address two separate issues: health and safety and the environment. Thus, it makes sense to separate these two issues in the organization. Previously, the HSE systems were staff functions, so to save costs on these types of functions, they were assigned to one employee. It is very important to understand that the work offshore regarding health, safety, and environmental issues is significantly more important than onshore.

This is not to say that onshore should be taken lightly, but the consequences of errors and the distance from incident to total loss are very minor if something goes wrong offshore, whereas onshore it is at least possible to create a distance and escape route if anything goes terribly wrong. The HSE work should therefore also be carried out offshore observing the following criteria:

- The positions in the company should be line functions, not staff.
- The work should be carried out as two independent positions.
- The coordination among health, safety, and environmental team members, QA/QC, and project management must be very tight, and information must flow freely among all functions.

These reasons take into consideration all of the rules and regulations that govern the work on ships, on barges, and in ports.

Health and safety reviews should be performed to the ISO Standard 18001 in order for the project organization to prepare the operations offshore and onshore. By systematically using ISO 18001, it is possible to implement, monitor, correct, and record all of the work that is going on in the project. It is also possible to communicate between all suppliers, contractors, clients, and authorities, even though the individual stakeholders may be from different countries.

The method of issuing and implementing the HSE system is also easier when the same system is applied. For various contractors and suppliers, it becomes very straightforward to deliver work, document safety records, and so on.

Environmental monitoring is done in compliance with ISO Standard 14001. The same principle applies for the environmental monitoring as for the health and safety work. The same benefits—that everyone understands and works according to the same principles—make sense because a common system governs all of the contractors, clients, and authorities.

So what is the common rule for environmental monitoring? The reason for this particular area of documentation is to make sure all processes and materials are done and handled in an environmentally safe manner. There is, of course, no logic in delivering safe renewable energy offshore if we mess up the environment we are working in while we are doing that. Thus, everything we do, install, bring, or consume during the process must be accounted for and be delivered in a manner that does no harm to the environment.

One example is when vessels are installed, including the processes and the materials brought in, assembled, installed, and consumed while doing the installation. The consumption of fuel by the balance of plant (BOP) is accounted for and calculated. The correct way is, of course, to use the least possible amount of fuel, lubricants, and so forth to install a megawatt of power offshore.

Furthermore, the principle of zero discharge applies. For example, in German waters, everything that does not stay fixed to the foundation must go back to shore, whether it be cardboard boxes, ballast, or gray water from the vessels. All of it must be delivered in port to a processing facility to be processed into manageable liquids or solids in a proper manner. Every kilogram of component packaging is to be accounted for before and after transport and installation offshore in order to secure that the entire amount of surplus material is brought back and discarded properly.

Even the onshore manufacturing processes are to be accounted for; this is unique in the sense that a "normal" production plant will have a system for discharging and processing waste but not necessarily a detailed plan for the actual number, composition, and processing of the various materials and liquids to be managed. In this way, the offshore wind industry imposes very high standards of engagement to use. If not requested by the authorities, the wind farm owner and the contractors normally request that a "better-than-average" philosophy be implemented throughout the organization and the entire construction setup. This is quite unique compared to an onshore construction site.

THE PROJECT EXECUTION PLAN

One of my favorite quotes from film is from Hannibal, the leader of the A-Team, who always said, "I love it when a plan comes together!" I always admired how the team looked at a problem and almost always decided it could be solved with a lot of explosives, welding together some steel plate, and a good beating inflicted on the bad guys. But the essence of the statement is that for every problem there is a solution.

The differences between the A-Team and the offshore wind farm project management team are, however, that we can neither use explosives nor beat up the customer (although I am sure there have been times when somebody wanted to do that out of frustration). That is another subject completely and certainly not covered by this book.

The offshore wind farm project will, however, have a natural flow of tasks that must be carried out consecutively if the project is to be successful. If all of the individual tasks follow the plan and are successfully carried out, you can hear Hannibal whisper those famous words in your ear.

The project execution plan is therefore the most important set of documents in the entire project planning phase. It determines whether the cost will be optimal and whether the program will be possible to realize during the construction phase. The project execution plan is the script of how the wind farm project is to be orchestrated from the delivery of the single component to the staging port, to the final handover of the entire wind farm.

The plan must therefore cover all aspects of the project, beginning with the permitting process. The project execution plan is a road map of sorts that lists all the tasks involved in the project, from environmental impact statements, stakeholder processes, offshore wind and geomapping of the site, the specific measures taken to protect the environment, what will be done to protect neighbors from emissions of all kinds, right up to the actual set of permits that must be obtained before construction can begin.

The process is lengthy, and often the final permits are given just prior to release of the tender documents and, in some cases, only after a tendering process for the entire scope of the offshore work has taken place. This is, of course, not ideal, but the nature of the industry is that because this is a relatively new market for almost all stakeholders and suppliers, the process is not yet set in stone and therefore variations do occur. In fact, the process of applying for a permit to install a wind farm varies from country to country, and thus the owner and/or developer will have to apply a new set of laws— in effect a new game plan—for each country in which an offshore wind farm is planned.

The project execution map also must contain the overall construction plan. But since we have just seen that this part of the plan may actually be firm only when the final tender has been completed, this, of course, comes with some degree of uncertainty. Also, since some of the offshore wind farm projects come with equity and bank financing, this is undesirable.

Financiers—whether banks, investors, or both—want a project that has very little financial or technical risk. Frankly, the best project to invest in is the one that is already built and has been operating for 20 years. At least, this is how the individuals who are going to be financing the project look at it. People who have made money know that the worst thing that can happen is that they invest in something and the investment loses money. Therefore, anyone who receives taxed investment funds will be very reluctant to release funds again and certainly not if there is a perceived risk of losing that money.

Therefore, to get a nonrecourse financing portfolio, the project owner must de-risk the entire process to as close to the absolute freezing point as possible. But the uncertainty of the installation of the wind farm is a major concern. And if the project execution plan does not give certainty of the BOP and the building permit is not firm on points like this either, it will become very difficult to finance the wind farm.

This is therefore a challenge that the owner or developer must overcome, and to do so, he or she will try to provide as many details as possible about the manufacture of the wind farm components—whether turbines, foundations, cables, substation, export cable to the shore connection, SCADA system for monitoring and surveillance, and so on—to secure the program and thereby financing. So the project execution plan will specify the building program, the BOP involved, and the sequence of the events that will happen; for example, the turbines will be manufactured, delivered to a specified port, transported, installed, hooked up, and finally commissioned.

This can seem to be like a jigsaw puzzle, and to get the pieces to fit together, the owner or developer will start the dialogue with all possible stakeholders in the project at a very early stage to determine whether the components and services for the program can be obtained. This process takes several years, and in the meantime the owner or developer and a team will revisit the plan many times to adjust it as necessary. This is an iterative process in itself. Here is an example.

●●●──

Sample Project Execution Plan

Assume that a project lease is given by the crown estate in 2006 to allow a developer to install a 200-megawatt (MW) wind farm offshore of the British east coast. The owner is, of course, delighted to have the seabed lease and the permit to harvest the wind and sell it to consumers as green environmentally friendly power through the electrical outlets in their homes. However, once the initial euphoria has evaporated, he discovers the following:

- The seabed is not as firm as was thought, and he needs much longer monopiles to get a sensible foundation.

- The metocean data are such that he needs a very long installation period in order to be able to install the foundations, cables, and turbines due to the fact that only poor equipment is available in the market where he intends to install in 2008.
- The biggest available turbine in the market is "only" 3 MW, and this in the end means that the internal rate of return (IRR) on his investment is less than 10 percent over 20 years.
- The banks and investors are therefore only remotely interested in the project, but they want a final investment business plan to help them make up their minds.

So, armed with the sad news, the developer begins his initial planning. He intends to use the new 5-MW turbine that Joe Bloggs Wind Turbines is developing because he can only deliver the IRR that attracts investors by using that turbine. But this turbine is not proven yet, and therefore the bar gets raised by investors immediately. So the IRR is now right, but the risk-teething problems with the turbine are higher, and we all know what that means. (Vestas had to take down 80 turbines on the Horns Rev project to repair them onshore and put them back up. That was very costly indeed.)

To add insult to injury, the marginal installation equipment available on the market cannot lift the Joe Bloggs turbine because it is too heavy and needs to be lifted higher than the existing 3-MW machines due to the larger rotor. The investors are not happy because this is yet another risk. But time flies, and two years later the first Joe Bloggs turbines are spinning onshore on a demonstrator project. Now the developer has de-risked the project to some extent, but the issue of installing the project offshore with the available equipment still prevails. So, to get rid of this problem, the developer decides to hire a prospective contractor who can build such a vessel.

Problem solved—except now there are the problems of delivering the vessel on time and financing it (which, by the way, at some point was as expensive as a 100-MW offshore wind farm). However, other projects are interested in hiring the vessel as well, so it is finally built. The project is now de-risked to an extent that all of the main project execution plan components are reasonably firm, and the IRR is now well over the magical 10 percent. The project can finally be planned and built!

This is, of course, only a very basic example, but nonetheless it does describe the development in the industry over the first ten years of commercial installation of offshore wind farms.

Components, such as higher feed-in tariffs and incentives, are also issues that can make a marginal project attractive over time. However, projects have been canceled for reasons such as poor ground conditions. So size and IRR are not the only showstoppers.

But let us get back to the project execution plan. The detailed version of the preceding statements is, of course, far more complicated than what we

36

just saw. Project execution is a process that can last several years, from the first components being manufactured in individual production plants to when the final components must be hooked up to the grid and taken over by the client.

Several main tasks have to be considered when developing the project execution plan. The main events that determine which phase the project is moving into or leaving have to be determined. The reason is that most of those events will trigger a set of milestones that normally constitute at least two important issues—namely, the handover of responsibility and payment for work rendered.

For any company working in the construction industry, this is very important since these events also mark the release of both warranty and bonds accordingly. It is also very important for the developer. Releasing warranties or bonds too early puts the developer at risk when errors, breakdowns, or damages occur.

So what constitutes a project execution plan? Here are the main items:

- The permitting process, where several milestones have to be achieved and passed; for example, the environmental impact statement, the geotechnical examination, the wind and metocean data gathering and processing, and the preparation of the park layout and main parameters.
- The financing of the wind farm. The power purchase agreement with the utility company—wherein the feed-in tariff and the agreement on number of terawatt hours to be purchased is determined—is included. In addition, of course, a financial closing date is determined in which the financial structure and agreement to invest the money into the wind farm is stated.
- The tender process for purchase of the components.
- The tender process for the BOP to be used for the installation.
- The start of production of the many components to be installed and the delivery date and place for all of them.
- The setup of the construction site.
- The actual transport of the components and installation of same.
- The hooking up and commissioning of the wind turbines.
- The handover to the client after the trial period of production.

Under each of these items, however, several subactivities are also carried out. The following sections describe the individual phases of the project.

START OF PRODUCTION

The start of production of the foundations, turbines, cables, substation—if applicable—and all other hard as well as soft goods and services that go into the offshore wind farm is initiated by the main project milestone: the contract being awarded to the suppliers of the preceding items.

Production can start as early as two or three years before installation of the final components and commissioning of the wind farm, and this will become more common in the future since offshore wind farms are becoming very large, with multimegawatt installation programs over several years in order to deliver a complete site. For example, the Forewind project on the Doggers Bank is 9 gigawatts. This project would have been impossible to install in only one or two seasons, simply because the size and number of units required are larger than the annual production and installation capability of the suppliers.

The start of production is the natural extension of contract negotiations and the award by the client and the possible suppliers. This is also a process that can last for an extended period of time due to the size and complexity of projects. The overall duration of project preparation, combined with the lengthy permitting process and environmental monitoring, as mentioned earlier, will therefore make the actual project take at least half a decade under normal circumstances.

But in the project execution plan, the first and most important items for this part of the process are the production dates from start via delivery to finish, the arrival of raw materials or components to be built into the final product, the duration of the fabrication processes, and the monitoring (HSE and QC of the actual production).

Documentation of these events also serve as fix points for all the processes to be carried out afterward when the components are delivered for transport and installation. Therefore, strict accuracy in determining when a raw material—say, steel—can be expected in the process for milling a monopile foundation is crucial to the project. This is because the manufacturing activity is already known down to the number of hours it takes to weld the plate into cans, which are then welded into tubes.

By knowing the process this thoroughly, the manufacturers of monopiles can give project managers an exact timetable for delivery Ex Works (i.e., a place of delivery and passing of risk) for a specific number of monopiles and provide details about the speed and size of transport vehicles and how many vehicles will be used to make transport efficient and to keep the production line running.

This summarizes a "just-in-time" (JIT) manufacturing principle, but the problem is that the capacity is fairly low. For example, a plant can only produce two or three foundations per week, so in this case the JIT method of manufacturing is not practical. Several factors must be taken into consideration when determining the production time and the delivery schedule.

So, for example, for a project where 80 turbines are going to be installed, the manufacturing time for the foundations would make it necessary to create a buffer of, say, 45 foundations. The reason is, of course, that the transport and installation can be done in one or two days, so the production will be too slow to follow the installation capacity of the BOP offshore. This can only be avoided by creating a significant buffer of components prior to starting the installation.

This actually creates another problem, which we will discuss later. The port facility often may not be able to accommodate 40 to 45 foundations at the same time. Thus, the port, the manufacturing sequence, the transport, and the installation time all affect the project and mostly contradict one another.

The Logistics Setup

The individual project owner chooses her own strategy for delivering the wind farm to the site and commissioning it. Therefore, a number of strategies can be developed to do so. The range in the past has been from DIY (do-it-yourself) to EPIC (engineer, procure, install, commission) delivery of the project. This depends on the internal skills and resources of the individual project owner. It is the same way for the logistical setup of a project. The project owner will look at internal capabilities and capacities and the track record of delivering such projects before deciding which strategy to use to install the project.

In general, the four variations are form of the contract, project options, DIY delivery, and EPIC contracting.

Form of the Contract

It is fairly obvious from Table 3.1 that a small organization with no track record should be avoided. The risk element is simply too large for the project to be financed and for the organization to carry out a cost-effective and

Table 3.1 Contract Strategy Matrix

Typography	No or limited track record	Long track record
Small organization	EPIC	Multicontracting
Large organization	Multicontracting/EPIC	DIY

commercially viable project. This type of project or wind farm owner is rare in the offshore industry simply because of the magnitude of the projects. However, small organizations with reasonably large competencies and track records in developing projects do exist.

There are some quite successful companies in the offshore industry today that are very capable of developing projects to either complete installation and sale, or that can develop projects to a stage where the commercial value has been established and the owner subsequently can sell off the project to a utility or a larger wind farm owner.

Project Options

The number of options that remain are the multicontracting, EPIC, or DIY of a large organization that has either a limited or comprehensive track record. The track record will thus determine what route is taken in terms of setting up the logistical solution for the project.

If the company is relatively inexperienced, it will be most advantageous to opt for either a multicontracting (develop packages of goods and services to be delivered in order to execute the project and then manage these packages) or to choose an EPIC contractor that can take care of all interfaces toward the individual contractors and consultants to the project.

Historically, this has been the case, and the results have largely been that the multicontracting route has proven to be the more cost-effective one. The reasons for this are, among others, that the EPIC contractors did not have very much experience in executing this specific type of project in the early years.

The engineer, procure, install, and commission projects have been very costly to the project owner and/or the suppliers who chose to go the EPC (engineer, procure, construct—not to be mistaken for the EPIC contract) route instead of multicontracting the work and supplies. Basically, if you do not know what the work is about in detail, you would normally apply safety margins on all aspects with which you are unfamiliar, thereby driving the price upward but without adding value for the money you are paid.

The fact that we were not very successful with the EPIC, or EPC, contracting strategies has led to the general consensus that the project owner should be large and experienced enough to be able to cope with this type of project from a DIY perspective. In effect, this means that the project owner will have to develop an organization that is capable of planning and executing a project from start to finish, in-house! This is now the case. Today all major utilities are developing in-house organizations that over time will be able to execute a large offshore wind farm project on time and on budget.

Do It Yourself

DIY in this case means handling the project from the decision to build through to the commissioning and operation of the wind farm. The design of foundations; the layout of the wind farm; the purchase and installation of foundations, turbines, cables, substations, and so on will all be done using in-house resources and competencies. The sensible third-party validation of designs and so forth will, of course, be necessary to keep a critical eye on the project and the product that results from these efforts.

This, however, requires a very comprehensive organization that has experience in this type of project execution, and such organizations are rare, although a number of European utilities now have designed and built a number of offshore wind farms. These utilities have thereby acquired an increasing level of competence, so the tendency to deliver projects on a DIY basis will become more common, thereby eliminating the market niche for individual and smaller consultancy companies that specialize in parts of the entire project delivery. This type of business will gradually be drawn into the project organization.

EPIC Contracting

But what, you might ask, went wrong when we tried to go down the engineer, procure, install, and commission route? The risk element of such a contracting strategy is very low. The EPIC contractor holds all responsibility until handover of the project when it has been successfully installed and commissioned. For the investor this is a dream. Here is a company with a balance sheet large enough to satisfy the bank's or financier's requirements, and if something is not working, the owner only has to say to the EPIC contractor, "Go fix it!" If the repairs are not made, the owner can claim damages from the contractor without having to establish who is at fault. It couldn't be any simpler—or could it?

Usually this is where it starts. The owner sees the EPIC contractor's responsibility and balance sheet as a guarantee that the project will get built. But you cannot build a wind farm offshore with money and insurance policies. You build it with skilled personnel who use the proper equipment. And without blaming anyone or pointing fingers, it has been my experience that all construction work—onshore or offshore buildings, ships, power plants, rockets, and so on—headed by an EPIC contractor often turn out to be a rehearsal. The contractor builds in a large risk margin and tenders for the lowest acceptable units that can be found at the lowest possible price for the work.

Further, it is my experience that the risk is handed down 1:1, if possible, to the subcontractors who have the specialist knowledge of, say, operating a jack-up. This is not necessarily the main competence of the EPIC contractor; he is the coordinator and the facilitator. This is acceptable for what it is worth, but in a very young and immature industry, this is a recipe for disaster.

The EPIC contractor probably does not know the risk of the project itself, certainly not in any great detail. This leads to even higher prices and greater risk aversion. The smaller specialty contractor will not have the firepower to fight back and will in the end have to accept a contract that is less than fair and far from optimal. So in this case, it will be everyone against everyone and only looking out for oneself. If one contractor can get an upside from the mistake or shortcoming of the other, this will be the result.

The EPIC contractor, who has limited specialist knowledge anyway, will not readily accept the blame because it would cost money, and usually the contract is prepared in a way that the EPIC contractor will have the opportunity to place blame and recover the loss before paying out. This has often been the case and is not really what the wind farm owner or developer wants in the end.

But is engineer, procure, install, and commission contracting bad? No, not as such. In a mature market, the EPIC route can avoid the financing conundrum just described. The contractor can take over the risk and quantify the cost of it and thereby make the project happen. I believe that in the near future the EPIC route will reoccur but in a slightly altered version.

In the new form, it will be seen as an EPIC contract for the transport, installation, and cold commissioning of the foundations, turbines, and cables. The reason is simply that these areas are within the competence of a number of large offshore contractors in the world. Contractors that have entered into the offshore wind sector during the past couple of years are used to this type of split in the scope of work.

Clearly, it is also evident that the marine package comes with a marine contractor and not with a project management company with a large balance sheet. The difference is that the risk is being handled by the people in the project who understand the risk and are thereby capable of quantifying it and pricing it right. Hopefully, this will be to the benefit of the project and, in the end, the end user—namely, the consumer who taps the environmentally friendly power out of the outlet in the living room.

TENDER AND CONTRACT STRATEGY

The decision on how to build a wind farm will also be reflected in the contracting strategy. Regardless of what the project and tendering strategy will be, it is important to make the decision at the very beginning. The decision to outsource all activities, or most of the activities, is reflected in the EPIC strategy and, as stated previously, this would be the preferred route chosen by the small or the inexperienced organization. The number of contracts will remain largely the same as within the DIY strategy, but the EPIC contractor will be managing them and all related interfaces. This is a big help to both the owner and its organization.

The tendering strategy will reflect the EPIC or EPC strategy in the same way that the owner of the wind farm project will seek out the various interesting general contractors in the market. They can effectively undertake the entire scope of work for the project, moving from planning, permission, and forward to the final handover.

The owner will then have to develop the entire work scope for the project, and this can be done using a fairly small but knowledgeable organization or even a consulting company that specializes in this type of work. Many companies consult on the development of the scope of work for an offshore wind farm, and they are capable of delivering the required level of documentation with the required detail for the EPIC and/or EPC contractors to effectively bid on the project.

The owner will use the data developed to carry out two tender processes:

1. A prequalification of EPIC, or EPC, contractors that are interesting and considered capable of delivering the scope of work.
2. An invitation to tender (ITT) for the jobs on which the owner wishes contractors to bid.

It is worth noting that the difference between EPIC and EPC also constitutes the perceived difference in the scope of work the owner wishes to give out. EPC—engineer, procure, and construct—is a limited scope of work compared to the previously mentioned EPIC—engineer, procure, install, and commission.

With EPC, the commissioning part is now covered by the owner. This in effect means that the final responsibility for the wind farm being operative now lies with the owner. The EPC contractor is now only responsible for the correct installation of the parts. Whether it works as originally intended is, strictly speaking, not the contractor's problem.

The EPIC contractor is not relieved of this responsibility because the company's job is also to commission and hand over an efficiently running plant to the client. This has to be considered when deciding the contracting route to take. It is also important to consider the commercial impact of the strategic tendering decision, since any obligation on the contractor's part will be met with a claim of payment. There is no such thing as a free issue in the contract.

Basically, the two forms of contract produce the same commercial result. The risk in the contract is handed over to the contractor by and large. The difference in the level of responsibility—whether the plant is functional after the handover—is not that significant. There will be a markup on the commissioning obligation, but the contractor will scrutinize the documentation thoroughly before pricing and entering into the contract. This can therefore actually be a benefit to the project and the owner in general.

But for the owner, the element of passing the risk on to the contractor is important. Within a small or inexperienced organization, it is highly likely that errors will be made, and they will no doubt be costly. Therefore, the cost of passing the risk onto the contractor is appealing, even though one knows there will be a premium to pay for the privilege. The cost is known and quantified, and for financing the project, this is important as well because financially de-risking the project will be crucial to lenders.

For many wind farm owners, passing risk is a rehearsal that is frequently carried out in the way that the risk should be taken by the contractor, but the cost associated should not be reflected back to the owner in the price. This is a no-win situation for the owner. Again, there are no free issues. If the contractor is to assume, say, all weather risks of offshore operations, the known amount of downtime is automatically priced in as the percentage of weather downtime that is to be expected. This is normal.

The owner is, however, often under the impression that the contractor must gamble on the unknown amount (i.e., the number of weather days above average). This can be significant in many places in the North Sea, for example, but it just is not possible. The contractor either possesses the equipment or rents it from a sub supplier who will charge for every single day the equipment is working. End of story.

Now, the client may think it has rid itself of the risk, but in effect it is simply priced in as an added margin on cost of plant (COP). This is logical. I often explain this to clients using the following example:

> If you can persuade my bank to only charge payment from me when the equipment is working, I will only charge you in the same way.

This case is equivalent to the situation where the farmer only gets paid for slaughtering the pig if I actually eat the roast pig afterward. In other words, I must deliver my part of the contract without necessarily being paid for my services if conditions prevent it.

The problem is, of course, that this is not feasible; therefore the concept of getting rid of both risk and payment isn't either. In the end, the client always pays for the full set of services; it is only a question of how the bill looks. If the installation contractor is asked to assume all the risks, the price will reflect that accordingly. This makes sense since the notion of working without payment is not really that popular.

QUALITY ASSURANCE AND QUALITY CONTROL REQUIREMENTS

Project delivery will be subjected to a set of requirements in order to document the proper quality of all products and services. The two issues can be briefly defined as follows:

> **Quality assurance:** The process of verifying or determining whether products or services meet or exceed customer expectations. Quality assurance is a process-driven approach with specific steps to help define and attain goals. This process considers design, development, production, and service.
>
> **Quality control:** The process employed to ensure a certain level of quality in a product or service. It may include whatever actions a business deems necessary to provide for the control and verification of certain characteristics of a product or service. The basic goal of quality control is to ensure that the products, services, or processes provided meet specific requirements and are dependable, satisfactory, and fiscally sound.[1]

Therefore, quality assurance (QA) is monitoring whether the customer's needs and requirements are met when the design is planned and taken to the production stage, and quality control (QC) is the process whereby the product is examined after production, during transport, and after being installed in its final position.

---

[1] Courtesy of wisegeek.com.

Thus, quality assurance is carried out mainly in the design stage as a sanity check on the product that is being designed, whereas quality control is the physical act of checking the product when it leaves the production line or, after it is installed, checking the product before payment is rendered or to facilitate the payment.

The preceding definitions will apply throughout this book. The QA/QC tasks in an offshore wind farm are executed according to those definitions to determine the quality of the product. In other words, quality assurance and quality control are performed when the wind farm is completed, and the work of the contractors, designers, and employees is approved.

HUMAN RESOURCES FOR INSTALLATIONS

As stated earlier, the organization behind the planning and execution of an offshore wind farm is crucial to the success of the endeavor. If no experienced personnel are in the top level of the organization, as well as some of the most important profiles in both planning and execution, the chances of success will diminish quickly.

The offshore wind installation company needs to have a complete organization to function efficiently. But the industry is, in general, very young, and it is therefore a major problem to attract skilled personnel, educate new staff, and retain the human capital once the organization is functional.

This is a serious topic since the demand far outweighs the supply. Plus, the tasks are both onerous and complex, which further puts pressure on an offshore wind installation company. So how does its organization look? There are many ways to develop and run a company, and the development part is the key to success.

When I started in this business 12 years ago, there were no human resource professionals who had any experience with the installation of large offshore wind farms commercially. The task of developing the company, A2SEA A/S, was therefore unknown to me, and the goal was to become operational in one year.

The following anecdote doesn't describe the way to develop an organization, but it paved the way for an understanding of what is required to start, develop, and run a competent management organization that can install offshore wind farms. This means whether you are involved on the supplier side or as the EPIC contractor that will take on the entire scope of work of engineering, procuring, installing, and commissioning the offshore wind farm.

●●●

Creation of A2SEA

The start of A2SEA was a coincidence. A good idea was in desperate need of money and human resources to solve the many problems and challenges only a start-up company will have. To me it was no different. When I began, I was the managing director, the project management, the design department, the business development manager, and the funding department all in one. This is, of course, not ideal, but it is the reality of any new company. So the first task was to define what the main focus should be in order to answer the many questions asked by all of the stakeholders around the company.

We initially counted three and a half people: myself as CEO; a project manager who was in charge of all practical issues relating to the technical development of the installation vessel we designed; a marketing manager who was in charge of everything from booking meetings, accounting, and developing the marketing material for the company; and finally, a retired naval architect who was our consultant and technically competent person to deal with the specific issues about design of the vessel.

Of course, this was only a skeleton crew, but the interesting point is that the organization—regardless of how small it is—will always deliver what is required to carry out business. In the A2SEA case it was no different. But what about competence?

Well, there wasn't any. So we defined what was required and determined along the way what was needed to deal with a specific issue. This was, of course, exciting but not ideal. It is obvious that a small organization like this cannot delegate a whole lot of assignments, and this was not the case either. But since the delegation was small and the number of people even smaller, it quickly became clear that sharing information would be key. If we did not communicate all the time, information was lost or forgotten, or not where it was supposed to be when we needed it.

Therefore, it taught every one of us that sharing information makes everyone smarter, more agile, and more capable of seeing the potential and the dangers, even if it was not the actual responsibility of the individual who discovered the problem. So, information shared is most certainly information gained. This should be the basic principle of any organization.

Furthermore, whoever picked up the phone or opened the mail was the owner of the problem or challenge or task, whichever suited the situation best. But this corresponds well with the preceding statement that all information should be shared. The fact that we knew at least a bit about what the others were doing made it simpler for us to act and be effective when someone was handed a task.

Finally, during the process of converting the installation vessels and putting up the first turbines in 2002, we learned through the teething problems of the vessels—and there were a few—that panic should not live in our neighborhood. No matter what happens, you should never rush to solve a problem, especially if you are working offshore.

You *must* assess the problem, evaluate the options, and choose the safest and the most effective solution. And you must do this the first time. If you choose to do something now because you are then being seen as doing something, this may come back and bite you very quickly.

A small organization is not in itself desirable, but it is very effective. When we installed the first project, 12 of us were doing all the work. A look at the organization of any offshore contractor today will reveal a much larger number of members. The job of installing wind farms has not gotten more complicated, but the details and documentation behind it have. This is the main reason why an organization like the one we started out with would probably not be sufficient now.

This work did lay the foundation, however, for what is considered the standard organizational structure in a company working in the offshore wind industry. Logically, the organizations and tasks have become larger, more complex, and more in-depth than what we could deliver in the early years. In essence, however, they remain the same.

●●●
_____

Related Images

Project planning starts at the desk figuring out who, what, where, and when.

Consumables and various components being loaded in port before departing for the offshore site.

Equipment that is not self-propelled is slower and more cumbersome to move in and out. Although the day rate seems appealingly low, the added cost of tugs and so on will drive the total cost up. When the extra downtime due to towing restrictions is added, the cost per turbine or foundation installed will rise.

Load over of the turbines from the transport vessel to the jack up. This work is, of course, an extra step in the installation process and is only carried out due to the lack of larger vessels that can transport more units at the same time.

CHAPTER FOUR

The Basic Organization

The basic organization is not very different from so many others. The top management layer must take care of coordinating and delegating tasks to the lower layers. Figure 4.1 shows the basic setup that I recommend. Details are discussed in the following sections.

As can be seen in the chart, the tasks to be handled by the top management are indeed many and complex, but they are all necessary to run an efficient installation company. The main difference to a "normal" organization is that the quality assurance (QA) and health, safety, and environmental (HSE) department (or departments) are line and not staff functions. This is not the case in an organization working within the onshore environment. Here, the QA and HSE departments advise and consult on their technical resources. Offshore, the departments are line functions and have the capacity to overrule, instruct, direct, and stop work if they deem it necessary.

Therefore, the HSE department in particular will have an extremely important role to play. This specific topic is discussed in Chapter 8, but for now it should be noted that the main task—besides installing the wind farm offshore—is to get everyone who goes offshore safely back on shore again after carrying out the installation work.

Let us go through some of the departments one by one to understand how they contribute to the organization and the installation project. It should be noted that the organization chart in Figure 4.1 is pretty much the same whether you are working as the installation contractor or the supplier of turbines, foundations, or other parts of the "hardware solution" for the project.

The head of offshore or the CEO of the company primarily controls the functions of the various departments, coordinates the information stream on the top level, and makes all the final decisions that will be communicated to the participants outside the company and to the shareholders. This is the normal routine. The role of the CEO is to be the final decision maker and to inform the outside world what the company has decided to deliver and how they will deliver it. As for the internal organization, the CEO manages and coordinates all functions, controls and forecasts economic decisions, and manages the human resources department, whose personnel will ultimately deliver the product or services the company is providing.

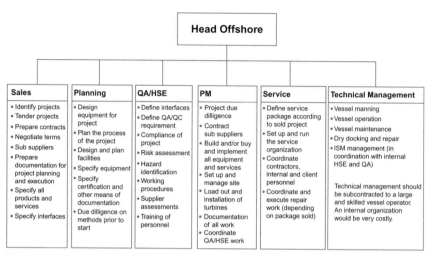

Figure 4.1 Organization chart and the responsibilities of the top management.

SALES

The sales department is, of course, responsible for identifying projects and/or service or repair opportunities that will accommodate the capabilities of the company. The sales executive therefore must focus on the areas in the organization chart. Let's go through some of them.

Identify Projects

It is logical for any company that wants to sell goods or services to the offshore wind sector to know what the market is. However, the good sales representative will immediately carry out a screening of which projects will suit the capabilities of the company and the equipment and services it can provide.

Furthermore, it is important to consider the capacity of the company. What can be provided and in what time frame? Does the project fit the operational availability of the company's equipment? If so, does the equipment fit project requirements?

Tender Projects

Tendering of projects is, of course, a very important part of the activities in the sales department. The tender process is long, complex, and costly, so the screening of the market will allow the sales department to concentrate on the projects that are of primary interest. Hopefully, tendering the more marginal projects can be avoided, since it's very disappointing to not be awarded a project after investing a lot of money and resources in it. The last item is possibly the most dissatisfying of all. Because a tender is a sweepstakes where the winner takes it all, coming in a handsome second is as frustrating as trying to eat soup with a fork.

The disappointment of not winning several projects will be demoralizing for the employees as well, and if this is occurs many times, good employees will probably try to find employment elsewhere. This must be avoided since competent personnel are rare in this business.

It is also vitally important that the tender documents include a draft contract. The sales department must scrutinize this document to identify the goods, services, and obligations that must be delivered to the client by the project management after the contract is awarded.

Prepare Contracts

Preparing the contract is actually part of the tendering process. However, it should be noted that during the tender phase, the draft contract should only be read. If you are participating in several tenders, the cost of employing lawyers to review contracts can become exorbitant. This is another reason to screen the market for projects in which you wish to participate. Let the legal department screen the contract before submitting the bid, and plan to do the contract reading when you know you have gotten past the first threshold in the tender process.

Reading the contract, however, is only part of the process. The project department will require the sales department to identify all the problems, challenges, and specific agreements that are unique to the contract you wish to enter. This must be done as part of the tender negotiations, and as we said before, this is crucial for a seamless takeover by the project department after the contract is awarded. If this is not done, the damage can be substantial when things don't go as predicted. In large, complex projects, such as installing an offshore wind farm, things often don't go as predicted.

Negotiate Terms

When a tender is successfully carried out, the next item is to negotiate the terms that are part of the draft contract. The terms—if ever I saw a draft contract for a tender—are owner-oriented, and often one will have to negotiate from a position nailed to the wall. This is annoying but nothing new. The problem is that starting negotiations on that basis means the process takes a lot longer and makes it more aggressive. But since the owner normally needs nonrecourse financing from the bank, he or she will therefore draft the contract so it looks appealing to financing institutions and investors.

The main problems are normally the warranties, the penalties, and the bonds to be delivered. The warranties are always onerous because the owner wants a wind farm without any risks involved, again due to the financing situation. So a warranty of five years is almost always imposed on all suppliers. So if you are only delivering a service, such as installing turbines, make sure the warranty is not part of the contract.

If you are in engineering and/or are producing goods for the project, and possibly installing them afterward, a five-year warranty is normal. But make sure that the dates of commencement are as early in the project as possible.

Hire Sub Suppliers

Handling the sub suppliers is also important. The contract will most likely rely on them to perform a variety of services. Therefore, at an early stage, the sales department should make sure that the sub suppliers are all accounted for; the binding offers have been given; and that offers are valid, clean, and back to back with the main proposal up until contract signing and project execution.

Prepare Documentation for Project Planning and Execution

As we said before, the project management will require complete documentation of the project, their obligations to the owner and the sub suppliers, and their rights and duties during the project.

Specify All Products and Services

The contract handover must clarify all points, products, and services. A specification of every one of the products and services is also included.

Specify Interfaces

This topic is dealt with in detail later, but to secure the full delivery of the goods and/or services owed to the owner, it is important to determine which interfaces are relevant to the existing contractor on the project. The reason for this is twofold: (1) to secure the full delivery of goods and services without missing or omitting your responsibilities and (2) to make sure that no more responsibility is taken on than what is specified in the contract.

This is extremely important. Imagine you turn over work that has flaws without mentioning them, and as fate would have it, the defects are detected after handing over to the next contractor. The takeover documents will show that the flaws appeared when you were in charge of, or responsible for, the components. Thus, remedying the problem becomes your responsibility, even though you didn't actually impose the defect or flaw.

Therefore, the identifying interfaces and allocating responsibility, as well as the documentation requirements, must be developed and implemented in the contract documents. The interface definition is discussed later in this book.

PLANNING

The planning department carries out all of the preproject engineering, determines the equipment that is necessary in order to carry out the work connected to the contract, and creates a large amount of the technical documentation that is required by authorities, owners, and suppliers and sub suppliers preceding contractors and the contractor to which the work scope is handed over after delivery by the company. In the following we discuss a few of the most important tasks in this section to give an understanding of what the planning department should do first and foremost.

Design Equipment for Project

It seems logical that the individual project will have unique technical challenges and therefore require solutions that are also unique to the project. It could be anything ranging from seafastening of the components on the contractor's vessel to providing method statements for performing the work or developing the logistical solutions for this particular project.

Here it should be noted that creating the method statements for the individual work processes is a team effort where the QA and, in particular, the HSE departments participate. This is the first event that shows how important it is to have an HSE department as part of the line organization. The development of method statements can only be done in cooperation with the HSE department, simply because this is the only department in the company that has the capability of collecting and interpreting the various national laws and compliance documents relating to health and safety. Therefore, their input is crucial to the process of preparing the project.

Plan the Process of Project

The planning department is in charge of developing a project process chart, which takes all of the relevant factors, such as weather, equipment used, distance to the site, number of components, and the contractual time schedule, into consideration. The project plan is the most important guiding document for the project management, so significant effort is put into this preparation work. A flawed plan will lead to delays, and delays lead to huge cost overruns, since the balance of plant (BOP) is extremely high for an offshore project.

Design and Plan Facilities and Specify Equipment

We will discuss the site layout later in this book, but for now it is enough to state that the correct allocation of facilities also lies within the planning department. It is very important to specify equipment as early as possible; determine what type, size, and operational envelope is required for all equipment.

Specify Certification and Other Means of Documentation

The equipment and methods developed by the planning department must be documented and certified by the relevant certification bureaus and authorities. This includes, but is not limited to, items such as the following:

- Method statement for the entire work scope
- Plan implementation and certification by local authorities in cooperation with the HSE department
- QA plan implementation and certification by local authorities in cooperation with component suppliers, the owner, and sub suppliers, if required
- Certification of all equipment
- Preparation of transport plans for review by the authorities
- Defining and coordinating the process with the other project contractors and sub suppliers for approval by owner's engineer

Perform Due Diligence on Methods Prior to Start

The methods, equipment, and services developed by the contractor prior to starting the offshore work must be subjected to a trial run first in order to detect any problems or flaws in the methods. The costing factor is a good indicator of why this is important.

When a flaw is detected during the planning or designing stage—say, in a drawing—the cost factor is 1:1. But if the flaw is detected while constructing, say, the seafastening after the drawing has been approved, the cost factor is around 5:1. The big problem arises, of course, if the flaw is not detected until you go offshore. If this is the case, the cost factor will be 10:1. So if the cost of the flaw is 10,000 euros in the design stage, the cost of not detecting it until offshore work has started will be 100,000 euros. This is just a rule of thumb.

The cost can easily be higher if, for example, the vessel can only transport 6 instead of 8 turbines in an 80-turbine project. The cost for the extra trips can be developed like this:

- 8 turbines per trip equals 10 trips
- 6 turbines per trip equals 14 trips—the last trip with only 2 turbines

If the cost per trip is 24 hours each way (the net installation time will be the same, but the weather will worsen and thereby further increase the cost), the 4 extra trips will add 8 net days to the project. If the transport cost is 150,000 euros per day, the cost is now 960,000 euros.

So the cost factor will be very high indeed, especially because the sales department could possibly have negotiated the extra trip into the original price for the work. This is a significant issue. Cost of equipment is the number one factor in an offshore project. The careful planning of the operations and processes required to install the wind farm offshore are therefore played out again and again during the planning, tendering, and contracting processes to get the cost and interfacing just right.

The worst thing that can happen in an offshore project is waiting for materials (Figure 4.2). Waiting on weather windows is a given that we have to work around, but enjoying the sun and having a chat while no foundation is in sight are not good for the project economy. Moreover, the weather window will be lost, and the possibility that another unexpected one will show up is not very likely.

QA AND HSE

Quality assurance (QA) tasks should be familiar to everyone, so for now we will only briefly describe what is involved. The QA role is mainly to ensure that the agreed-on standard of work is delivered both to the company from previous contractors and passed on to the next contractor in the project

Figure 4.2 Waiting for the next component to arrive at the pile guide.

without any damages, defects, or flaws in the product or the service. To ensure this, a set of handover documents must have been developed and constantly updated and fine-tuned in order to secure the perfect interface management. We discuss this more later.

The health, safety, and environmental (HSE) department has a significant role to play in the planning and executing phases of the project. The HSE officers and their crew will be safeguarding the entire process from the starting point until successful delivery of the entire scope of work. Some of their key functions are as follows.

Define Interfaces and Requirements

The interfaces are not exclusively a QA/QC matter, they are also a question of when the next contractor takes over responsibility for the employees on the project. To secure this, the HSE documentation thoroughly describes what is required in order to deliver the work—from a safety point of view—when the work starts and at what stage it ends. In this way, the employee is aware of what he or she must do, what triggers a task, and what ends it. Furthermore, any deviation from the job description must be recorded, and changes or corrections must be made to avoid repeating an unplanned event.

Compliance of Project

The project must comply with a large number of rules and regulations that don't always correspond to one another. The rules could be maritime, construction regulations, port working statutes, and so forth. Often, these rules and regulations are international in some cases and national in others. This can cause some confusion, and therefore the HSE department should develop a set of site-specific regulations that comply with the building permit and national, international, onshore, and offshore regulations. This is not an easy job.

Rules may actually clash, and what seems perfectly sensible in some countries or cases is considered madness in others. A good example is using two cranes to lift a load. In some countries this is considered unacceptable, whereas in others, it is done all the time. Imagine, however, if you go to a country to install a wind farm and part of your loading schedule includes upending of the turbine towers before loading them on the installation vessel.

Even though the towers may not be too heavy for the crane to lift, the upending possibly will require the combined lift of a crane on the pier and one on the installation vessel, as shown in Figure 4.3. This is actually a standard operation in Denmark and other places, but it was not particularly

Figure 4.3 Upending a tower section using a shore crane and the vessel crane in Esbjerg, 2002.

preferred in the United Kingdom because there is an inherent danger of things going wrong if the crane operators and banksmen don't have the exact figures on the loads they are lifting, the proper crane load curves, and a good understanding of the physics of the lift itself.

Risk Assessment

Oddly enough, this leads us directly into risk assessment. How bad is the example of using two cranes, actually? For the professional crane operator and banksman or rigger, this is a run-of-the-mill operation. But for the HSE department to carry out root-cause analysis of what can go wrong, it is the start of a long statement of possible accidents. This is, of course, not the case, provided that everyone adheres to the method statement developed for this specific purpose. If properly managed, this is a safe, fast, and efficient way to conduct an operation that occurs every day in the industry.

Logically, this is only the case because it is actually possible to make a risk assessment that addresses the possible dangers of the operation. Solutions are presented and detailed, and finally it is established that everyone involved in the process is properly trained and understands the requirements in the method statement.

If an unskilled, untrained person develops his or her own makeshift solutions to the situation, it can quickly develop into a potential catastrophe, even for seemingly simple tasks such as shown in Figure 4.4. The potential dangers of this procedure are very apparent. See how many you can think of. List them, and then suggest countermeasures or correct procedures for carrying out the task.

Hazard Identification

The documentation of possible dangers connected to an operation is called a hazard identification document, and it should list all of the perceived hazardous operations, products, materials, waste, and actions that will take place or result from the work. When listed in a readable format in a paradigmatic way, the reader will be able to understand what the dangers are and how to mitigate them before starting the work. This is a very important task of the HSE department, and it should be carried out offensively rather than defensively and only impose more restrictions that are actually hindrances. (Unfortunately, the latter seems to be the norm these days.)

Figure 4.4 Even standing on an aluminum ladder in the water, this workman doesn't seem concerned.

Working Procedures

Working procedures are debated elsewhere in this book, but in general they describe the specific assignment that has to be carried out in detail, considering the legal and safety issues related to the assignment. This must all be written in a straightforward way that a worker on deck can understand.

Supplier Assessments

Supplier assessments are also important. While you may have a very safe, cost-effective, and streamlined production process, this is useless if your suppliers or sub suppliers do not perform to the same standard. Unfortunately, the lowest common denominator will be the guiding one in this case, regardless of whether that is how you want it or not. You will be judged by the total performance of the team you and your suppliers put together to deliver the scope of work. Therefore, it is necessary to assess suppliers and be sure that the client will do so as well.

Training of Personnel

This is fairly obvious, but it is still important to discuss it. Training your personnel will reduce the number of accidents and near misses. This is also a significant task for the HSE department. The personnel will provide all of the employees in the project with the necessary training, be it offshore survival, height rescues from the turbine by means of hoisting down injured persons (or themselves), helicopter rescues, and so on.

The list of skills required to work offshore on wind turbine installation vessels is long and ever growing. This is healthy because offshore you rely on the person next to you if an accident happens. And if a fire breaks out on a ship, it is everyone to the rescue! There is nowhere you can go, so if you cannot be an asset and assist in putting out the fire, you quickly become a liability to the rest of the crew. Therefore, everyone who is responsible for a specific job on the vessels and turbines has specific training requirements. That way everyone can be an asset in the relevant situation and avoid being a liability when there is no current requirement for the skills possessed by the individual.

PROJECT MANAGEMENT

Project management (PM) is, of course, the company engine room and bridge control during an installation project. The project manager or managers have full responsibility for the project and therefore also need to have full knowledge of the scope of work (SOW) to be performed. Thus, it is necessary to include the project management once the tendering process becomes serious and documentation and resources have to be developed to finally win the contract.

So when it comes to the most important tasks of the PM department prior to and during project execution, you can get a good idea of what is necessary, who should know certain details about the project, and how the project is controlled as you perform the SOW. The following sections discuss the tasks listed in Figure 4.1 in more detail, focusing on the major issues.

Project Due Diligence

We touched on this subject earlier in this chapter, but the previous due diligence discussion concerned methods. So what type of due diligence must the PM department carry out?

Well, as we already said, the project manager must be able to deliver the SOW. But is the service that the salespeople sold also the service that can be provided in real life? And if not, what are the changes, alteration add-ins, and omissions that need to be done to get to the goal line? What is the time consideration and the cost implication of changing what was sold? And does the project manager buy in to the project at all?

These are some of the issues the project management, and the project manager in particular, must consider and clarify. The project manager has to have the practical knowledge, equipment awareness, and ability to assess whether the proposed methods, machinery, materials, and manpower will do the job.

If she cannot do the job using her skills and knowledge of the processes, she will have to back out of the project. This is why it is important to consider involving her in the process once we expect to win the project. She can deliver the final intelligence that is needed to get the SOW right and also to get the price tuned properly and, finally, to win the confidence of the client. But as we said, before starting the project for real, the project manager will do a complete sanity check of the job, normally during the handover from the sales and planning department to the PM department.

Contract Sub Suppliers

The sub suppliers have, of course, been contacted by the sales and planning departments in order to receive a binding offer for their products and services. But after winning the project, the project manager must secure the availability of the goods and services, plan them into the project in detail to get the sequence right, and establish the project parameters and communicate them to the sub suppliers in the form and order necessary to handle the project both in the practical sense and the documentation of the SOW as required by the customer.

Build and/or Buy and Implement All Equipment and Services

The planning department, along with the sales team and the customer, will have specified a number of items to be produced specifically for the project in order to deliver the wind farm offshore. One item could be seafastening for the turbine components. This item is specific to the turbine and to the ship on which it goes—at least until today.

The PM department inherits the design and methodology that has been developed and must now finalize it, tender it, and construct and install it on the installation vessel. Since the project manager works in the real world, she is also faced with the burden of trying to save money while doing this. It may well have taken half to a full year to tender and win the project, and in the meantime prices and delivery times may very well have changed dramatically. This is a problem the project manager will face throughout the entire project. But in the end, this will be her job, and this is why a project manager is the specialist in bringing all the details together and turning it into a fully installed wind farm.

Set Up and Manage Site

How difficult can this be, you may be thinking? It's on the water, isn't it? Well, there are a number of things to consider in this respect. First of all, the site is not one but two or more locations. Odd, yes, but you have to consider that we do work onshore in a staging port, where we prepare all of the components for loading and transporting to the site, and offshore, where of course we install the foundations and turbines.

Now this should be straightforward, but the components we install are very large in volume, very heavy in weight, and certainly very delicate to handle because of their surface treatment of either gelcoat or paint. This makes the issue of setting up the site quite challenging.

A staging port with 30 or 40 foundations and/or turbines is a staggering sight. The sheer volume of the components is impressive, and apart from their size we have to move them from the transport vehicles to the storage area, from the storage area to the preassembly location—which is normally on the pier where they are then loaded—and from the preassembly area on to the installation vessel to sail them to the site offshore.

A going figure in the industry has been for a staging port area of the size of around 60,000 to 70,000 m^2 where it is possible to store the components coming into the port, preassemble a full load of turbines, and then load them on the installation vessel at the same time. This number is just a guide, and nothing can be taken as the fixed value.

A number of issues will influence the choice and the size of the staging port:

- Layout of the area; if it is very irregular in shape, you need more space.
- Pier layout; whether you can easily access the vessels.
- Road or sea access; if you need to sail components in from another country, you must consider the traffic of the vessels going in and out.

(Transporters deliver components and installation vessels go out to the site with preassembled units. They should not obstruct one another.)

- Paving of the port; whether you need to pave the area with blacktop or concrete in order to store turbine components.
- Load-bearing capacity; whether you can move the heavy equipment—and components—on the site.

These are just some of the issues to consider; so, as you can see, the right port location and access and usability of the offshore site are much more complicated than it may seem to the spectator.

And this is just onshore. Offshore the question is whether or not you can access the site with the equipment you have chosen, whether you should stay out there with the installation crews or go in and out from port, and which traffic constraints are posed by the many vessels out there. Vessel traffic control, which is dealt with later in this book, gives one an insight into this particular problem. This will be a daily struggle for the project manager who cannot actually see what is going on.

Load Out and Installation of Turbines

The project manager has been handed an installation concept by the sales and planning departments that at best is workable. Of course, they know what the job requires, but they are not as deep into the details as the project manager is, and if she changes the setup—even slightly—the implications on the installation program will be significant.

Imagine the project manager realizes that the vessel can only hold 5 instead of 6 turbines. For 80 turbines this means an extra 2 trips with the installation vessel (16×5 vs. 14×6). This alone is a cost of at least 500,000 euros in vessel expense alone. Or, even worse, she realizes that the best port available can only accommodate 4 turbines per load out instead of the 6 that were offered to the customer. Now it is 20 trips instead of 14, and the cost has just risen another 2 million euros.

The tradeoff can, of course, be that the chosen port will give some upsides that another better-designed port would not. So in all fairness, this part of the job needs much care and attention from the project manager.

Documentation of All Work

In the past, we just did the work, and no one really cared about how we got to the finish line. Those days are over. Today, particularly in offshore wind projects, there are so many interfaces and such high costs involved that the proper delivery in all links of the chain must be documented. Normally the last 5 percent of the project costs are withheld by the customer to secure the necessary documentation that must be presented to the certification body and the authorities to get the wind farm released for production.

Coordinate QA/HSE Work

Even though the project manager does not physically do the documentation himself, he must ensure that the QA and HSE departments, along with the crews he manages, document all of their actions and all their handovers in order to relieve himself and the installation company from any liabilities and to get paid.

The project manager will be a very busy person during the execution of the project. He must be skilled in the management discipline, but he must also be able to think on his feet when unforeseen events occur, good as well as bad. A good event, such as unforeseen excellent weather periods offshore, presents an opportunity. But if it is the weekend and he must call in extra staff or speed up the delivery of components to the preassembly site because he might run out, the project manager must be able to tackle this situation quickly and efficiently.

A bad event is, of course, if things break down or if accidents or incidents happen. He is ultimately responsible and must be able to take action, get to the right people and authorities night and day, and keep a calm head. This is a rare combination, and successful project managers are few and far between, especially since the offshore wind industry is relatively new with a limited track record. We deal with project preparation in Chapter 5 and project execution in Chapter 6.

SERVICE DEPARTMENT

A service department will be needed in order to manage all of the work that the turbine and, to some extent, the foundation supplier will have to carry out after commissioning and delivery of the wind farm to the client. For the first five years—normally—the turbine supplier will warrant the turbines, the availability, and the service and maintenance of the wind farm. Onshore this is usually no problem. Offshore, however, it can become more difficult, even though in general the principle is the same.

It is far more costly to send an employee to the turbine offshore because the transport boat is subject to strict rules and regulations. Also, the cost of running a vessel that is capable of sailing out to the wind farm, which is normally quite some distance away from the shore, is extremely high. This problem is discussed in more detail in Chapter 12, but for now it is important to know that an entire organization will be dedicated to this particular job.

Finally, it should be noted that the trend for offshore service—and commissioning for that matter—is increasingly carried out with the personnel available on site in "hotel vessels" to cut down on transport time to and from the individual turbines.

TECHNICAL MANAGEMENT

As mentioned before, it is preferable for technical management to be outsourced to a skilled and experienced vessel management contractor, unless your company has a large fleet of installation and/or service vessels. The reason is that the operational organization to run a vessel, regardless of type, will be very costly, and since all the skills are needed whether you have one or ten vessels in your fleet, the cost of adding this part to your organization will be extremely high. However, there are also those who claim that installation vessels, in particular, are so complex that no ordinary fleet management company can run them satisfactorily.

This is partially true, and therefore this part of the operation of the vessels should be kept in house. But the manning, purchasing of goods and services, dry docking, and so on will not be specialized to a degree where a normal management company couldn't sort it out. This is why technical management should be divided into two areas.

One area should cover project-related vessel services such as preparation for projects, demobilizing, maintenance programs, upgrades, crew training, and so on. In short, you could say that it is everything that has to do with specific tasks concerning the work that the vessel does.

Another area of technical management is related to services such as manning the vessel in compliance with regulatory frameworks; supplying fuel, lubricants, and provisions; arranging ports outside of the project area (inside they are normally determined by the client); dry docking; and transporting spare parts and other supplies to the vessel. This is something for which it is difficult to create added value for the individual ship owner with one to five ships in a portfolio.

Since the costs are extremely high for operating such vessels, pooling them in large clusters will create a possibility for discounts and rebates during

the operation phase. For this purpose, a number of operators, such as V.Ships
Ship Management in Germany, Navios Maritime Holdings in Greece, ASP
in Australia, and others, specialize in work like this. There are many other
suppliers of this type of operation; these three were the first a quick search on
the Internet revealed. What they have in common is that they specialize in

Figure 4.5 Loading of transition pieces in port.

operating merchant ships for clients all over the world, pooling solutions on everything related to ship operation on behalf of the owner.

For the specialized vessels, such as jack-ups, the number of management companies is significantly smaller, but still this service can be provided by several companies worldwide. Figure 4.5 shows the loading over of transition pieces in port prior to moving the jack-up barge to the offshore site. Even though the feeder system is not the best solution from a logistics perspective, the method works. In a situation where there is not enough equipment available, this becomes a viable way of carrying out the operation.

●●●

Related Images

The last blade has been lifted out of the transport frame and is on its way to the final position on the turbine.

A fixed gangway from the installation vessel to the turbine is crucial to keep a high level of safety during construction. The gangway can also serve as support for power, air, and hydraulic hoses if needed.

Fitting the hammer with anvil on the pile prior to driving. The hammer alone weighs more than 200 tons, so heavy equipment moving on the vessel is a daily procedure. However, when work becomes routine, the risk of accidents rises due to the crew's reduced level of awareness of danger.

Working with the pile fitted into the pile gripper. The size of the components and the equipment is impressive.

CHAPTER FIVE

Project Preparation

Project preparation is crucial to the seamless execution of both onshore and offshore work, so it is imperative to plan and describe all of the activities that will be performed during the project. Furthermore, the health, safety, and environmental (HSE) requirements and the quality assurance (QA) and quality control (QC) documentation must be developed in detail in order to follow, monitor, and correct behavior, activities, and deliverables during project execution.

Therefore, thorough planning of the operations and the identification of interfaces and responsibilities are extremely important. The installation of wind farm components offshore happens so fast that often changes will be implemented at a stage that will prove to be too late, simply because the time to implement a new method takes longer, due to design and approval, than the installation process for a foundation or a turbine.

DEFINE PROJECT PARAMETERS

Once the project has been handed over to you from the sales department—much as just described—the project manager and her team will have to define the outline of products and services that are to be delivered, the boundaries of the project, the legal framework, and the number and kind of activities to be performed to carry out the work that has been sold to the client.

The project parameters are mainly defined by the size, location, and layout of the wind farm itself; the building permit that has been awarded to the client; the legal framework in the country where the wind farm is to be located; the location onshore from where the project will be staged; and, finally, the type of turbines, foundations, and the equipment that will be installed or used onsite during the project.

Once the parameters are defined, the project manager can start building up his team and basically choose between two types of organizations to do this. The first type is the matrix organization, where the project manager "borrows" resources from the various departments in the various phases of handling projects, as shown in Figure 5.1. This means that the same resource manages more work but not as efficiently.

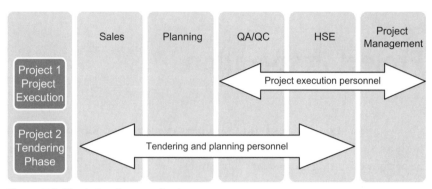

Figure 5.1 Matrix for the organization.

This method works well in a small organization that has several projects in various stages running at the same time. However, this also creates some confusion in the team because different people will assume the responsibilities for different functions throughout the duration of the project, and team members will have to work on several projects at the same time.

The second option is the project team basis (Figure 5.2), where a firm team with a number of fully allocated members will work closely throughout the entire project. This is very good for the project because the team

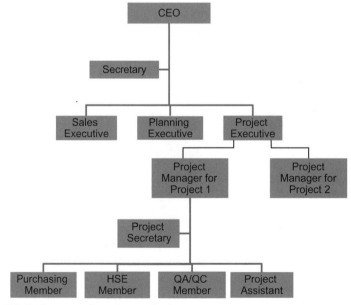

Figure 5.2 Project team-based organization.

members are generally attached to the project and will not work on others during installation of the specific project. The downside is that it requires a larger organization, and since personnel are scarce in this industry, team members will be forced to take on other assignments or, worse, will transfer among job offers to get more lucrative positions.

Finally, it should be noted that entering and exiting such a team can be very difficult. Many believe that team-based project management (PM) is significantly better since the team members know one another and the strengths and weaknesses of their colleagues. This will lead to a team integration that over time will prove superior to the matrix organization. In Figure 5.2, the organization is taken out of the main organization as a team, and the project manager for Project 2 has her own organization buildup.

CONTRACTING PRODUCTS AND SERVICES

As we said before, project planning and preparation also include contracting of the products and services that the sales and planning departments have included in their tender and project specifications. The proper way to secure continuity is to have the sales department make sure that all of the offers from sub suppliers—whatever the type of goods or services have been sought— will be valid through the tender period and through any possible extensions of validity after the company has been chosen as the final supplier of the project deliverables. The sub suppliers should still be able to deliver according to their original offers to the company.

In this way, PM will be able to finalize negotiations with the sub suppliers and can then secure what was offered to the client as the solution chosen for project execution as originally intended. This is very much like a normal tender process, where the scope of supply for the sub supplier was detailed in a formalized tender document in order for the sub supplier to fill out the scope of supply (SOP) and thereafter for PM to be able to compare and choose the proper supplier for the work.

For this work, a planning and engineering resource will always be necessary, since this department originally developed the SOP and the proper solution for the SOP, as well as the chosen sub suppliers from which the company requested a bid for the work. Table 5.1 shows what the process would look like.

As shown in the table, the sales and project management departments interact in the final phases of the bidding and contracting process. When

Table 5.1 Planning and Contracting Process

Phase/ Stakeholder	Tender	Project Takeover	PM Negotiations	Outcome
Sales	Produce and submit bid	After awarded, hand over to PM		Succesfully finalize tender and contract the project
Project management	Assist in final stage	Participate in takeover from sales	Negotiate sub supplier bids	Take over and execute project
Sub supplier	Submit offer to bid	Hold and/ or validate offer	Bid according to PM tender and negotiate contract	Enter into contract as sub supplier and deliver SOP

the contract is finalized, the sales department exits and thereafter PM is responsible for the process and the project moving forward. In this way, the project manager can work out his own schedule based on the master planning that the sales and planning departments have made and possibly negotiate better terms with the sub suppliers before starting the work. In that way, the project manager will be able to get some financial and/or time savings, since he can focus on the detailed parts of the project with every single sub supplier and negotiate in more detail before entering into a contract with the sub supplier.

DEVELOPING PROJECT HSE PLANS AND PROCEDURES

The HSE department will have made a plan for the HSE work specific to the project. However, this plan is not as detailed as it will have to be during project execution. For the installation contractor, it will be a matter of demonstrating that the HSE system is there, in working order, and suitable for project execution. However, the building permit is very onerous in terms of the specific project requirements, and the "general" statements that have been made during the sales process will have to be developed into project-specific plans, procedures, and working instructions. This will have to be done by the PM in cooperation with the HSE department.

First and foremost, the HSE plan will have to be detailed in such a manner that the responsible authorities can evaluate whether the company is capable of working in the offshore wind industry without risking lives and limbs of the workers. Then a set of certificates, ranging from vessels certificates on trading, safety on board, environmental policies (e.g., waste management, oil spills, antipollution certification, and method statements of the vessel(s) that will be used) must be demonstrated.

The building permit will have stated the client's requirements at all stages of the operation. These requirements will also be the set of HSE boundaries within which the project manager will have to work. Is this feasible? Can he operate the vessels, crews, and components in an order where he can safely and effectively finish the project on time?

The HSE plan and method statements, such as a safety handbook for the project, waste management plans, and so forth, will be developed in order to respect the rules and regulations, as well as the specific requirements in the building permit. This can be a reasonably complicated process, where things such as bird migration and marine mammal breeding areas come into play. These are also part of the environmental boundaries under which the project must be built. They will have to be coordinated with the weather windows that make working offshore safe and the time schedule that has been sold to the client.

If this is not the case, then money will be lost, one way or the other—as a result of time wasted because of inclement weather or other limitations or due to cost overruns from unforeseen or badly handled incidents that occur. The bottom line is that the buck stops at the project manager; this is why she has to be very experienced, sometimes cold-blooded, and multiskilled in order to see the problems and challenges of the project as it is spread out in detail.

DEVELOPING QA/QC PLANS AND PROCEDURES

The QA/QC plans and procedures will have been part of the tender, if for no other reason than because the single component passes through a large number of hands before it is fitted on its final position offshore on the site. The client will always want to ensure that any damage to a component is addressed immediately and the responsibility is made clear and claims issued to the responsible party.

As with the HSE plan, the QA/QC plans have to be developed into a sensible set of interface documents that clearly define what is checked, what the handover state of the component was, who checked it, how the component was handled, and so on. Chapter 6 discusses in detail the what, who, where, and when of each operation, so we just touch on the subject here. However, it is a very important, if not the most important, part of the work done by the QA/QC department since this work will directly affect the bottom line in the project.

If the QA/QC methodology and interface management are not detailed properly, it will lead to more cost and aggravation since the client will require that any damage be covered. Even though a construction all risk (CAR) insurance policy is always taken out for the entire project, the insurer will have regress toward any contractor or supplier who delivers defective work or products. For this particular reason, QA/QC documentation is carried out.

DETERMINING METHODS AND REQUIRED EQUIPMENT

Surely the sales department has been in touch with the project management to determine what equipment is necessary to carry out the various parts of the work in order to deliver the scope of work (SOW) for the project. And surely the sales department has been informed of the resources and capabilities of these. But putting it all—the rented resources and equipment from the sub suppliers and the same for in house—into one contract and making it work together is more difficult and requires detailed planning. This is the project preparation that the project manager will have to do to be able to fulfill the contract SOW.

First, a set of guidelines under which the work must be carried out needs to be prepared. In this instance, the guidelines set out the main operations: which vessels are actually being used, how many components will be transported, how the components will be reassembled and stowed onboard, and how they will be installed offshore once the installation vessel arrives onsite.

Of course, this work is much more detailed, but once you have agreed on the principle, you have a solid base to work from. And just to be clear: What the sales department has developed for the bid may not necessarily be the solution that is chosen in the end. On more than one project we sold one solution, and then during the project planning, PM changed it because there was a better time schedule to be made or time to be saved if we went for an alternative solution. But once the main guidelines, or parameters, for the project have been established, the next two items can be determined: the balance of plant (BOP) and the final project time schedule.

DEFINING THE BALANCE OF PLANT

As we mentioned before, the BOP will be determined during the tendering and contract negotiation process. But the changes in the project execution planning as they were stated when the methods and required equipment were determined will be included in the final BOP. This is because the project manager ultimately has the responsibility for the transport and installation of the components, and therefore she must decide which equipment is to be used and how.

So in order to be responsible for the process of installing the components offshore, the project manager needs to ultimately decide, along with the client, who may very well be a component supplier. Normally, this would be the case if the company is installing foundations, turbines, cables, or other items on the wind farm offshore.

The BOP is huge for an infrastructure project like an offshore wind farm. The number of vessels transporting to and from ports; installing offshore; guarding the offshore site; transporting installations; commissioning crew, marine surveyors, client representatives, and spares; working on the preassembly site onshore; monitoring traffic; coordinating vessels, trucks, and cranes; and managing people is impressive. On a site today, between 500 and 2000 people would likely be working for a shorter or longer period. It is the job of the project manager to keep track of everyone, the project's progress, and the safety and quality of all the work going on.

The BOP consists of three parts:

- *Hardware,* such as vessels (installation, crew, hotel, guard, tugs, survey, cable, dredging, surveying), cranes, trucks, areal platforms, forklifts, and so on, and all the consumables connected with operating them
- *Infrastructure,* such as port(s), access routes, offshore site, preassembly site, and offices
- *Human resources*—the personnel who fit into all the various listed units mentioned in the two first items

So even if the sales and planning departments have done a lot of work, the remaining part—detailed engineering in this case—will have to be carried out by the project manager prior to commencing the onsite work. This part of the work will be very time and resource consuming.

The preparations for an offshore wind installation project will start at least one year before. Usually the vessels will be contracted first, since they are key to the installation process. If they are not secured immediately, the project may be jeopardized. The key piece of equipment is always the installation vessel, mainly because it has the capacity to do so but also because the time schedule is related to this particular piece of equipment more than anything else. So if this fails to materialize, the project will most likely not finish on time.

The remaining BOP will also have to be contracted—or employed at the earliest possible stage after the contract is awarded. Human resources will have to develop all of the method statements, work order descriptions, interfaces, safety procedures, and so on in order to receive a permit to carry out the work once mobilized on site.

The balance of plant is, of course, more than just hardware. Personnel and planning are the key factors to a successful installation program, and the careful planning of processes will determine whether the project finishes on time or will suffer delays.

CREATING THE FINAL PROJECT TIME SCHEDULE

The second important part is the results from the decisions on methods and BOP—the project time schedule. Once the BOP and method of installation have been chosen, the project manager will look at the installation times for all parts of the project relating to the process from preassembly to final onsite cold installation. The hot installation or commissioning is always carried out by personnel working for the turbine supplier since that supplier will have to warranty the turbines for the first five years of production.

So, for this purpose, the project manager has recruited a number of personnel who will either work on the installation vessel from cold commissioning to delivery or from hot commissioning and also to delivery to the wind farm owner (Figure 5.3). For the project manager, in this case only, installation until cold commissioning will be considered. There are four items that, in general, determine the overall installation time schedule.

The Choice of Equipment

What will the vessel carry? How fast can it go? What is its jacking and installation capacity? We discuss the specifics of vessel characteristics in Chapter 12, but for now these questions are what we would consider.

Figure 5.3 Loading Siemens 2.3-MW turbines on *Sea Power*.

The Turbine Type

In a sense, each type has a specific preferred method of preassembly and in-stallation. If the turbine can only be transported in a large number of parts, the installation time will take longer, and therefore this will have a great deal of influence on the timetable for the following reasons.

Installation Time

Normally the number of components to be installed offshore should be as low as possible. So, the optimum is to prepare as much as possible onshore before loading up the vessel and going out to the site.

Necessary Weather Window

The installation will need appropriate weather conditions and the amount of possible downtime should be considered. The longer it takes to assemble the turbine, the higher the installation vessel operational parameters have to be. If the weather window needed is around 48 hours to install the tur-bine, but the vessel will require the waves to be less than 1.2 m Hs, the amount of downtime on an exposed site will increase because the longest statistical weather window of this Hs is only 36 hours. This means that the vessel will have fewer good weather periods available to install the turbine. There are therefore only two options: increase the length of the installation period or contract an installation vessel with a higher weather criterion for operation.

Onsite Metocean Conditions

As we said before, metocean data onsite will influence the program and may result in downtime offshore. If the combination of waves, wind—which of course is the main reason why the wind farm is installed in the particular location—currents, and tides is unfavorable to the installation spread, the installation program will take longer. So the installation vessels must be suitable to the conditions onsite, or the program of installation will suffer due to the lack of suitable weather windows. The only way to resolve this problem will be to hire more rugged vessels; but, they may not be available. If they are, they will cost significantly more.

Distance and Navigation Time

The distance to the staging port and navigation time are challenges. The vessel loading and transit capacity will result in a set number of trips to and from the site with components. As an example, a vessel that can carry 5 turbines will have to travel to the site a minimum of 16 times to install 80 turbines. This means navigating the route between the port and the site at least 32 times. If the distance is 50 nautical miles (NM) and the speed of the vessel is 6 knots, the time consumed will be $32 \times 50/6 = 267$ hours, or 11.1 days. If the speed is 10 knots, however, the time spent is only 160 hours or 6.66 days. This will be reflected in higher costs, but more significantly, the weather window must be somewhat longer if the vessel is slower because it will take longer to reach the site.

For example, 50 NM can be transited in 8 hours and 20 minutes with the slower vessel, and in 5 hours with the faster. This means that the weather window must be at least 6 hours and 40 minutes longer (the vessel has to go back again). Therefore, the distance to the staging port is a significant parameter.

Figure 5.4 shows the principal setup of a detailed time schedule. Once the loading, transporting, and installation sequence of one load of turbines—in this case, three turbines per trip—has been developed, it is possible to create the overall installation time schedule by repeating the sequence, adding the appropriate weather contingency to the plan as you go through the project.

7	Installation of turbine #1-3	7.54 days	Sat 04.09.10	Sat 11.09.10	
8	Jacking and preload harbor	1 hr	Sat 04.09.10	Sat 04.09.10	5
9	Loading of 3 x 2 tower sections, 2 nacelles, 2	18 hrs	Sat 04.09.10	Sat 04.09.10	8
10	Preparation for sea transport	2 hrs	Sat 04.09.10	Sat 04.09.10	9
11	Jack down	0.5 hrs	Sat 04.09.10	Sat 04.09.10	10
12	Sea transport Nyborg - Baltic I (112 sm)	15 hrs	Sat 04.09.10	Sun 05.09.10	11
13	Positioning	1 hr	Sun 05.09.10	Sun 05.09.10	12
14	Jacking and preload	1 hr	Sun 05.09.10	Sun 05.09.10	13
15	Preparation for lifting WTG 1	1 hr	Sun 05.09.10	Sun 05.09.10	14
16	Lifting and installation of tower section(s)	5 hrs	Sun 05.09.10	Sun 05.09.10	15
17	Lifting and installation of nacelle	5 hrs	Sun 05.09.10	Mon 06.09.10	16
18	Lifting and installation of rotor	4 hrs	Mon 06.09.10	Mon 06.09.10	17
19	Jack down	0.5 hrs	Mon 06.09.10	Mon 06.09.10	18
20	Repositioning	2 hrs	Mon 06.09.10	Mon 06.09.10	19
21	Jacking and preload	1 hr	Mon 06.09.10	Mon 06.09.10	20
22	Preparation for lifting WTG 2	1 hr	Mon 06.09.10	Mon 06.09.10	21
23	Lifting and installation of tower section(s)	5 hrs	Mon 06.09.10	Mon 06.09.10	22
24	Lifting and installation of nacelle	5 hrs	Mon 06.09.10	Mon 06.09.10	23
25	Lifting and installation of rotor	4 hrs	Mon 06.09.10	Tue 07.09.10	24
26	Jack down	0.5 hrs	Tue 07.09.10	Tue 07.09.10	25
27	Repositioning	2 hrs	Tue 07.09.10	Tue 07.09.10	26
28	Jacking and preload	1 hr	Tue 07.09.10	Tue 07.09.10	27
29	Preparation for lifting WTG 3	1 hr	Tue 07.09.10	Tue 07.09.10	28
30	Lifting and installation of tower section(s)	5 hrs	Tue 07.09.10	Tue 07.09.10	29
31	Lifting and installation of nacelle	5 hrs	Tue 07.09.10	Tue 07.09.10	30
32	Lifting and installation of rotor	4 hrs	Tue 07.09.10	Tue 07.09.10	31
33	Jack down	0.5 hrs	Tue 07.09.10	Tue 07.09.10	32
34	Return to harbor (112 sm)	15 hrs	Tue 07.09.10	Wed 08.09.10	33
35	Contingency	3.12 days	Wed 08.09.10	Sat 11.09.10	
36	Weather (35%)	2.38 days	Wed 08.09.10	Fri 10.09.10	34
37	Cable contigency (7.5%)	0.36 days	Fri 10.09.10	Sat 11.09.10	36
38	Technical (6%)	0.28 days	Sat 11.09.10	Sat 11.09.10	37
39	Turbine supplier contigency	2.5 hrs	Sat 11.09.10	Sat 11.09.10	38

Figure 5.4 Example of a time schedule.

In this way, a very reliable time schedule can be developed, and the program should be solid and able to be carried out in the allocated time.

So when balancing out the various pieces of equipment and resource allocation, the project manager and his team are facing a giant puzzle where all of the parts are variable. If one part is changed, it will make an impact on the other parts. Such changes are not necessarily intended but are a consequence of changing parameters. This exercise creates some interesting problems that must be dealt with.

Auditing the Contract Suppliers

Once all of the contracts with suppliers and sub suppliers have been finalized, the project can commence. In order to start up the project, an audit of all hardware, software, and human resource must take place. This means that the project manager will have to physically inspect all of the suppliers and their equipment prior to starting the project. This is normally done 6 months to 2 weeks before the contract in question becomes effective.

For the installation vessels, in particular, an "on-hire survey" is carried out. This survey is ideally a combination of certificate control and an inspection of the vessel to see whether it is fit for the purpose, in good shape, and no outstanding repair or maintenance is needed. This is important because you do not want to stop in the middle of the project in order to repair defective equipment. Surely the vessel will go "off hire," but you have the entire spread of equipment, personnel, infrastructure, and components coming into the staging port and it will start piling up.

This is very costly indeed, and even though the repair may be only a few thousand euros, the downtime cost is more than a million per day. So remember to carry out a thorough inspection of the equipment you rent because if the shortcomings or damages were not noted in the on-hire survey, it will be a part of a claim when the project is finished and the vessel is "off-hire" surveyed.

The same principle applies for all matters in the project, but the 80/20 rule applies, and 80 percent of your problems will arise from 20 percent or less of the plant, starting with the installation vessels, should they not function properly. We discuss on- and off-hire surveys in more detail in the following section.

IMPLEMENTING PLANS AND PROCEDURES FOR SUPPLIERS AND CONTRACTORS

The contracted supplier, contractors, and subcontractors will have to be made aware of the plans and procedures of the project that they are about to enter into. Even though all of the information that the wind farm owner has gathered in the form of permits, data, and so forth is made available, the end result will be a contracted scope of work. Here, the individual contractor and supplier will have to adjust to a final project framework where some of the contractor's requests and ideas will be accepted. In general the individual company has to adapt to the overall program that the project management puts together.

The final program is made by the project manager at the project's leading company. As an example, the project could be multicontracted, with a number of individual companies delivering the completed work in order to install the wind farm; the program will be determined by the owner.

If, instead, the wind farm owner decides to go down the EPIC route of contracting, the EPIC contractor will define the program and plan everything according to the main schedule of the owner. Usually this is determined at the time of the tender offer or agreed to at the project award date to the EPIC contractor. He will then coordinate all of the activities moving forward and certainly make all of his subcontractors and suppliers aware of what their duties will be during the project, including the program and the plans; that is, HSE, QA/QC, and other important items to complete the work in a safe and cost-effective manner.

PREPARING ON- AND OFFSHORE CONSTRUCTION SITES

Preparation begins with contacting the owners of the onshore site; generally it will be the port where the components are unloaded and the staging port for the project, provided they are not the same. The offers made for the project bid during the tender process must be made firm and the area must be contracted with a set of deliveries that will satisfy the wind farm owner, the authorities, and the contractor who rents the area—onshore.

Preparing a construction site is normally a straightforward job: measuring out the area, making a plan for activity areas, dedicating the various purposes of work, pointing out storage areas, defining site roads and access routes, and so on. Until now this is the same for a staging port for an offshore wind farm project. However, there are some issues that make the work here different from other construction sites, including the following.

Security

The staging area and the port of unloading are indeed PORTS! Since 2001, the security levels in ports all over the world have increased dramatically, and this is somewhat problematic to deal with for the project manager in the staging or unloading port because she must fence in the entire area and account for all personnel who are working in the port and on the project. With possibly 500 people or more, this is not an easy task. People enter and leave, but not always through the front gate.

As discussed before, ports are subjected to the International Standard for Port Security (ISPS) code, and every person entering or leaving the port—or a vessel in the port—must be accounted for with personal details, company references, and the type of business to be carried out. So consider the following.

A person enters the front gate of the staging port to go onboard the installation vessel to go to the site offshore and carry out work. After finishing the work offshore, she must return to the port, but the installation vessel stays out there to finish other small jobs in order to be able to hand over a section of the work. The person in question, however, is offered the chance to go to the port in a crew transfer boat, which she does. But the crew transfer port is not the staging port. (Typically the staging port is a large port in the area where the wind farm is installed, but the crew transfer port is the closest possible small port where a 20-m crew boat can dock.)

So now one person left the port that she entered from the shore but leaves at another port, possibly to enter the staging port the next morning to work. This means she entered twice without leaving, so the record of the port security will state that she exited, but there will be no record of entry because she entered via the staging port.

To satisfy the requirements of the authorities, this creates a massive necessity for control. It is now dealt with through a very broadly designed software tagging system that records the whereabouts of all personnel using chip identity cards and monitoring points at every entry and exit possibility. But this is an interesting procedure, where there are many opportunities for mistakes.

Ground Preparation

Once the port—whether staging or unloading—is chosen, the suitability of the area must be documented and established. In some ports the ground-bearing capacity is not sufficient to handle the axle loads of the trucks and trailers moving the components around. So for that matter, it may be the closest port to the project, but it requires attention to its infrastructure, such as bearing capacity. This is normally dealt with by repaving the surface, a contract cost that can be considerable.

Piers and Waterfronts

The pier where the unloading and preassembly of the components takes place must be capable of carrying the components and the machinery handling it. This is normally also dealt with by finding a port with sufficient bearing capacity or upgrading the port of choice. Occasionally, a port is not capable of delivering either option. Previously, at least one project had last-minute changes due to the lack of pier bearing capacity. Instead of using the installation vessel crane to load the turbines, two very large onshore cranes had to be contracted to load the turbine components onto the installation vessel, as can be seen in Figure 5.5.

Figure 5.5 Crawler cranes offloading equipment to the transport vessel. The installation vessel is waiting on the right side of the photo.

The reason was that the pier could not support the weight of the trucks—not even the truck that bunkered the installation vessel, and this truck was street legal. So the careful selection of port and a detailed description of the required infrastructure capabilities should not be neglected. Notice the distance to the two onshore crawler cranes offloading the transport vessel. The pier looks very nice, but it could not carry the weight of the components. The cost of the loading operation in this case went into extremely high figures.

The Seabed in the Port

This is an important consideration because the installation vessel will normally jack up (if fitted with legs) in port to load components onboard. In this way, the loading goes a lot faster because the vessel will only have to worry about stability after completing the loading procedure. If afloat, the vessel will constantly have to move water ballast around in order to compensate the shifting of loads. Ballasting takes much longer and is very complicated.

The jack-up can determine where all the load has gone before jacking down and thereby determine the loading conditions and the related stability. But a port is not the ideal place to jack up. Normally, the sheet piling that forms the steel barrier to keep the ground in place is only driven far enough into the seabed to hold the pier in place. But the jack-up will make holes in the ground where the legs and their spudcans penetrate as a result of the high pressures.

In the beginning, this may not affect the sheet piling, but when the legs are retrieved, there is no ground in front of the base of the sheet piling, at least not where the legs penetrated the ground. This means that the sheet piling will move forward at the bottom and the entire area of the pier will sink. It might not sink immediately, but when heavy loads are placed on top of the area where the sheet piling was undermined, it will sink.

This is exactly the case for the staging port. Heavy loads will repeatedly be placed in the same area, and if not addressed, this is an accident waiting to happen. So normally two precautions are taken:

1. The installation vessel moves away from the pier before jacking in order to create a minimum distance to the sheet piling to avoid undermining it.
2. A stone cushion is dumped on the seabed to create a hard, firm base to jack the vessel on.

So in short, onshore preparation at a minimum should address the fencing, security, paving, preservation, and stone dumping issues prior to starting work, but the main issue is still missing: creating the work layout for the staging area to receive, store, preassemble, and load the components. This is done by designing a site plan as shown in Figure 5.6. This plan is for unloading the wind turbines for a small project. But even if the number of turbines is not significant in future terms, the staging area is no less impressive.

Preparing the offshore site is in some ways a bit easier. The main issue is, of course, that it is situated on the water, so fencing in will mainly be placing navigational buoys so as to surround the exclusivity zone, which is normally the construction site circumference plus an additional 500 m outside this. In addition, notice must be given to mariners that something unusual is going on with navigational restrictions that they will have to observe.

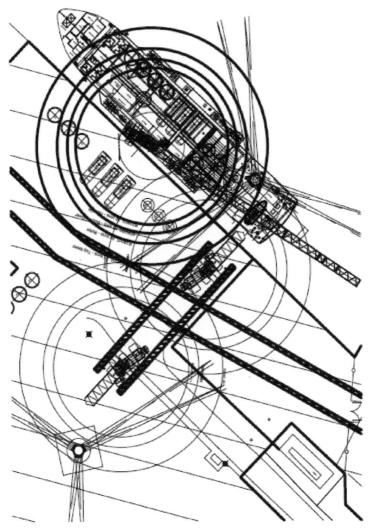

Figure 5.6 The layout of a loading pier for an offshore project. Notice how crane tracks and positions of preassembly areas are laid out to facilitate fast and easy access to the installation vessel. *Courtesy of EnBW, Baltic 1.*

●●●
Related Images

Lifting the third blade for a 3-MW Vestas nacelle vertically in order to insert it into the hub.

Exchanging a gearbox as part of a repair campaign in an offshore wind farm.

Lifting a 3-MW Vestas nacelle during installation.

Lifting components, such as bolts and consumables, onboard the M/V *Resolution*.

CHAPTER SIX

Project Execution

With all the preparations discussed in Chapter 5 and time to the starting point getting short, project management (PM) must make sure that the activities can and will start smoothly, as planned. A number of items must take place at the beginning, and we have already touched on a couple of key items: auditing of suppliers and subcontractors and on-hire surveys. These activities will take place prior to the onsite startup in a pyramid type of organization.

AUDITING

The wind farm owner will audit the main suppliers with whom she is directly contracting. They in turn will audit and survey their contractors and so on. This is illustrated in Figure 6.1. On-hire surveying of all suppliers is carried out both on their premises, if they are supplying materials and/or components, and when arriving onsite to deliver components or services. This is required because the supplier or contractor can be presented best in his own factory yard.

When arriving onsite, however, the practices and equipment delivered may be less impressive or simply not suitable for the project. Thus, the owner audits the main contractor to determine whether the company actually delivers what was agreed on in the contract and that the delivered equipment, components, and resources are actually fit for the purpose.

Necessary Documentation

The performance of an audit is extremely important for the installation vessel. Therefore, the owner or main client will access the vessel with a team of marine experts in order to control the following concerns.

Certificates

These can include trading permits, dry docking, last special surveys, class records, MARPOL—the antipollution plan, SOPEP—the anti-oil spill plan, and so on. They should be complete and valid for the entire project period if

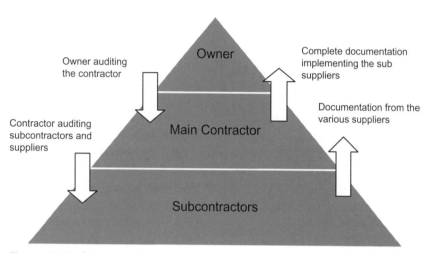

Figure 6.1 Auditing procedures.

at all possible. If not, you may have to take the vessel out of service for a period in order to update these certificates, and this would create a delay for the project.

Operation Manual

This is the general operation manual for the vessel in order to operate as the type of craft it is hired for. For example, the jack–up vessel must have a manual stating how it is to be operated in all aspects of the offshore work it can be subjected to.

Project-Specific Method Statement

This is the document that describes the specific working methods of the project it is hired to carry out. It is necessary to have all approach routes for the individual turbine locations planned, the anchor patterns if they are required, jacking operations (the specific soil data for the site will determine how the vessel is to be jacked up), how long the preload times are, and what the thresholds for canceling the operations should be.

Further, you must describe the lifting operations; the installation procedures that the vessel will undertake, such as lifting loads, seafastening loads, and sailing with the components on deck; the maximum weather criterion the vessel can manage with the components on board; and so on. This method statement should be developed well in advance of the project start and handed over to the authorities by the client or the wind farm owner, whoever is the charterer, for approval and returned to the vessel where the officers and supervisors should familiarize themselves with its contents.

Health, Safety, and Environmental Plan

This is of paramount importance. If the vessel does not operate with a valid and approved HSE plan, it will not be permitted to work offshore at the project. As said before, the client and the wind farm owner will develop a project-specific HSE plan based on the building permit that sets out the criterion for the plan and work onboard the vessel. The HSE plan must be fully implemented, and the crew must be familiar with it.

Working Instructions

The crew on deck must know exactly what type of work they will have to perform within the project. This should be explained in a simple and straightforward manner in the working instructions. They must be present at all work stations and easily navigated by the crew in order to quickly and effectively understand their duties in a specific working situation. Special attention is always given to hot work and work in confined spaces, where there is an increased risk of accidents (Figure 6.2).

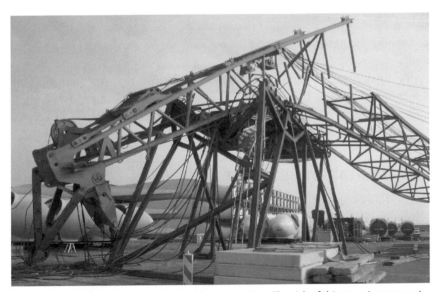

Figure 6.2 Accidents can happen anywhere, anytime. The risk of things going wrong is, of course, reduced when using detailed working instructions. *Courtesy of P. Pedersen.*

General Condition Surveys

This is a view of the general condition of the vessel. It should be functional, clean, and tidy; have all the agreed-on equipment; and all dents in the hull, paint damages, and so forth should be documented. If this is not done, the vessel owner could claim that damage happened during the charter. If it is not recorded prior to delivery, it will be difficult to prove when any damage occurred, and the vessel owner could claim compensation.

Fuel and Lube Gauging

It is very important to measure the contents of fuel and lube oil onboard the vessel at the on-hire survey. The fuel tanks could easily contain 200 or 300 tons of oil and at a cost of 700 to 900 euros per ton, fuel cost is a significant item. So if the tanks are not gauged on arrival, there is a risk that the charterer will pay for more than what is used.

Special Equipment

The vessel has a crane, jacking equipment, and seafastening onboard, which carry a very high dollar value. This equipment is crucial for the operation of the vessel within the project and must be scrutinized carefully for defects and/or damages. Furthermore, the seafastening is almost always paid for by the client and therefore should be in perfect working order. It should be noted that the marine warranty surveyor (MWS) will be present during the on-hire survey, and some of the main areas of expertise and interest are the special equipment, the certificates, and the general condition of the vessel.

Normally, the on-hire survey is done within a two-day period. This seems short, but the number of tasks that must be performed are planned well in advance, and the focus is on the areas where you would usually—as the experienced surveyor—expect to find problems.

PROJECT STARTUP SEQUENCE

Once the on-hire survey is properly carried out and the punch list (i.e., points identified in the on-hire survey that have to be rectified prior to project start) is worked through, the project can commence. It is logical that the on-hire survey is applicable for all contractors and their scope of work, so the fact that one contractor—in this case, the installation contractor—is ready to start working does not in itself mean that the project can begin. First, when all of the relevant contractors (relevant to the work scope—in this case, installation of components) are cleared and ready to start working, the project can commence.

The commencement of the project must follow a proper sequence, which is given by the overall time schedule for the work. If this is the installation of the foundations, the time schedule will start when all of the equipment relevant to the scope are surveyed and approved to start.

The first work to be carried out is the loading of the initial components—in this case, the foundations. The first complete load of foundations must be ready on the pier for unloading in order to avoid waiting time for the installation vessel. Therefore, onshore preparations will have to run ahead of the survey of the offshore equipment, because if they are not, you will experience waiting time for the first load. If this happens, the onsite installation contractor can make a claim for waiting time immediately; this should, of course, be avoided.

Shore-Based Preparations and Progress

To get started without waiting time after the mobilization and on-hire survey of the installation equipment, the first load of foundations—or turbines, for that matter—must be prepared on the pier beforehand. The agreed-on number of components will be placed on the pier in a setup that allows the crane onboard the installation vessel to lift them without having to move around too much or to relocate the vessel.

Care and attention should be taken to make the operation run smoothly so that the installation vessel can load, seafasten, and exit the port in the shortest possible time. As we saw in the detailed time schedule in Figure 5.4, the time for loading is already preset in the plan. It is therefore crucial that the installation vessel does not have to wait for components to be ready. If this is the case, the contractor will immediately make a claim for delays.

This is important because the contractor will surely encounter bad weather conditions at the offshore site at some point. And if there are waiting times in port for loading or preparing components that have to be loaded up, two things will happen:

1. The contractor will offset the waiting time incurred in port against the downtime offshore.
2. The contractor will claim that usable weather windows that were missed due to delays in loading the vessel were the cause of the delay and therefore not part of any claim for delays the client or wind farm owner has against the supplier.

In fact, the claim for loss of usable weather windows is very likely, particularly if the project is carried out in marginal weather conditions, which is often the case.

Logically, the cost involved in such cases is high, and therefore the minute calculation of downtime for delays and/or weather is carefully noted. Chapter 13 is devoted to the port operation, the layout, and the work to be carried out during the project.

Offshore Site Preparations and Progress

Prior to commencement of any work, the offshore site must be prepared. The area should be clearly designated with navigational aids such as buoys and other markings. A notice to mariners must be issued, stating that the area is a construction site for a period of A to B and that it is closed off to commercial as well as leisure crafts and vessels. This is all part of what the authorities set out in the building permit.

The requirements are quite onerous and specific. This is because most of the areas in which wind farms are installed offshore are close to onshore areas with reasonably dense populations, and marine traffic will also be heading for these areas. One example could be Hamburg, Germany, where the Elbe navigational channel ends in the southern North Sea, and the majority of the German wind farms offshore are in the area of very busy shipping lanes in and out of northern Europe. If you start putting up a wind farm without notifying the mariners in that area, you will quickly get into trouble with unplanned activity from some large container ships.

In the case of, say, German waters, the national or regional maritime authorities—Wasser Schiffarts Amt (WSA) or Wasser Schiffarts Direktion (WSD)—have to be notified, and all your activities will have to take place according to the guidelines contained in the permit. This makes sense because they are the main authority for seagoing traffic in and out of Germany.

The next task you will be required to do, or have done, is a seabed survey to determine the issues that follow, not only for operating the equipment during the installation but also, as described in Chapter 11 for foundations. The seabed survey is, of course, carried out first to give designers a set of parameters to determine the right type and size of foundations. But in the project execution phase—and prior to that, the tendering phase—seabed surveys will give the installation contractor, regardless of which type of job is being carried out, the necessary information to develop plans and procedures.

So to work offshore on the installation site, you must have access to and evaluate the survey data described in the following sections.

Seabed Scan

Normally, this is a side scan to a depth of around 5 m, which will give you, as the contractor, details about the top layers of sediment. Furthermore, it will tell you whether there are any hidden objects in the top layers that a jack-up or cable plough could hit during the installation. This is very important because the contractor might hit an old wreck or, even worse, unexploded munitions when either plowing the cables through a wreck or jacking a leg down on top of a bomb previously dropped from a warplane. So, in general, a seabed scan is required in order for the insurance company to let the contractor work on the site.

Cone Penetration Test

The cone penetration test (CPT) is where a pilot hole is drilled into the seabed and a special cone is then driven through the hole into the ground. The cone is calibrated against various layers of sediment; this exercise can tell you more about the density and composition of the various layers in the seabed. This will give you a very good idea of whether the vessel can stand on the ground and jack up or, if it penetrates the seabed, to what extent it will do so.

Core Drillings

These are actual core samples drilled out of the seabed and taken to a laboratory for examination. This type of testing will give the contractor valid data for the design of the foundation and also an exact description of the various soil layers where he will be working.

The preceding three types of seabed investigations or surveys are the most common, so it should also be noted that each has a different cost. The seabed (side) scan is the least expensive, but for most cases, this is valid for installation contractors, since they work in the upper layers if they are firm.

The CPT is more expensive because it generally requires more time and a jack-up to carry out the test. However, the results are far more informative, and combined with the side scan sonar, they usually form the major part of the design basis of the project. Also, combined with the core drilling, which is work-intense and costly, the side scan sonar and CPT should be all you need.

For the procedures of carrying out the three different types of investigations, it should be noted that normally the pre-tender documentation is made up of a comprehensive side scan report of the complete site; a number of CPTs done at locations scattered over the entire site; and, based on this, a small number of core drillings where the CPTs have shown major changes in the seabed boundary layers.

For the post-tender and actual offshore work, the contractor must have access to a core drilling for every location being worked—that is, for every foundation position on the site. If he does not have that, the contractor may risk a punch-through of the seabed because he is working from inaccurate data.

The punch-through is when the seabed gives under the pressure of a jacking leg on the jack-up barge or the jack-up vessel (Figure 6.3). This is

Figure 6.3 Punch-through of a jack-up barge. In this case nobody was hurt, and the jack-up was recovered later. *Courtesy of P. Pedersen.*

very serious and the worst-case scenario the offshore contractor can be faced with. For that reason mainly, the offshore contractor needs to have as much and as detailed information about the seabed as possible.

Therefore, with the properly marked construction site in place, and with a set of approved jacking and access plans, you can start accessing it with installation and other vessels required to install the wind farm. For project management, however, it is important to constantly monitor and update what is going on at the offshore site.

MONITORING THE ACTIVITIES

When work begins offshore, progress must be recorded at every stage. This is true for a number of reasons, of course, but it is mainly in order for the various contractors to be paid for their work. For PM, and ultimately the wind farm owner, stages need to be known to start the wind farm in sequence; to inspect the work done to secure proper delivery of the wind components, intact and in working order; and/or to penalize contractors who are late in their deliveries.

The project monitoring program is therefore set up in a number of parallel sequences, all focused around interface management, QA and QC procedures, and the commercial departments of all stakeholders in the project, including banks and financers, to state what has been done, what state the work is delivered in, and whether the project is on time or not.

Again, the time schedule is very important because it is the guideline for the entire work process. PM will gather and validate all information on the project progress to develop a real-time, accurate image of the status of the project. (This should be done as stated previously in Chapter 5, Figure 5.4, with regard to the pyramid type of documentation, where the various layers of project stakeholders report upward to their main clients and ultimately to the wind farm owner.) Once this has been done, project managers can evaluate the status compared to the ideal scenario—the overall time schedule—and thereby report to the owner and issue payments or penalties, whichever is required.

Furthermore, project management can issue warnings, claim penalties, or ask contractors and/or suppliers to speed up the work where appropriate. For this, PM must be well suited, and the set-up of this part of the organization is extremely important, since this is the only relevant place and body where decisions within the project can be made outside the normal scope of work for individual contractors.

PROJECT MANAGEMENT SETUP

The setup of PM is important, as previously stated. Therefore, it is of some concern how this part of the organization works. As with everything in the world, everyone has his or her own preferences, but the following organizational setup has previously worked well. The reader is invited to form his or her own opinion on the subject.

Referring to Figure 5.2 from project manager down, it can be seen that the project group is a dedicated one that does not carry out any tasks other than those related to the single project and the single project manager (Figure 6.4).

The organization can be much bigger, depending on the service or product offered for the project. Figure 6.4 would be the typical organizational layout from the contractor's side. The wind farm owner would require this type of organization for every work package and then add an extra layer for senior project management, as shown in Figure 6.5.

For ease of understanding, the complete organization for the project has been left out of Figure 6.5. The main principle is that a complete top layer of management is allocated to the project in order to handle the many different scopes of work required to complete the project. And it doesn't really matter whether it is the EPIC contractor or the wind farm owner who is depicted in the organization chart because in the real world, the two organizations will look identical. However, PM is now a complete scope of work for all the activities in the project.

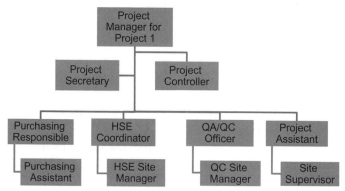

Figure 6.4 The project organization as seen from the project manager's position.

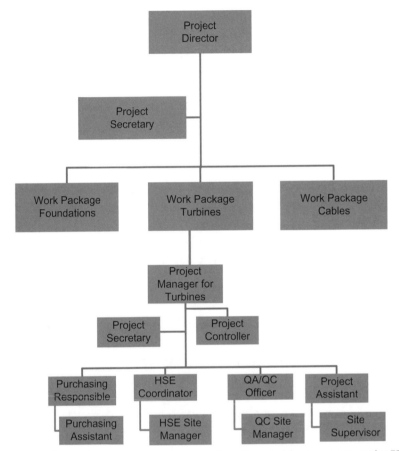

Figure 6.5 The project management as seen from the wind farm owner's, or the EPIC contractor's, side.

This, of course, puts a lot more strain on the entire organization, and the requirement is for a large number of people who can interact in the project without losing sight of the common goal: to install the wind farm offshore. The management task itself becomes extremely complex in this case, and if you do not work out the chain of command properly, two things will happen.

- The decision-making process and the ability to decide on behalf of the project become unclear. This means that whoever feels he or she is competent enough to make a decision will do so—or abstain from doing so. This has consequences because, in both cases, information is lost or distorted, which is devastating to the progress of the project.

- Since the progress line in an offshore wind farm project is very fast, the information has to be available almost instantly to have any value. Consider the case where the weather forecast from yesterday becomes available today at 0700, where the work has been discontinued due to the lack of information on the applicable weather window. This means that an opportunity is lost, and claims for delays will be sent to everyone and his dog—as long as they are part of the project.

A lost day in an offshore wind farm installation program can easily cost 1 million euros, so whoever is left with the bill will feel pretty bad and try to get rid of the expense. This is why monitoring the project in almost real-time updating is crucial. And even though the example is half bad—normally the weather forecasts are very reliable—the consequences of even small mistakes or slips will be very expensive.

The decision to organize the work as just shown means that the individual work package managers will have to be equipped with a clear set of instructions in order to establish:

- *Chain of command*—who can decide what. It is not up to the work package manager to decide whether something should be done in a specific way. The chain of command must be very clear in this case.
- *Span of command*—what can the individual work package manager decide and what has to be carried to a higher level.
- *Coordination*—daily coordination between the work packages must take place in order to ensure information sharing and updating.

The final issue here is that the top management level must be very charismatic and strong. In previous projects, it has been the case that top management was not equipped with a clear mandate from the line organization; this led to a vacuum in decision making that haunted the project throughout its entire duration.

The really bad problem was that PM of the project hired in a large number of work package managers at the same time, without defining the organizational placement of the individual work package manager, his duties, and his span of command. This led to disaster: every work package manager tried to position himself or herself as second in command altogether, and this was not the intention.

The final result was a project that was poorly carried out, a large fluctuation of personnel that came and went, and cost and time overruns that were very serious. Therefore, the organizational behavior, the span of command, and the project management should be carefully planned and, when hiring personnel, carried out by strong, available, and well-prepared top management. This, unfortunately, is not always the case.

●●●

Related Images

Loading monopile foundations in port prior to sailing to site. The turnaround time is crucial to saving overall project time. Therefore timely planning of the process is required before startup of installation work.

Preparing the next nacelle at the front of the pier. Note that the installation vessel has only just departed.

The feeder concept in practice. The components are loaded over from the transport vessel to the jack-up barge.

Towers stored on the pier. Notice the concrete blocks acting as ballasts to keep the foundations firmly on the ground.

CHAPTER SEVEN

Interface Management

What is an *interface*? It is, in principle, the point where one contract party takes over or hands over parts of or the entire work scope to either another contractor or to the wind farm owner. If you scour the Internet, the only valid explanation you will get is that an interface is a tool that allows two components to interact or correspond with each other. But this is not what we are looking for here. In our case, an interface is the borderline between two scopes of work, and this is how we will use the term in this book. An interface is also the point where the responsibility for a product or service passes from one contractor or work package to another.

THE MAIN INTERFACES AND ACCOMPANYING RESPONSIBILITIES

For a typical installation process, the interface identification is usually named and listed in an "interface matrix" (see Table 7.1) that indicates who is responsible, who is to be notified, and who is to carry out a task. The table shows the typical interface matrix for the installation of the wind turbines offshore, but the one here only lists who is responsible for carrying out a task. However, the next task in the line will possibly be carried out by another contractor, and this constitutes a handover in the sense of responsibility changing hands. Please note that the contractor who hands over is also the possible receiver at a later stage. In this matrix, the turbine supplier hands over the turbine on the pier but receives it back on site.

As Table 7.1 shows, a large number of interfaces are required just for fitting the turbine on the offshore position and hooking it up to the grid. A great number of these events constitute a handover to another contractual party, and it is logical that this must be managed in order to safeguard the deliveries of the individual contractors and, of course, the wind turbine generator (WTG) components themselves.

Table 7.1 Typical Interface Matrix

Task	Turbine Supplier	Employer	Installation Contractor
Onshore Storage and Preassembly			
Transport of all WTG parts and related tools and support equipment to preassembly site in XXX	X		
Provision of the area for storage, and preassemble the WTG components inclusive surveys		X	
Supply turbine supplier with the area loads given by the harbor of XXX, including quay loads. Additional load calculations for the quay if requested by turbine supplier		X	
Harbor dues and taxes		X	
Required civil and electrical work within the preassembly area (potential quay modifications exclusive)	X		
Connection for telephone, Internet, power, water, sewage		X	
Provision of office and messing facilities for all turbine supplier personnel throughout the project—white collar		X	
Provision of office and messing facilities for all turbine supplier personnel throughout the project—blue collar	X		
Provide a web-based, weather monitoring and forecast system available for turbine supplier throughout the project		X	
Storage, handling, preassembling, and testing of all WTG components: 1. Up-end tower sections and Premont internals, cables, and "power unit" 2. Prepare nacelles with "met mast," aviation lights, etc. 3. Assemble rotors exclusive of spinner nose	X		
Provide frames to support upended tower sections at the quay side	X		

Table 7.1—Cont'd

Task	Turbine Supplier	Employer	Installation Contractor
Move WTG components to the quay (within reach of the vessel crane) in sets of three, and prepare for loading (inclusive mounting of lifting brackets and covers for towers)	X		
Supervision, hook on, and unbolting from temporary frames on the quay during loading of the installation vessel	X		
Remove WTG-related return goods from the quay	X		
Inspection of the individual components before loading		X	
Offshore Installation			
Vessel coordinator interfaces with relevant authorities and facilitates registration of personnel offshore		X	
Supply installation vessel inclusive of crew, crane operators, access arrangements to the components onboard and the foundation, generators, winches for line holders, and accommodation facilities for turbine supplier personnel		X	
Free issue of seafastening as it was at the completion of the Rödsand II project (no repair, maintenance, transport, or handling inclusive)	X		
Transport, handling, repair, modifications, and mounting of seafastening		X	
Supply flange measurement report before start of installation of the WTG		X	
Supply of foundation flange including measuring report		X	
Bolts and nuts for flange connection	X		

(Continued)

Table 7.1 Typical Interface Matrix—Cont'd

Task	Turbine Supplier	Employer	Installation Contractor
Ladder arrangement forms internal tower base platform to internal foundation platform	X		
Service lift including fittings	X		
External stairs from tower door to external platform	X		
Mounting door stop	X		
Removal of all protective parts from WTG and foundation when first arriving at foundation	X		
Supply specialized lifting equipment to install the WTGs	X		
Rotor lifting and transporting to quayside (water front)—turbine supplier's responsibility	X		
Rotor loading and seafastening to installation vessel; onshore crane supplied free of charge by turbine supplier			X
Loading and seafastening of WTG components to installation vessel			X
Transfer, positioning, jacking, preloading, and placing the gangway to the foundation			X
Lifting planes and lift studies (drawings and descriptions of the respective lifts)			X
Method statements for the crane work related to the installation of the WTG			X
Method statements for the installation of the WTG	X		
Signaler (turbine supplier foreman) responsible for giving signals to the crane operator	X		
QA control of the individual component before lifting	X		
Unfastening from seafastening before installation lift on site			X

Table 7.1—Cont'd

Task	Turbine Supplier	Employer	Installation Contractor
Supervision of unbolting from seafastening and installation of the individual turbine	X		
Delivery of all tools required to install the WTG	X		
Lifting and installing WTG components at the destination			X
Bolt connections and dismantle lifting equipment	X		
Electrical power for yawing the turbine, light, and tools		X	
Torque bolts and prepare turbine for transport	X		
Deliver and operate the required crew change vessels		X	
Toilet facilities on board vessels		X	
Cable Works			
Deliver and operate the transfer vessels		X	
Offshore transfer time and associated costs (ready for offshore transfer → on the individual foundation)		X	
Prepare "crane" on WTG foundations and load the required tools and parts	X		
Power supply for Racon in one WTG	X		
Power supply for Met station in one WTG	X		
Power supply for and connection of DAVIT	X		
Supply 33-kV cable from switch gear to sea cable termination point		X	
Premount 33-kV cable on switch gear	X		
Connect and test 33-kV cable from sea cable termination point until switch gear in WTG		X	

(Continued)

Table 7.1 Typical Interface Matrix—Cont'd

Task	Turbine Supplier	Employer	Installation Contractor
Supply power for dehumidifier, light, lift, and electrical tools for electrical completion and commissioning of the individual WTG		X	
Supply all tools and parts to complete the cable works and commissioning lift	X		
Complete all cable works inside the WTG and commissioning the lift	X		
Connect earthing cable between foundation and tower	X		
Supply Ethernet switch and VoIP phones	X		
Supply and route fibers, Ethernet, and power cords	X		
Provide power supply for LAN, VoIP	X		
Commissioning Works			
Deliver grid connection no later than 4 weeks before start of actual installation		X	
WTG commissioning			
Commission Ethernet switch and VoIP phones	X		
Commission fiber, Ethernet, and power cords	X		
Commission power supply for LAN, VoIP	X		
Commission 230 VAC/15A in 1 turbine for turbine supplier's radar	X		
Commission 230 VAC/15A in 1 turbine for turbine supplier's Met station	X		
Commissioning work and turbine start-up	X		
QA inspection and handover of the WTG	X	X	
Test on completion	X		
Snagging	X		

There are basically two approaches to the handling of interfaces and responsibilities in relation to the transport, assembly, and fitting of wind turbines:

1. Contract out to an EPC contractor
2. DIY or multicontracting, where the owner takes over the responsibility as soon as possible and handles the interfaces personally

Both scenarios have their pros and cons, as we will soon see. To explain the various pros and cons for both scenarios, it is the intention to cover:

• Project examples
• The amount of work involved
• The economy of the project-related interfaces

CONTRACTING TO AN EPC CONTRACTOR

As examples of projects where the client chose to let others handle the responsibilities during the project period, we will use Barrow and North Hoyle. We must assume that the client's reason for contracting that part of the project was either that he did not have the competence in house or that he wanted to minimize the risk as much as possible. This can be a very attractive solution because we may assume the client has enough work, too little resources, and not enough experience to handle the interfaces during the project period.

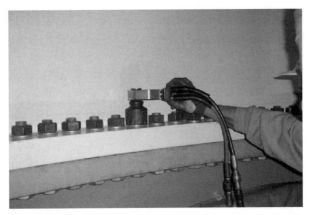

Bolting the turbines together and controlling the work leads to the handover interface between installation contractor and turbine manufacturer who should verify the documentation of the work.

From this point of view it will seem a practical solution to let people with the appropriate experience handle these interfaces, since there are several and the handling is very important in relation to the successful completion of the project. However, by choosing this solution, the client also renounces influence during the project period. Should the project be delayed, the client has very little influence on the outcome, and furthermore we may assume the suppliers will contest the client's right to day penalties.

Both of the previously mentioned projects were severely delayed, with serious repercussions for the client's economy for the wind farms as a result. The delays may also have had a negative impact on offshore wind farms as such and given fuel to the critics of this kind of energy extraction. During the delays, the client had no influence whatsoever as to the management of the project, the solutions, or the contractors chosen. And with the chance of one or more of the contractor's risk of bankruptcy as a direct result of the project, the likelihood of getting the appropriate day penalties paid were reduced. Therefore, letting others handle project interfaces and responsibilities has its risks.

Obviously, by choosing this solution, the amount of work during the project period is dramatically reduced as the client in principle will not have to deal with anything until the turbines are operational.

DIY or Multicontracting

As examples of projects where the client handled the responsibility himself, we will use Horns Rev and Nysted. The clients in these cases chose to handle the responsibilities themselves because, in part, they had the manpower necessary and because of their own organizational structure as a project organization. In other words, the clients in these cases believed they had the knowledge and time to handle this major task during the given project period.

Both of the preceding projects were finished on time and with no extra cost to the client. That meant the economic prognosis could stand firm. As a result of the client handling the project management herself, she had very close contact with all suppliers and contractors and, therefore, also at all times, had an idea of how the project was progressing. When a delay appeared in one area, she could adjust the rest of the project flow immediately. The same, of course, goes should one area progress faster than anticipated. The owner's PM team could adjust to the new situation without having to liaise with others in any significant way.

The close contact to the project and the daily "hands-on" would also give the client a unique possibility to keep ahead of possible delays and thereby also the ends and means to minimize possible technical and economic impacts that may arise.

THE ECONOMY

If the supplier and/or contractor is asked to assume responsibility during the project period, most will raise the cost per turbine not only according to the increased responsibility but also due to the extra workload to be assumed. It is estimated that the suppliers or contractors will add another approximately 30 percent per turbine for handling the project interfaces and responsibilities.

Once the nacelle is bolted and the rigging gear is released, control is passed back to the turbine manufacturer.

It should be emphasized that should an external contractor assume responsibility during the project period, he will almost always make provisions in his contract. Thus, the client will not be completely relieved of responsibility. So even if the client does contract this area to sub supplier(s), the responsibility cannot be completely handed over. The economic consequence of this will be that the client pays to make others assume the risks; even so, however, he must assume part of the risk himself.

So, depending on the client's economic estimates for producing, assembling, and erecting the turbines, either solution may be economically viable. However, the bulk of offshore projects so far have struggled with economic issues, mainly because of risk allocations made to EPC or EPC equivalent offers, thereby delaying or even stopping the projects.

INTERFACES AND HANDOVER DOCUMENTS

During big and complicated projects such as Barrow, London Array, or others, there are, in principle, an infinite number of things that can go wrong. Therefore, the importance of strict project control cannot be stressed too much.

Taking control of components in the port also forms part of an interface. Once lifted, the hub is deemed to be in proper condition. Any claim arising after this point, until the next handover, is the responsibility of the installation contractor.

A crucial part of project control is an analysis of the interfaces the components go through from the moment they roll out of the production site until the turbine is ready for production on site.

It is very important to have a complete overview of all the interfaces that occur during the project to ensure that every contingency is covered and nothing is forgotten. An interface occurs every time the responsibility for the component goes from one contractor to another. The fact that the responsibility shifts also indicates that something is about to happen with the component. It will be handled in some way or another according to component and project progress.

The potential of something happening to the component during the actual transport (e.g., an accident) or while the component is sitting on the pier waiting for the next phase is another potential problem. Components that are left sitting on a pier are exposed to the elements. They can also easily be damaged when they are moved from the ground to the truck, the truck to the vessel, and the vessel to the final site.

These situations must therefore be described in detail every time they occur. To that end, a handover document must be written for every component; it should describe the exact component, the exact interface, the issues of this particular interface, any comments, and so on. The document must then be signed by the two parties of the interface or their representatives.

The finished foundation with a level top flange is the main interface between the two scopes of work: foundation installation and turbine installation. The contractor checks the quality of the work prior to continuing installation of components.

In the case of defects, a drawing of the component may come in handy, since it is much easier to show the defect on a drawing than to describe it in writing. It is also possible for a component to be damaged several times during transport, and it will be easier to establish at what point the damage occurred and who is responsible if the damage has been indicated on a drawing.

The handover documents for each interface and for each component are then collected by the project manager so she can decide what to do if damage to a component occurs. The handover documents will serve as evidence as to when a certain damage was done and also who had the responsibility for the component at the time.

Before transport of any component begins, it should be decided who has the final say if something goes wrong during the transport. For example, there may be a minor scratch on a tower from transport. The scratch will have to be repainted, but the defect will not affect the overall timetable. However, more serious defects may occur, and an immediate decision must be made as to how to proceed. Also, the defect may affect the timetable for the whole project, so other procedures will have to be set in motion so that certain parts of the project can be delayed and others may be hastened along. These decisions will have to be immediate in order to reduce the economic impacts such incidents can have on a project.

Delivering components: In this case, a tower in the staging port also forms an interface.

The interfaces just described may be changed according to how the final transport, assembly, and mechanical completion contracts will be negotiated. It does not, however, change the fact that the following phases still must be completed; only the sequence may change.

Interface 1

General Description

The first interface occurs when the component is to be transported from the supplier to the pier by the transport contractor. This transport will normally be by truck and may at a later time be by vessel. A vessel will be chosen when the supplier is far from the pier, and transport by vessel is usually more convenient and cheaper than transport by road.

In the case of vessel transport at this stage, an additional interface must be described. The first interface issue relates to the quality of the component delivered. For example, is the quality at the expected level? Are there any defects? Can the component be unequivocally identified? Is the surface treatment (e.g., paint) as it should be?

Control must be according to the specifications for this particular interface. They are taken from the contractual specifications between the client and the supplier and are described in detail in the handover document (see also *http://mgeps.files.wordpress.com/2010/04/incoterm2000.gif*).

The component supplier must deliver the following:

- Total number of components
- Turbine type and turbine brand
- Weights, measurements, and volume of each component

The transporters must deliver the following:

- Total cost of the transport, but if relevant the separate prices must appear. Incoterms FAS (Free Alongside Ship) is an asset of internationally agreed terms defining when responsibility for cargo passes from buyer to seller.
- Which components are to be transported on which vehicle
- Where and when the components are to be picked up, as well as ETA (estimated time of arrival), both where and when
- Costs in case of delays, as well as other costs that may occur as a result of delays; other costs are extraneous
- Documentation of all necessary permits
- Documentation for all vehicles' capabilities in relation to conducting the transport
- Copies of all vehicles' registration documents
- Copies of all chauffeurs' driver's licenses, certificates, and so forth; no driver can be substituted without prior approval of the client
- Documentation of all lifting and lashing equipment, with the expiration date no sooner than one month after the expected finishing date
- Documentation for the elected route from supplier to pier; route description to be enclosed
- HSE and Q plans for transport
- Method statements

Handover Responsibility

The client must decide who has the onsite authority to say go/no go in case the client is not present personally but uses a representative; in which case, the local person must contact the project manager for further instructions.

When the handover document has been signed, the component can be lifted on the truck for further transport, providing the component is deemed acceptable according to specifications. The responsibility has now been transferred from the manufacturer to the transporter.

Damage and Impact on the Time Schedule

In the case of damage to the component, it must be documented in the handover papers, and the responsibility for the damage must be established. The client or his representative reserves the right to deny loading of the component should the damage be extensive.

Depending on the extent of the damage, the component may not be delivered on schedule, and as a consequence the entire project is in danger of delay. The project manager must immediately be informed of the damage, any potential delay, and the expected new delivery date.

Document Control

The original signed handover document should be given to the project manager, and a copy should be given to the supplier of the component. All other documents concerning the transport must do the same: original to the project manager and copy to the transporters.

Payments

Payment for the component can proceed when:

1. The component has been loaded with no damage
2. The handover papers have been signed
3. The responsibility for any damages is established and the cost of repair is settled

Safety and Security

The component supplier and the transporters must document their safety organization and ensure that the chauffeurs have had the relevant safety courses. They should also supply contact information for their safety manager and quality manager.

Interface 2

General Description

Interface 2 is the interface between the truck and the pier. The issues here relate to whether there has been any damage to the component during the first leg of the transport. It is not uncommon for the surface of the components to be slightly damaged either from the lashing or from the transport cradles.

This is most often due to the surface treatment not having thoroughly hardened before transport usually because of tight time schedules. However, it is very important to look for any damages because the environment of the turbine's final destination is very harsh, and proper protection is needed to ensure the projected lifetime of the turbine.

As previously mentioned, a drawing of the turbine can be very helpful in this case. A circle or a cross indicating the exact location of the damage is easy to make and easy to understand. In most cases damages like these will not delay

Workers in the process of lifting a nacelle onboard an installation vessel in port. At this point, it is released for transport by the client, the installation contractor, and the marine warranty surveyor.

or in other ways influence the timetable. Repair work can easily be done when the component is sitting on the pier waiting for the transport to the final destination. Even when fitted on the foundation (providing the damaged area can be reached), repair work can be carried out successfully.

When transporting blades, damage sometimes occurs due to the fragile construction and the difficulty of protecting the entire blade during transport. Blades will do what blades are meant to do: They catch the wind and therefore must be handled cautiously when moved. Blade damage (other than surface treatment damage) must be dealt with carefully because blade condition will affect the turbine's output.

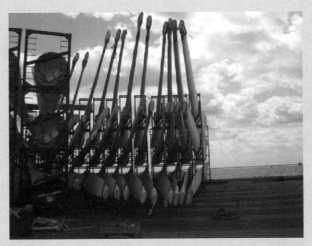

Blades waiting for transport in port.

Handover Responsibility

The client or the client's representative, together with the transporters or the transporters' representative, must sign the handover documents. When the component is sitting safely on the pier and the documents are signed, the second interface is completed. The responsibility is now transferred from the transporters to either the client or the company hired to do the pier assembly (depending on how the project is to be implemented).

Damage and Impact on the Time Schedule

In the case of damage to the component, it must be documented in the handover papers, and the responsibility for the damage must be established. The client or a representative reserves the right to deny unloading of the component should the damage be extensive.

Depending on the extent of the damage, the component may not be delivered on schedule, and as a consequence, the entire project is in danger of delay. The project manager must without delay be informed of both the damage, the potential delay, and the expected new delivery date.

Document Control
The original signed handover document should be given to the project manager, and a copy should be given to the transporters.

Payments
Payment for the transport can proceed when:

1. The component has been unloaded with no damage
2. The handover papers have been signed
3. The responsibility for any damages is established and the cost of repair is settled

Interface 3

General Description
The various parts for the turbines will arrive at the pier individually. Depending on the choice of foundation, the turbines, tower, blades, and nacelle may all arrive separately. This is because the parts are produced at different manufacturing sites and because of transport. For example, it is not possible to transport a fully assembled tower by road. The length in itself is a problem, but the weight can also pose particular problems on certain types of roads, especially in the summer when the sun can soften the asphalt.

Depending on the pier facilities and the vessel used for the final transport of the turbine to the site, part of the turbine assembly will be completed before putting the parts on the vessel. It is a good idea to do as much of the assembly on the pier as possible; the working conditions are much more controlled on land than at sea, and the assembly cost will be cheaper on land than at sea.

Towers being preassembled prior to load out. It is clear that damages can occur during this process as well; thus, there is a requirement to document the proceedings and hand over the work as part of a proper interface management process.

The handover documents of this interface will describe in detail which parts are to be assembled on the pier and to what degree the assembly should to be done. This description must be taken from the manufacturer's manual, as well as from the contract scope. It is very important for this scope to be described accurately to avoid misunderstandings and also to ensure the quality control that must be done in relation to the handover to the next interface.

The contractor doing the fitting on the pier must do the following:

- Document security plans for the working areas
- Document all relevant courses for the fitters
- Prepare HSE and Q plans
- Compile method statements
- Document calibration certificates for relevant tools

Depending on how the project will be implemented, there is a possibility that instead of assembly on the pier, assembly will be done on site. In that case the components will, of course, be lifted from the pier and onto the vessel without assembling.

Handover Responsibility

The handover documents for this phase must be verified by a representative for the fitting contractor and the client's representative with responsibility for harbor logistics. The signing for this phase will verify that the turbines are ready to be lifted on board the vessel for transport to the site. Responsibility is transferred from the fitting contractor to the vessel when the component is lifted from the pier.

Damage and Impact on the Time Schedule

In the case of damage to a component, it must be documented in the handover papers, and the responsibility for the damage must be established. The client or his representative reserves the right to deny acceptance of the component should the damage be extensive.

Depending on the extent of the damage, the component may not be ready on schedule, and as a consequence the entire project is in danger of delay. The project manager must immediately be informed of the damage, any potential delay, and the expected new delivery date.

Document Control

The original signed handover document should be given to the project manager, and a copy should be given to the fitting contractor.

Payments

Payment for the fitting can proceed when:

1. The component has been fitted with no damage
2. The handover papers have been signed
3. The responsibility for any damages is established and the cost of repair is settled

Interface 4

General Description

From the pier the component is lifted on board the vessel for the final leg of transport to the site. The components must be placed on board according to the loading plans. From the moment the lifting is started, the responsibility goes from the fitting contractor to the vessel.

The responsibility stays with the vessel until the time the components are placed on the foundation. The vessel contractor must deliver the following:

- Total cost of the transport but if relevant, separate prices must appear
- Loading plans, handling manuals, and instructions
- Description of lashings and securing plans for the components
- Vessel organization, including the HSE plan
- CVs (i.e., work histories) of all vessel officers and crane drivers
- Vessel specifications and specifications for vessel crane(s)
- All relevant certificates other than compulsory vessel certificates
- Specifications of pier crane(s)
- HSE and Q plans
- Method statements

During this phase of the project damage may mainly occur in connection with the lifting. Often there is not much room on the deck, and the components are placed without enough space between either the components or the vessel. And, because most of the time lifting cannot be done in completely calm weather, swaying of the components can cause damage to surfaces.

When on board the vessel the components are secured with the appropriate sea fastening, and the journey to the site can begin. After arriving at the site, the components are lifted from the vessel to the foundation. Depending on the project, the towers may come as parts and be fitted on site, or they may come completely assembled.

Loading onboard itself requires documentation. Normally, the turbine manufacturer handles the components but the contractor seafastens them. This cooperation requires documentation of "who did what."

The nacelle may come as a "bunny ear" or separate with the three blades to be fitted when the nacelle is in place. The handover documents for this interface must describe these lifts according to the lifting plan.

The details for the lifts are very important because lifting several hundred ton in a difficult environment naturally may be the cause of high risk. The wind speed is perhaps the most important factor here. Different components may be lifted in various wind speeds, with the blades as the most crucial component. For every nacelle there is at least one blade to be lifted and fitted on site, and this particular lift is a bit tricky because the blade must be lifted from a horizontal position on the vessel deck to a vertical position to be fitted to the nacelle.

The safety of the crew and the fitters must also be taken into consideration. Working the guy lines during a complicated lift can be a physically difficult task, and only experienced fitters should undertake it.

Handover Responsibility

The handover documents for this phase must be verified by a representative for the vessel and the client's representative with responsibility for vessel logistics. The signing for this phase will verify that the lifting has been completed without damage to the components. The responsibility is transferred from the vessel to the installation contractor when the handover documents are signed.

Damage and Impact on the Time Schedule

In the case of damage to the components, it must be documented in the handover papers, and the responsibility for the damage must be established. The client or a representative reserves the right to deny acceptance of the component should the damage be extensive.

Fitting a blade is the most delicate part of the entire installation; therefore documentation of the process needs to be detailed. The task requires a lot of focus due to the risk of damage. It would be difficult to repair a blade after installation and would require a vessel to stay in position longer—at a very high cost.

Depending on the extent of the damage, the component may not be fitted on schedule, and as a consequence the entire project is in danger of delay. The project manager must immediately be informed of the damage, the potential delay, and the expected new delivery date.

Document Control

The original signed handover document should be given to the project manager, and a copy should be given to the vessel.

Payments

Payment for the vessel can proceed when:

1. The components have been successfully lifted from the vessel
2. The handover papers have been signed
3. The responsibility for any damages is established and the cost of repair is settled

Interface 5

General Description

At this interface the mechanical completion of the turbine and the final handover to the client is accomplished. After arrival at the site, the specific foundation, the tower, nacelles, and blades have to be fitted. The fitting of the components and the mechanical completion must be described in detail.

This is a crucial matter because the client must ensure that the turbine is delivered as expected. The fitting of the components are pretty simple: The tower must be erected, and the nacelle with the blades must be assembled on top. However, the scope of work (SOW) for mechanical completion can be another matter altogether.

The SOW for mechanical completion must describe all the parts in detail, including components and consumables, that will be installed on the site. It should be noted that scaffolding and other means of equipment necessary to fulfill the SOW should also be included.

The contractor doing the mechanical completion must deliver the following:

- Relevant certificates for all fitters
- HSE and Q plans
- Method statements
- Document calibration certificates for relevant tools

The handover documents for this interface must describe the entire scope of work.

Handover Responsibility

When all of the handover documents have been signed, the client accepts final responsibility for the turbine. Therefore, a verification of mechanical

Fitting towers offshore.

completion according to the SOW is crucial. The client or a trusted representative must verify that the mechanical completion is according to the scope of work. Once the handover documents have been signed, the responsibility for the turbine is transferred to the client.

Damage and Impact on the Time Schedule

In case of damage during the mechanical completion, it must be documented in the handover papers, and the responsibility for the damage must be established. The client or a representative reserves the right to deny acceptance of the turbine should the damage be extensive.

Depending on the extent of the damage, the turbine may not be ready on schedule, and as a consequence the entire project is in danger of delay. The project manager must immediately be informed of the damage, the potential delay, and the expected new delivery date.

Document Control

The original signed handover document should be given to the project manager, and a copy should be given to the fitting contractor.

Payments

Payment for the mechanical completion can proceed when:

1. The mechanical completion has been successfully completed according to the SOW
2. The handover papers have been signed
3. The responsibility for any damages is established and the cost of repair is settled

●●●───

Related Images

Fitting a full rotor on a Siemens 2.3-MW turbine is tough work. The 5- and 6-MW turbines are almost twice as large and therefore present an enormous challenge. Installation of a wind turbine from M/S *Sea Power* at Lillgrund Wind Power Plant, 2007-08-04.

Lifting a 3-MW Vestas turbine on deck and checking to be sure the bolts are all removed.

Personnel checking the components on the seafastening frame for the Vestas 3-MW turbine.

Lifting a 3-MW nacelle, hub, and two blades—-the "bunny ear" way—from the deck and placing it on top of the tower.

CHAPTER EIGHT

Health, Safety, and Environmental Management

What exactly is health, safety, and environmental (HSE) management? To fully understand it, we must first define what is involved in HSE work. Referring once again to the Internet, we see that one definition of HSE is an act taken "to prevent people from being killed, injured, or made ill by work."

A more down-to-earth explanation could be the proactive work to plan and execute operations in order to prevent damage to property and accidents to people and to create a safe and healthy working environment. This is, of course, easier said than done, since the environment where offshore work is carried out is by its very nature dangerous. The oceans have claimed lives for thousands of years, and they still do. But as stated earlier in this book, the overall objective is to bring everyone back safely from the excursion to the offshore site, and if a wind farm can be built during this excursion, it is an added benefit.

WHY HEALTH, SAFETY, AND ENVIRONMENTAL MANAGEMENT IS IMPORTANT

We rigorously carry out HSE work for many reasons. Here are some of the most obvious ones:

- We want to safeguard our employees from any danger that might arise from unsafe work practices.
- We are subjected to an environment that is inherently dangerous, and therefore every sensible measure must be taken to accommodate the challenges we face when working in the offshore environment.
- We want to safeguard the environment we work in by ensuring that all activities are safe, environmentally friendly, and leave no pollutants behind.
- The wind industry is a renewable one, and environmental protection is paramount to its stakeholders.

Popularly speaking, we should not attempt to solve the carbon problem by creating dangerous working environments or to try to replace the carbon savings by working with methods and equipment that are not state of the art and, as a result, create more problems.

The Most Important Single Activity in a Project

Nothing can justify the loss of a human life or the contamination of a marine environment just because the benefits are anything from cheap electricity to air conditioning. The belief is, of course, that the twenty-first century working environment is supposed to live up to a minimum set of standards, such as safe working conditions and a clean working environment, where we use the best practices of employment to protect both employees and the marine environment.

Finally, the law protects all of the preceding standards, and this is what HSE management cares about. But how is this work organized?

The Health, Safety, and Environmental Organization

Setting up the HSE organization is a task that must be taken seriously. In most, if not all, countries where offshore wind projects are carried out, either clear regulations that govern how this is to be done are already in place or regulations that will apply to the project are being developed.

The structure of an earlier project is used, mainly because it originated in-house in AOS, so neither the final version of how things should be done nor the theoretical exercise of how to model the organization has been established. The HSE organization takes care of two issues:

- The health and safety of the workers and the surrounding society.
- Environmental care, which involves not polluting the environment as a result of doing the project. Pollution, in this case, should be thought of in the broadest possible sense—for example, noise, garbage, obstruction, or disturbing marine life and the onshore environment.

Therefore, the organization should be set up to deal with both issues at the same time.

Figure 8.1 shows one way to organize the HSE; the structure is actually very simple. The management of HSE is divided into two areas: back office functions and onsite organization. The back office is the part that will organize and structure all aspects of the HSE work according to rules, regulations, building permits, and client requirements.

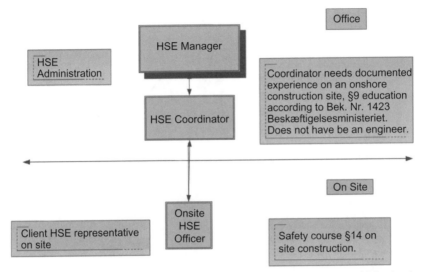

Figure 8.1 The Health, Safety, and Environmental organization as it would be implemented. *Courtesy of Advanced Offshore Solutions.*

The onsite organization, which is represented by the HSE officer, or officers, will carry out the work and enforce the HSE system as defined by the preceding stakeholders. It is important to note that, again, the HSE organization and HSE functions are not staff functions. Unlike so many other organizations or project structures, the HSE organization is a line function with the authority to stop work everywhere at any time.

The reason for this is simple: If unsafe work practices or a risk of environmental pollution is latent, the task of the HSE organization is to stop the situation from developing into something that can be dangerous, careless, or create a lot of unnecessary cost or aggravation. And it very often happens that a situation that seems harmless onshore can quickly evolve into a life-threatening situation offshore. This must be avoided at all costs and using all efforts possible.

THE HSE DOCUMENTATION STRUCTURE

There are, of course, several ways to arrange the HSE documentation structure. Figure 8.2 shows a common method that uses three tiers. In general, the three tiers represent the level of need-to-know for all the parties involved in the project. The top level shows that the managers of the project, the owners of the wind farm, and residents of the surrounding area will

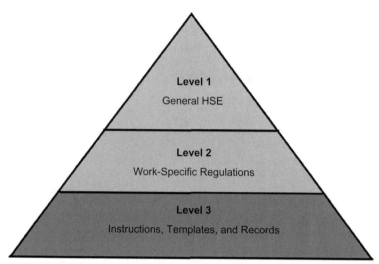

Figure 8.2 HSE document structure that defines which personnel are responsible for certain elements of the project. *Courtesy of Advanced Offshore Solutions.*

assure all project stakeholders that they will adhere to all of the requirements that are enforced throughout the duration of the project.

A cynical person might point out that this level is where the toast speeches are made, the bragging about all the goodwill and best endeavors. That person is partially right. But the top level is also saying that the top management of the project promises that all aspects of the project, the processes, and the behavior of the employees involved will be dedicated to protecting human life, the environment, and the values created in order to supply electricity. Yes, this sounds a bit emotional, but this is what the top management must do.

In some countries this pledge will ultimately land the responsible person in prison if the HSE work is not performed according to the requirements. This is potentially the case in the United Kingdom, where the managing director will, at the very least, be questioned by the police and possibly charged with neglect if a major accident occurs. The managing director will not necessarily be imprisoned or actually charged, but the possibility is more apparent than, say, in Denmark, where gross negligence would possibly have to be determined before such drastic actions can be taken.

Level 2 is where things start to take shape. Here, the actual rules and regulations of the work in question are outlined, the national HS rules are described and clarified wherever required, and the project-specific guidelines for the health and safety work are established. Level 2 is also where the HSE documentation is unfolded in order to address the specific work package and the requirements. In this way, the individual work package can search for and pinpoint exactly what is relevant to them without losing oversight of the entire project's HSE plan. Thus, level 2 is for the project management and the work package managers in particular.

Level 3 is where the working instructions are fleshed out in detail. The individual task is described and the specifications for the job, not only from the HSE point of view but details of the actual working instructions, are defined. This is done in such a way that each employee can read about and understand exactly what the task involves, what the health and safety concerns are, and what he or she must do to perform the job safely.

A special section is dedicated to high-risk work—for example, hot works, diving operations, working in confined spaces, and working at great heights. Since these tasks require special training, such items should be specifically addressed in individual documents. This establishes that if it is not in the HSE standard for the work you are allowed to carry out, you cannot do the work. When training the offshore crew, this logic would then be implemented and the statement should be taken literally in the sense that if it is not strictly allowed, it's forbidden.

Monitoring and Reporting

Since the onsite HSE officer is part of the line organization, he or she will have a lot of control over when work starts and stops. This means the officer will constantly monitor the work and write all the necessary reports.

As we said before, work must go on 24/7/365, and it's the same for the HSE work. Special attention is normally given to working at night, since the risk of accidents increases as a result of two factors:

- Lack of daylight makes operating the machinery more difficult.
- Fatigue due to working at odd hours is a latent danger, and the chance of a worker paying less attention because he or she is tired, can lead to accidents very quickly.

So why not just work during daylight hours?

One reason is because the wind tends to decrease at night, and it is relatively certain that the majority of both foundations and turbines have been installed during the night or early-morning hours in the past. Therefore, working around the clock and monitoring the activities are crucial to the HSE department. Figure 8.3 shows a crew working at night fitting slings to the gravity foundation. Here a number of things could go wrong.

Figure 8.3 Notice the poorly illuminated ladders and the person sitting on top of the partition in the foundation. *Courtesy of Advanced Offshore Solutions.*

Auditing and Correcting Actions and Methods

Since the offshore wind industry is still young, one will obviously encounter methods and procedures that were considered genius when they were created in the office. This is the case for all activities in the industry. But when the procedure, work instruction, or method statement is implemented in the project and the offshore crew (or onshore crew, for that matter) have worked with the procedure, instruction, or method statement, it often becomes clear that additions or changes must be made to this documentation, and possibly complete cancellation.

It is for this reason that the onsite HSE officer must be present and aware of what work is going on. The officer will have to monitor and evaluate the work as it unfolds during the project. As soon as a requirement for a change or an update of a document is necessary, it is the job of the HSE officer to detect, record, describe, and establish the change requirement for the work instruction that has been monitored. This is a constant and ongoing process in order to make the working environment as safe as possible for everyone involved in the installation.

THE HSE AND NATIONAL AUTHORITIES

Today, wind farm installation is carried out in many countries in Europe. The installation contractors, the service suppliers, and certainly the component manufacturers are also from Europe. This creates certain problems in that the various HSE regulations are not the same, so the HSE regulations in one country that are valid for one stakeholder are not relevant to the other.

Further, a working procedure that is deemed safe in one country under the relevant HSE rules can be declared unsafe in another country for another project. This creates more problems because the method statements and working instructions will vary from project to project. Therefore, adapting the HSE documentation to suit all projects is imperative. This, unfortunately, is a very labor-intensive process because all documents must be validated against the various national rules and regulations.

As an example, using two cranes to lift and upend materials is a common practice in many countries, but it is not done in the United Kingdom. So what happens when you have to upend the components in Denmark or Germany, transport them to the U.K., and then place them horizontally on the pier? Well, in a case like this, you need two sets of method statements for the same job that must be changed to accommodate both ports. Until we figure out an easier, less expensive way to do this, it will mean adding time and cost to projects.

Related Images

The loading of the vessel is always a tight fit; all space is used. This, however, makes operations potentially hazardous. Therefore a risk assessment of work, vessel loading, and mitigation of safety hazards is always carried out.

A tower section makes its way to the final destination on the offshore foundation.

The North Sea highway! The gangway from the installation vessel to the turbine is a safe and effective way to transit people and small tools while work is ongoing. It also serves as a safe escape route from the turbine in case of an emergency.

Securing the crane hook is part of the work carried out prior to leaving port and/or installation sites.

CHAPTER NINE

Work Vessel Coordination

The need to coordinate vessels is the result of the requirement of managing the traffic in the confined space of a building site offshore. A unit is set up to deal with the movement of material, machinery, and manpower. Why is this necessary?

When you look at a map of any given offshore area and draw a boundary around the part where you want to work with the installation of an offshore wind farm, you will come to the conclusion that, at face value, there is a lot of empty space, even when the turbines have been plotted in. So why would you need to regulate the traffic and the number and location of work vessels at all? It would seem that the area is big enough to host a large number of vessels going about their business without even a remote chance of interfering with one another. Can this be true? Or is there a good reason to enforce a specific set of rules and regulations for the activities on it and the transit area to and from the site? The answer is, of course, yes.

The best way to demonstrate the need to regulate traffic and vessel behavior onsite and to and from port is to draw the construction site boundaries and start to plot in the various vessels with their anchors deployed in the water. When you realize that the vessel's anchors and the mooring lines actually stretch out over the seafloor for several miles when deployed, you also understand that even though the area looks vast, it very quickly fills up; that is, when some vessels are moored, some are hindered in their ability to navigate, and, for example, the cable work has to be carried out in lines that cross the transit or mooring paths of the other vessels in the area. For this, a small but specialized organization of experts is set up: traffic coordination.

ORGANIZATION SETUP AND FUNCTIONS

The traffic coordination organization is an actual location with marine crew such as master mariners, navigators, or other skilled personnel who can survey the entire traffic in the area in which the wind farm is to be located.

Furthermore, the experts monitor all of the traffic going in and out of the staging port and determine whether they have any business related to the installation work—after installation and during maintenance—or whether they are just supposed to pass by the wind farm without interfering with any activities related to the area and the wind farm itself.

The organization will normally consist of a two- or three-person group working around the clock during installation and repair work, and mainly during operational hours—that is, daytime during operation and maintenance. The Vessel Traffic Coordination Center (VTCC) is equipped with the most up-to-date surveillance equipment, such as radar, AIS systems, and highly complex software programs, that can track and monitor in real time all of the activities onsite and going to and from port.

From the center, all daily traffic is planned and the routes of transit and intersite movement are marked on the computers in order to see whether or not one activity interferes with another. This is very important since, for example, laying cables should not take place in areas where other vessels are anchoring or carrying out subsea activities such as dumping stones as scour protection or, even worse, interfering with the diving work that could be going on simultaneously.

It is also important to know how many individuals are onsite. Therefore, all of the vessels going in or out will have to declare the number of people on board (POB) in order for the VTCC personnel to know how many are at the site; en route to or from port; or, in the worst case, missing. It is therefore crucial to have a system that can track every single person working offshore.

In case of an emergency, the VTCC personnel should know who is where and doing what. If a rescue operation is initiated on the basis of a missing person, the rescue crews will need to know where to look, for whom, and what this person is supposed to be doing when he or she is finally found.

The worst thing that can happen in an emergency operation is if the rescue teams spend a lot of time searching for an apparently missing person who is not actually offshore. Valuable resources are lost, as well as the time wasted searching for the person.

Therefore, it is vitally important for the VTCC personnel to have a POB tracking system and an accurate count of the number of people on board and who they are. Several systems are available that can track people, and they have been used on offshore projects over the past several years with good success.

VTTC Rules and Regulations

The International Maritime Organization (IMO) has developed a global convention and set of rules for the regulation of marine traffic that all vessels must obey. They are the result of hundreds of years of navigating the seas and are updated frequently by the IMO convention in order to always operate a set of navigational rules that match the types of vessels and traffic that corresponds to them, their work, and the density of traffic in coastal areas.

THE OPERATIONS CENTER AND THE WORK CARRIED OUT

The operations center is basically an office that is hooked up to radar and other navigational aids, and its role is to observe the traffic going on in the construction site and the approach area, including the ports and navigational channels. Furthermore, the operations center has a central role in planning for future procedures:

- Which activities are going to happen in port, in the transit channel, and onsite for the next period of 24, 48, and 72 hours?
- Which vessels are going to be there?
- Which activities are they going to perform?
- What will be installed in the wind farm itself and how many vessels are going to be involved?
- In which areas of the wind farm will they be working?
- Are any of the vessels going to interfere with one another?

The operations center plans and the coordinates all of these activities. They should be matched up against upcoming weather forecasts, and if the weather is marginal for an operation—say, for cable laying, which normally is subjected to lower weather restrictions than, for instance, the foundation installation—the planning should take this into account.

If the cable layer is located near a turbine installation position and the installation vessel for the turbine is positioning itself by means of anchors, this should be observed and planned for accordingly. However, if the weather is marginal for the cable layer, the possibility of the vessel not working should be accounted for, giving the turbine installation vessel some more degrees of freedom when working. On the other hand, if the weather changes and there is an opportunity for the cable-laying vessel to work, the turbine installation vessel will be subjected to the original restrictions, and the planning procedure will need to be adjusted.

Traffic Coordination

Naturally, in a large project, such as an offshore wind farm, a lot of vessels will be at the building site at the same time. Therefore, a well-functioning traffic coordination policy must be put in place in order to have the right personnel; the right equipment at the right place; and, of course and most important, to avoid collisions.

For this purpose, a traffic policy and a corridor where the traffic can be controlled must be established in coordination with all parties involved. This can be made with two lanes or with separate lanes for in- and outgoing traffic, and it must be tied to a point at the wind farm location. Then the traffic can be controlled safely, and it will make it possible to secure the traffic zone because the area can be declared an exclusive zone during the building period.

Traffic rules for the building area must be made in such a way that the vessels that cannot maneuver themselves have exclusive rights. Because this can mean extra cost to the contractors, it must be taken into consideration during negotiations, and then it can mean extra waiting hours and, again, a delay in the contractor's timetable.

Traffic Control

Control and supervision of the traffic situation in the harbor, on the way, and on the offshore building site must be performed by a coordination group. Therefore, a watch group will be responsible for coordinating the transport of personnel, vessels, and equipment during the entire project, 24 hours a day, 7 days a week.

It is also important for this group to coordinate the building traffic with the normal traffic in the harbor because when foundations or turbines must be transported out, other vessels will have problems coming into the harbor. And when transport vessels are having difficulties maneuvering themselves, all of the other vessels must give them exclusive rights.

The coordination center must have radar surveillance, combined with a VHF-based communication system, in order to coordinate with all vessels when and how many personnel are on the way. In case the control center is far from the site, a satellite-based communication system will be necessary because the VHF system has only a limited range. This will make it possible for the guard and warning services to react in order to have sufficient vessels, helicopters, or other transport means standing by if an accident should occur.

This way, the operation center can be contacted quickly in the event of oil spills, and then the staff at the operations center will know exactly how many vessels are at the building site. Then the proper authorities can be informed quickly, and the area can be secured and the risk of an accident can be minimized.

Naturally these conditions also apply for search and emergency services. They must know exactly who is missing and where the person is supposed to be. This saves time in a search operation and also provides a much higher chance of finding the person.

Guard Vessel

The final item under traffic control is the guard vessel. In Germany, this is a requirement and is actually a vessel that has no other assignment to the project than that of looking out for vessels that are on a possible collision course toward the wind farm and/or the vessels working in and around the exclusion zone. The guard vessel must patrol the area 24/7/365 to secure the area.

ORGANIZATION OF SURVEILLANCE

Surveillance can be done by a coordination center, with a manager and six employees placed in the staging port. The surveillance center must be placed close to the staging area because of the distance to the building site, and it should be located near the project management offices, since they will have close daily contact and an overall requirement to coordinate the offshore work and, of course, the traffic going in and out of a port.

The usual recommendation is for the main center in the port to have three employees who will be onsite 24 hours a day to work as part of the surveillance on the jobs performed by the offshore subcontractor's personnel. Thus, the control center will have the necessary personnel, and the correct amount of handling equipment available for the safety of vessels.

●●●

Related Images

Crews often do "flying changes" on an installation vessel; that is, they are transported out to the working vessel along with other goods. This saves time and is possible because the onshore/offshore coordination work is closely monitored and managed.

Bridge personnel at work on a vessel while underway. It is important to have a relaxed atmosphere and good cooperation between all parties involved.

A jack-up barge being towed from port to site with a monopile onboard. This is a slow process and is much more weather-dependent than working with self-propelled equipment.

CHAPTER TEN

Logistics Solutions

What is *logistics* actually, and why is it important? When you google the word *logistics*, you get the following definition:

> Logistics is the management of the flow of goods, information, and other resources in a repair cycle between the point of origin and the point of consumption in order to meet the requirements of customers.

This means that an optimum relationship exists among the need, the availability, and the delivery of goods or a service from the point of origin to the final destination. The idea is also to avoid excess availability of any such goods or service anywhere in the process or logistical chain.

In layman's terms, logistics means that when you buy the last tomato at a farm stand, a new tomato plant has sprouted from the ground. In other words, no resources are wasted, and the required materials are exactly where they should be at exactly the time they are needed. This concept is also often referred to as the "just-in-time" principle, where a component moves just one position forward in the chain to take the place of a position that has been vacated.

This way of carrying out the logistical operation for an offshore wind farm is, however, too simple. There are a number of reasons for this, and mainly they evolve around the fact that it is not possible to deliver a promised number of foundations, turbines, and cables using the just-in-time principle.

The problem actually starts at the *end* of the process. The time frame for the installation offshore is limited due to weather and the capabilities of the equipment that is used. This is defined as *weather criteria*, which we discussed earlier, and that determines the actual operational envelope for the equipment and personnel on a specific site during a full year.

If the installation envelope only allows the vessel to install the components for five or seven months during the spring, summer, and fall, the entire construction program will focus around that. So if 100 turbines, 100 foundations, and 100 cables are to be installed during this period, there are only a few days open for each component to be installed.

The delivery of all of the components in the unloading port must then take place just prior to and during the installation time period. This creates a massive constraint, for obvious reasons. To unload and install, say, the turbines, a total of 100 days is needed, working from the assumption that the gross installation time per turbine is one day. However, all of the foundations, or at least most of them, must already be there because turbine installation is faster than foundation installation.

But the manufacturer cannot deliver 100 turbines in 100 days. The delivery and preparation for unloading take much longer. This means that there is a mismatch between the installation time and the manufacturing and preparation time, and this creates the requirement for a buffer storage of components. That means that the just-in-time principle can no longer be applied, but the logistical challenges just got much worse.

We will consider an example that demonstrates some of the issues the project management will experience. Let us say that all of the components are very big and heavy, and the port facility is not necessarily capable of storing many components at one time. Ideally, the average 80-turbine project needs around 65,000 to 70,000 m^2 of flat, well-paved, high-load-bearing surface.

This type of area is very coveted by port clients in general and therefore is the most precious land to rent out for the port. Typically, this is something a long-term tenant will qualify for. The direct access to the pier is an added challenge because easy access for any port customer means a very short turn-around time for their vessels and cargo. For the wind industry, it simply means not having to transport the components too far in order to load them.

Furthermore, the rotors, towers, and nacelles take up an enormous amount of space when preassembled and ready for unloading. This means that little or no activity can take place concurrently with the unloading of the turbines. Asking for this type of service for a relatively short-term contract is not the sort of business proposal the project manager can sell easily to the port. This is the first logistical challenge the project execution team will face.

So this means that the port facility has to reserve a very large, well-paved, and easy accessible area for the storage of components in buffer storage. However, ports are not so keen to reserve areas of the size and type ideally demanded by the wind industry, and there are many examples of ports not welcoming this type of business at all, let alone reserving an area of this size.

In 2001, the project manager of a client was almost thrown out of a port office for insisting that they were among the largest customers the port would ever have. The port captain pointed out the window at a ship loading oil and gas supplies for an offshore field and yelled at the project manager that the day she could bring in this magnitude of business for a 20-year period guaranteed, he would consider trading piers, but until then, the area he had specified was what the customer could get; "If you don't like it, go somewhere else."

Knowing also that most ports do not have this type of area readily available, the few ports in the interesting areas (for the offshore wind industry) that have them will have the various project buyers and project managers beating at their doors. This is another problem that ports would rather not have.

Therefore, planning for and renting a suitable staging area in a port that is favorably close to the offshore site are tasks that should not be underestimated. Usually this is part of the tender documentation. The wind farm owner will have to nominate one or more ports believed to be useable as staging areas for the project. However, the contractor and/or the project manager must decide, together with the tendering team, which port to use and how to execute the transport to the site; storage of components; and preassembly, loading, and transport to the offshore site. Figure 10.1 shows an example of a staging area in a port that was used for previous offshore wind projects.

Figure 10.1 Loading and storage in Lowestoft for the Scroby Sands project in 2004. Notice the confined space. *Courtesy of A2SEA.*

●●●
Related Images

Fitting a Vestas blade in place onsite. This is the most critical lift during the installation since the blades are delicate and the space in which to fit them is very tight.

Rigging and crane operations involve large equipment; what needs to be handled is heavy and potentially dangerous. Therefore, staff training and work experience is very important.

Shown here are some installation vessels in port.

CHAPTER ELEVEN

Commonly Used Installation Methods

The method of installation of both foundations and turbines influences the staging port layout, size, and location to the offshore wind farm site. Manufacturers and contractors have developed specific strategies and working procedures for all of the different types of foundations and turbines. Therefore, it is obvious that the individual project can be executed using various methods of installation, depending on the foundation and turbine chosen. The sections that follow discuss the main characteristics of the methods and possible future trends.

FOUNDATIONS

In general the industry prefers working with four types of foundations: gravity-based, monopile, jacket, and tripod.

Gravity-Based Foundations

Gravity-based foundations stand by weight on the seabed. These foundations, which weigh over 2500 tons, are very common in the industry. It is not possible to manufacture a large number of gravity-based foundations, because they are cast in steel-reinforced concrete using prefabricated molds. The molds are so large and the process takes so long to prepare rebar and the mold for casting; thus, the contractor typically sets up 6 to 8 molds and starts manufacturing the foundation at least 9 to 12 months prior to installation offshore.

This, however, means that a large number of foundations have to be handled and stored. Once again, there are two obstacles to be met, both relating to the port facility. The picture on the last page of this chapter shows the casting and storage of gravity-based foundations. The area for manufacturing and storing is quite substantial.

Furthermore, it is almost self-contained, but the sheer weight of the foundations will require a lot of ground-bearing capacity in order to withstand the pressure of the individual foundation. Therefore, it is common to wet store the foundations offshore near the installation site and thereby relieve the pressure on the staging port.

Installing Gravity-Based Foundations

The installation of a gravity-based foundation is largely carried out using three pieces of equipment:

1. A large floating crane. The capacity of the crane should be more than 2500 tons in order to actually lift and place the foundation on the seabed.
2. A very large barge that can transport and possibly store a number of the foundations onboard.
3. A tugboat or group of tugs that can tow either the barge or both barge and crane into position in order to set the foundation in the right location.

In addition to these pieces of equipment, there are, of course, several vessels carrying out different jobs to prepare the seabed.

This would include dredgers that can level the seabed in order for the dumping vessels to lay a stone cushion so that the foundation can be lowered onto firm and level ground. Also, there would be vessels laying out the scour protection to prevent the foundation from being undermined by seabed currents that will try to flush away material from around the foundation when it is installed.

As just stated, the fabrication of gravity-based foundations must start well in advance of the installation in order to have a buffer of foundations so the vessels—when started—will not run out of foundations and then would have to wait on standby. Typically the fabrication will start one year in advance of offshore installation; and around four to six months prior to installation, the seabed preparation work along with the cable trenching will commence. When the installation season starts—typically in March or April—enough foundations will be ready for the installation to end a few weeks after casting of the final foundation.

The seabed preparations will also be well enough advanced for the foundations to be installed, and thereby the cabling work can commence shortly after. This is a well-organized working plan and so far has been a very successful one. The wind farms in the Danish and Swedish parts of Øresund (the sound between Denmark and Sweden) and in the Danish area of the southwest Baltic around Lolland (Nysted 1 and 2) have all been installed using this well-proven method of operation.

The challenges are few, and the main problems that the projects in the Baltic have run into have been ice developing in the ports where the foundations were cast; for example, in Swinoujscie in Poland during the tough winters there.

Monopile Foundations

A monopile is basically a long tube that is driven into the seabed by means of a very large hydraulical hammer. The tube is made of steel and welded together from cans, which are in effect steel plates of varying dimensions rolled into round form and welded together using specialized machinery.

Monopiles are the most often used foundations in the offshore wind industry for several reasons:

- They are relatively simple to define for the designers.
- They are relatively inexpensive to manufacture.
- They are fairly easy to handle and store.
- They are simple to install and maintain.

Figure 11.1 shows monopiles loaded on a barge in two layers. The top layer has been installed, but the cradles for the piles are still visible for two of the

Figure 11.1 Seafastening of monopiles on a transport barge.

three piles in the layer. In this way it is possible to increase the efficiency of the installation process and to make the slow transporting by barge more cost-effective in general. The seafastening of the piles is, however, both expensive and engineering intensive, as can be seen in the figure.

There are more good arguments for using monopiles, but the preceding reasons are enough to persuade the owner of a wind farm, if it is at all possible, to work with a monopile foundation. This is the easiest, fastest, and most well-defined method of manufacturing, installing, and maintaining the foundations for the wind farm.

Monopiles are indeed easy to install, provided that the equipment, method statements, site preparations, and soil data are well defined and chosen. Surprisingly, monopile installations have been carried out with a very large bandwidth of success in the past. This is not necessarily because the monopile is difficult to install but because the challenges have not been taken seriously enough.

The monopile is, however, not the final foundation. The pile driving process actually, under the circumstances, will forge the top of the pile due to the constant pounding on the surface with the hammer. This leads to the metal becoming brittle and thereby unsuitable for any load bearing.

Therefore, the common solution is to fit a transition piece, which is another oversized pile on top of the monopile, so that the monopile is slotted into the transition piece over a distance of 6 to 8 m. The transition piece is then adjusted to true verticality, and the annulus between the pile and transition piece is filled with a high-density concrete, commonly called grout.

This is a good solution in order to construct the wind turbine, which is then installed on the transition piece. This piece therefore holds all access platforms, cable tubings, and other appurtenances to create a functioning access platform for the turbine. The main pieces of equipment that should be used for the installation of monopiles are as follows:

1. Installation vessel (discussed in more detail in Chapter 12); the preferred type of installation vessel is the jack-up that can load and transport the monopiles and transition pieces on deck.
2. Large hydraulical hammer for the pile-driving process. The hammer comes complete with power packs and control room for the monitoring and controlling of the driving operation.
3. Pile handling tool that is used for holding and positioning monopiles vertically before and during the driving of the pile.

4. Grouting equipment in order to cast the monopile and transition piece together.

5. If required, a drilling rig, which would be used if there are large boulders or hard ground underneath the pile. The drilling rig is a reverse circulation drilling system that will be placed on top of the pile, and the large-diameter drill will then drill internal relief so that the pile can be driven once the material has been removed.

For the monopile, the principle remains the same as for the gravity-based foundation. There are a number of other vessels and a lot of other types of equipment that are used to carry out the foundation work. There are scour protection installation vessels that—as for gravity-based foundations—will install a filter layer, which is a stone layer of small (10–20 cm) stones that will be used as a cover layer on top of the seabed to create a solid unscourable surface around the pile when it is driven.

Unlike the gravity-based foundation, no seabed preparations are made for the monopile. Other than laying out a filter layer, nothing is normally done. The savings are therefore significant in comparison. Both time and expense can be reduced dramatically, and this is another reason for using the monopile type of foundation.

The monopile will be driven through the filter layer, and once the cable has been fitted—via a special tube called the J-tube, the cover layer—stones from 30 to 60 cm will be dumped on top to lock the filter layer to the ground (Figure 11.2).

Today, the monopile has been used in the vast majority of wind farms and, interestingly enough, where the consensus three or four years ago was that the monopile was not cost-effective over 25 m or more, it now seems that projects with up to 30-m water depth are considering the use of the monopile. Furthermore, the ease of installation and maintenance makes it interesting for wind farm project owners to consider; the added material and cost of this, compared to the installation cost, and the proven ability to serve as offshore foundations is a reasonable tradeoff even at extreme water depths. This means that the monopile foundation will be a valid option for many years for offshore wind farm installation.

Jacket Foundations

A jacket foundation is a lattice structure made using the same principle as a lattice tower for a TV aerial or radio transmission mast. A square cross section with the structural members is fitted to four corner tubes to create a

Figure 11.2 This monopile foundation is ready for installation of the tower. *Courtesy of A2SEA.*

light and strong construction that will withstand large forces at light weight.

For the offshore jacket foundation, the idea is to have a small cross-section in the splash zone—the area in the water column at the surface where the waves are strongest—and thereby to reduce the forces created from the waves when passing the foundation. Contrary to the gravity-based foundation, the monopile, and the tripod, the jacket foundation—since it is made from small-diameter steel tubes—does not have a large surface that the waves can attack. The other types of foundations all have a large cross-section in the surface region due to the tubes protruding from the water. This means that the jacket can be built lighter, but there is a price to pay.

While the jacket is strong and lightly built, the nodes—the positions where the tubes are welded together—are extremely difficult to manufacture, and the cast high-tensile steel nodes are very expensive. Furthermore, all the weldings on a jacket are handmade. It has not yet been proven efficient to manufacture jackets using welding robots and other types of automation in the process. This is not the case for the monopile, which is largely manufactured using welding robots and a large degree of automation.

So while the jacket presents some interesting advantages for the owner in terms of low wave loads, high capacity to carry turbines, and the ability to stand in very deep water, the downside is, of course, the price and the complexity of manufacturing the jacket.

It is also important to note that the same principle, which dictates that the gravity-based foundation must be manufactured well in advance and takes up a lot of onshore storage space, is valid for the jacket (and the tripod) as well. The manufacturing time for a jacket is longer than one week, if not calculated in four to six weeks. This means that just-in-time manufacturing is not feasible. Furthermore, in order to stand by means of the slender lattice structure, the footprint of the jacket becomes very large. Usually the footprint is around 24 × 24 m.

Therefore, given that this type of foundation is used in very deep water, the length of 50 plus meters will prohibit horizontal storage due to an area's capacity. You must be able to move the foundations around in order to load them on the transport/installation vessels. So the obvious solution is to stand the foundations up for as long as possible. However, there is a limit to doing this as well: wind.

The wind will, of course, try to topple the foundation, and this will be a major criterion when deciding where to store the foundations. The challenge has not yet been addressed due to the small numbers of jacket foundations previously installed on one single site. The Alpha Ventus site has six jacket foundations and six tripod foundations, and storage was carried out at the fabrication yards: large ship/oil rig yards that had the capacity to store such numbers. It will be interesting, however, to see what happens when 60 units need to be stored for a project.

Because of the size of the foundations, they will probably be built in a shipyard, so it will be helpful to arrange for a storage area in the port. For example, if the foundations are supplied to a German project, they could be built by a local manufacturer such as WeserWind on the North Sea coast.

With a footprint dimension of 24 × 24 m, even a few foundations will take up more than 6000 m², and this area must be directly in front of the bulwark of the pier because transport of the foundations is very difficult, although it can be done with extremely large equipment, such as SPMT trailers or the like (Figure 11.3).

In case, for example, 80 repower turbines are chosen, the logistical challenge can only be met by using such equipment, since the need for storage space far exceeds the 40,000 to 60,000 m² that is normally required for foundation and turbine installation. Furthermore, a setup where at least 40 of the

Figure 11.3 Port transport of a jacket. *Courtesy of Talisman Energy.*

jacket foundations are manufactured early can be the consequence. Even more likely is the need to split the fabrication of the foundations between two or even more yards.

Installing Jacket Foundations

The installation of the jacket foundation is actually fairly straightforward. The principle is that the jacket stands on four anchor piles—one at each corner of the foundations. The anchor piles are fitted to the jacket using so-called pile sleeves, and the connection between the anchor piles and the jacket is established by means of grouting (or casting) the anchor pile in place using high-tension concrete (Figure 11.4).

However, the process is slightly more complicated, since we didn't mention anything about how the anchor piles got into the seabed or how the jacket was placed on the anchor piles. And, as always, the devil is in the details.

TRIPOD FOUNDATIONS

A tripod is a different type of deep-water foundation. While it has a relatively small cross section in the splash zone—due to only having one large tube, compared to the jacket, which has several smaller tubes protruding from

Figure 11.4 Pile sleeve on a tripod foundation. The principle is the same for the jacket, only the support structure is different. *Courtesy of renewable energy sources.com.*

the splash zone—the tripod also uses a number of anchor piles fitted through the base of the foundation (Figure 11.5).

INSTALLING FOUNDATIONS

Most projects can be accommodated by using one or more of the preceding four types of foundations. A monopile is suitable for shallower sites of up to 25 m. Some examples are the projects Horns Rev, Kentish Flats, Rhyl Flats, Barrow, and Greater Gabbard. The method of installing is simple and is done quickly, compared to the jacket, tripod, and gravity-based foundations.

Due to the size of the foundations, the wind farm owner should consider whether it will be possible for the manufacturer to deliver directly at the offshore position (Figure 11.6). This is, of course, particularly relevant if the supplier is from outside the vicinity of the project. An example would be if the supplier was from Denmark—say, Bladt Industries—and the project was in the United Kingdom.

If this were the case, would it make sense to transport the foundations to the onshore port facility in the United Kingdom, store them and reload them onboard the installation vessel, and transport them to the site to install them? Most likely not. The direct route is more cost-effective, and if the

Figure 11.5 Tripod foundation for the Alpha Ventus project. *Courtesy of the website renewableenergysources.com.*

Figure 11.6 Delivering from offshore. *Courtesy of A2SEA.*

supplier has enough storage capacity, it will be easier and less expensive to do a direct transit to the offshore site.

This becomes very interesting both timewise and financially because storage in a yard is normally much less than in a commercial port. The can be done by loading the foundation onto a barge from the supplier, and the barge will then be towed directly to the offshore position. Then, the foundation will be erected and placed by an installation vessel. Upending a monopile is both complicated and dangerous. Some piles have actually been dropped by inexperienced contractors. In this case, however, the A2SEA crew has everything under control (Figure 11.7).

All four types of foundations are so rugged that it is possible to mount them with a floating crane. This is already done in the offshore oil and gas industry and more recently with the monopiles on the Baltic 1 project in 2010. Therefore, this method can be used for the erection of the foundations as a good, safe, and tested method.

Considering that the foundations—in particular jackets and tripods—can be transported either standing or laying on the barge, it can be determined how far away from the site that production of the foundations can take place.

Figure 11.7 Upending a monopile on the deck of *Sea Worker*.

The wind farm owner should consider production in places where the cost is low because transport on a barge is cheap and reliable. Calculations on this point should be dealt with in the prequalification phase—that is, whether to manufacture foundations in places like the Baltic area, Spain, Portugal, or even further away, and then transport them directly to the offshore site for erection.

Transport for Foundation Installation

The transport for installation of the foundations can be done in two different ways. With a barge and a tugboat, transports of foundations directly from the manufacturer to the offshore construction site are easily done (Figure 11.8). Then, by using an installation vessel, the foundations are installed into the seabed. This method is considered feasible, safe, and reliable and is used already in the oil and gas business.

With a specially equipped crane vessel, the foundations can be picked up directly from the manufacturer. This will mean a shorter distance from the manufacturer to the construction site. Picking the foundations up directly from the manufacturer is preferable so that the crane vessel does not have to be under charter for too long.

Figure 11.8 Offshore transport of a jacket foundation. *Courtesy of Talisman.*

The foundations must be loaded in the shortest possible time and transported to the construction site. Upon arrival at the onsite location, the foundation is lifted from the crane vessel or barge and positioned and fixed with piles driven into the seabed. This also is a good and reliable, but far more expensive, method because the crane vessel is more expensive than a barge, tug, or sheerleg crane. In addition, the installation process must be carried out quickly in order to save money.

Both methods should be considered and as much information as possible should be obtained before the project is begun. Choosing the best transport and mounting solution will save both cost and time. The logistics of the project can change dramatically, however, if the Vestas turbine is chosen. Then it will be advantageous to move all of the foundations from the manufacturer in large batches directly to the construction site and to keep storage of piles in the Bremerhaven harbor.

The second solution gives very limited choices on manufacturers and a loss of possibilities if the prices from low-cost areas are not investigated because the quality normally is very good. The range of floating cranes in northern Europe is excellent, and a competitive price can be negotiated. Furthermore, experience has shown that installation can be done quickly and favorably (Figure 11.9).

Figure 11.9 Photograph of one of many floating cranes available. *Courtesy of Talisman.*

Furthermore a combination of tugboat and barge can be made, both in Germany and other European countries. Competition between the different contractors gives better transparency in the price calculation, which can give the wind farm owner a fair chance of saving money.

Securing Installed Foundations

When the jacket foundations are erected, they must be secured with the driven piles. This must be done at the same time as the positioning of the foundation, which means that a rather large barge is needed because the piles are 30 m long or more. All of the piles must be transported at the same time and driven with a hammer before the foundation is secured. The grouted connection between piles and pile sleeve in the jacket must be cured before it is possible to erect the turbines.

Because of the small diameter of the piles, it will be possible to choose between several subcontractors. Normally this is offered by offshore subcontractors and gives a seamless installation procedure. When the piles are driven into the seabed, the foundation must be leveled. This is done by preinstalled jacks on the pile sleeve. Then the annulus between the pile and sleeve is filled with a high-density silicate concrete and left to cure.

When the grout has cured, the hydraulic jacks can be removed and used for the next foundation. In comparison with the monopile, the erection of a jacket or tripod is far more cumbersome, but it certainly lasts longer. This must be considered when a time schedule is drawn up. However, if a Vestas turbine is chosen, a large-diameter monopile is needed as a foundation.

Only a limited number of manufacturers can be found in Europe, since rolling of thick plate dimensions and welding the tubes are work for specialists. One of the most experienced is the Netherlands-based Smulders Group, which is capable of manufacturing large numbers of monopiles and has done so for several offshore wind farms.

The pile driving is simple and straightforward. The pile is driven to the desired depth, and a transition piece is mounted and leveled in the same way as the jacket—by means of hydraulic jacks—and the annulus is filled with grout and left to cure.

IMPORTANT THINGS TO CONSIDER

All in all, the following must be considered when transporting and erecting any foundations.

Harbor

It must be possible to manufacture the foundations either close to the pier edge or by a manufacturer with unhindered access and ability to transport to the edge of the pier, or from another place of manufacture directly to the offshore site by using a barge or crane vessel.

Loading

In the harbor, it must be possible to either lift or roll the foundations on-board. The best solution is rolling when using a barge, and the cost of the transport process will be low. When using a crane vessel, it could also be done elegantly, but probably the cost will be higher.

The following must be defined:

- Access and criteria for loading on the vessel or barge
- Criteria for seafastening—that is, procedures and equipment
- Conditions for approval of seafastening

Transport

The foundations can either be transported by tug and barge to the construction site or loaded on a crane vessel. It should be noted that the "Nordschleuse" lock in Bremerhaven is only 35 m wide. This means that at least two, maybe three, of the preceding offshore contractors will be excluded because their equipment cannot pass the lock. It is extremely important to make bidders for the project aware of this in case Bremerhaven is chosen as the site harbor.

During and after Departure

It is important to pay attention to the items described in the following sections.

Installation Offshore

The installation procedure should, as an absolute minimum, include a detailed summary of the various manuals and instructions relating to both vessels and crews. This is required in order to assess whether there is sufficient capability and resources for carrying out the work as desired, without risking either equipment or personnel.

When fitting components, attention must be paid to weather limitations on the chosen equipment, and before leaving the harbor, it is essential to

Figure 11.10 This is an unusual photo of the North Sea and was taken when installing the Beatrice turbines. This is not how most of the days are in that area. *Courtesy of Talisman.*

have a recent weather forecast available. The importance of long-range and accurate weather forecasts is underlined by the fact that traveling distances are more than 100 km and at a speed of around 6 to 8 knots at best, traveling times last up to 10 hours (Figure 11.10).

Lying idle in the port or onsite without any possibility of working for long periods, and not being able to return to port, can be a disaster to the time schedule. Thus, the offshore contractor and the wind farm owner must jointly decide on a supplier of weather forecasts that can be accurate for up to 72 hours and provide fresh updates at a minimum of every 6 hours.

Positioning the Vessel and Wave and Current Conditions

When it comes to jacking procedures, how and at what wave heights and periods can the vessel operate? What about calculated penetration, preload times, and pressures? For this, a determination of methods, as well as specific instructions for the project, must be made.

Figure 11.11 This photo shows a more typical day in the North sea. The wind blows and the waves are reasonably high, if not very high. *Courtesy of Talisman.*

When considering lifting methods and lifting procedures, the capacity and type of equipment must be determined. Which wind force can the lifting withstand, and how long before the limit is reached must work stop in order to secure the crane and components from overload? Who does what during the procedure, and who is responsible for the correct lifting equipment and pile driving spread? (See Figure 11.11.) In addition, working procedures and instructions for the specific project must be made because a method that worked for one project might not work for another.

Pros and Cons of the Different Methods

So, why is one method better than another? And why does one method work better than another? The reasons are many, and they are worth discussing.

If you choose a monopile and install it using an afloat type of vessel, the timing of the process will be critical. The normal weather window in which this process can be made is lower than for a jacked-up vessel. But the isolated cost of the installation vessel can be lower. However, as we will see later, the cost of the entire BOP will be higher because the requirement is for more equipment and a longer duration onsite.

If you choose a different type of foundation—say, a tripod—then some of the installation vessels, afloat or jacked, will not be able to handle the foundation for different reasons. Here are some examples:

- The *Svanen* with more than 8000-ton lifting capacity
- The *Resolution* with 600-ton lifting capacity
- The *Sea Jack* with 1200-ton lifting capacity
- The *Swire Orca* with 1500-ton lifting capacity

So what makes one or more vessels capable of handling a tripod and not others? The reasons could be the following:

The *Svanen* cannot lift and place the foundation because the footprint of the tripod is too large to fit between the pontoons of the installation vessel. Even though the vessel has the highest lifting capacity of all the ones just listed, the component—in this case the foundation—will not fit where the crane hooks can lift and place it.

The *Resolution* may be marginal, even though the lifting capacity was recently increased from 300 to 600 tons. But this capacity lies close to the stern of the vessel, and again the footprint of the foundation may prohibit it from being able to launch it over the stern. This must in any case be carefully investigated before chartering the vessel.

The *Sea Jack* from A2SEA may be able to install the tripod provided the foundation is turned to the right side against the vessel. However, it will not—like the *Svanen*—be able to transport it out to the site and therefore automatically requires a transport barge. But this in itself may not necessarily disqualify the vessel for the process.

The *Swire Orca* was designed to be able to transport this type of foundation and will surely be able to transport and install it. But the *Swire Orca* is no longer available, so we have to find another solution.

The tradeoffs are therefore cost, availability, and capability of the vessel. The contractor may choose a less expensive installation vessel in order to get the job done. There might be no other alternative to chose from, so the weather criteria will be lower than the optimum solution. Furthermore, the cost of tugs, barges, and other support vessels will in the end be higher than the ideal case, but this is the reality the contractor has to work with.

This means that until the market for installation equipment is filled up with second- and third-generation vessels, the industry will keep seeing solutions that are the best possible alternatives from the available tonnage in the market. This will keep costs high.

How Does the Chosen Method Work?

It is, of course, completely individual how the chosen solution works. But let us for argument's sake, take the *Sea Jack*, a transport barge, and two or three tugs, and let us run through the sequence.

The first task is to load the tripod onto the transport barge. This is done in port where the tripod is manufactured, and here there can already be some challenges. Some of the ports where tripods are manufactured have navigational limitations. Height is an issue if there are bridges, such as Lowestoft in the United Kingdom, where previously the port area for wind installation vessels was restricted to the area in front of a bascule bridge. Width can be a concern if there are locks or narrow navigational channels, such as Bremerhaven, where the foundation manufacturer is located behind a lock gate with a beam restriction of 32 m.

First of all, the barge must be wide enough to support the tripod. The footprint is around 24 to 25 × 25 m; this is quite big. A reasonably large barge must then be mobilized to transport the tripod. Furthermore, the tripod is heavy and very tall. This means that the center of gravity is positioned far above the barge deck. This must also be considered when ordering the transport barge.

A naval architect or master mariner will be able to calculate stability based on the barge characteristics; an educated guess would be that a barge in the region of 28 × 75 to 90 m will be a good choice. However, it should be noted that the size can only be determined by a direct calculation.

Once the barge is chosen, the next item is to load it. This would possibly be done by a floating sheerleg crane in port or, if possible, as a rollout by means of heavy goods trailers—also known as self-propelled modular trailers or SPMTs. The loading could be carried out in either way. The sheerleg would possibly lift the tripod from the pier and reposition it in front of the barge and offload it onto the deck.

Once positioned on deck, the tripod will be seafastened by means of brackets and knees welded onto the deck and finally secured with heavy-duty chains. The marine warranty surveyor will oversee the entire operation, and when satisfied with the proceedings, the surveyor will sign off on the transport.

The barge will be moved out by means of two tugs for maneuvering in port and through the access channel. Once again, a set of challenges lay ahead. If the navigational channel is narrow with a fixed route, navigating the barge will be difficult. Other vessels in the area must either give way or the time of transport must be planned so that the tow does not interfere with other traffic. Otherwise, the other vessels must respect the towing of the barge.

Situations similar to this occur frequently for transports of wind farm components. In several projects, A2SEA transported three or four turbines with rotors fully assembled on deck. This meant that the installation vessel

was around 70 m wide. Transiting through Storebælt in Denmark, which is
the main access route to the Baltic Sea, meant that the vessel was a severe
hazard to other traffic. The planning of the transports were therefore carried
out very meticulously, and an added complication was that in one event
there was a sailing regatta with more than 2000 sailing yachts attending.

Figure 11.12 shows the enormous fangs that the vessel presented when
going through the Storebælt main traffic route, but even so, no incidents
occurred due to detailed planning and a good HSE plan for the transport.
So it is difficult to plan and to execute the transport of a load of these
dimensions, but with careful scrutinizing of all of the parameters that need
to be considered, even this type of freight can be safely transported over long
distances.

Once at the offshore site, the barge must first approach the area of entry
in order to get to the correct position. The tugs must position the barge and,
under normal circumstances, moor it against the installation vessel or anchor
it close to the vessel. The majority of installation vessels will not be able
to accommodate the weight of the barge. If the current or the waves press

Figure 11.12 Notice the enormous overhang of the rotor blades, only 6 to 8 m
above sea level. A sailing yacht would easily be demasted if it collided with the
vessel. *Courtesy of A2SEA.*

the barge against the installation vessel, the weight could endanger the structural integrity, and therefore the normal procedure is to moor the transport vessel or barge close to but not against the installation vessel.

This is difficult, and weather restrictions will be lower than the possible maximum of the installation vessel. This is due to the danger of one of the tugs losing grip or power, so the weather window will be considered to be lower than if the installation vessel was carrying the foundation itself. If this were the case, the weather window would be defined as the tradeoff between the vessel's capabilities and the foundation's maximum dynamic contribution to the lift due to the buoyancy effect when passing through the water column.

Dynamic effect is the next issue that has to be considered. When the installation vessel lifts off the foundation, the difference in movement between the two floating bodies—or the jacked installation vessel and the floating barge—will introduce acceleration forces to the load. This force is the dynamic effect, which will be significant.

Therefore, an offshore crane is designed different from an onshore crane. The offshore crane has the dynamic effect designed into the steel structure, the winches, the ropes, and all other structural parts of the crane. This means that the dynamic forces from lifting a load will be taken up by the crane—at least to a certain extent. This must be calculated carefully before lifting any load offshore.

The lifting offshore is therefore the most crucial operation of all. Planning and preparation must be defined, calculated, and described in detail before setting out on the offshore venture. The marine warranty surveyor will ask for this documentation before loading the foundation on the barge. The preparation of all these documents must be done by the installation contractor and the manufacturer of the foundation prior to loading because once the marine warranty surveyor arrives, that person will decide whether the load goes out of port or not. The detailed lifting plans, and if possible in 3D animation (a number of software programs can simulate the transport, lifting, and placing of foundations offshore), will determine whether this is the case or not.

But hang on! The *Sea Jack* has a Manitowoc M1200 crane onboard. This is an *onshore* crane! Onshore cranes are not designed for dynamic lifting. Thus, the derating when using them offshore for dynamic lifting or lifting from a listing, or heeling, vessel, and the onshore crane is only designed from a horizontal position. (*Listing* is the technical term for the vessel having an inclination in either the longitudinal or transverse direction due to wave movement.) Therefore, an inclining position will result in a derating of the crane of up to 50 percent of the lifting capacity.

So the lifting operation (*offboard*—the technical term for lifting a load over the vessel's side when either the load itself or both load and transport vessel are buoyant—must be considered a zero seastate, or less than 0.5 m significant wave height. This is so as not to demonstrate the structural integrity of the crane. Once lifted off the transport barge, the tugs can tow it away. The installation vessel—in this case the *Sea Jack*—will have the load hanging free in the hook and can start lowering the foundation toward the seabed.

Lifting offboard will be an issue once again when the foundation is lowered through the splash zone and placed on the seabed. If the tripod is lowered too fast, the closed chambers formed by the tubular structures will give the foundation buoyancy. This has happened on previous projects where the crane operator has let go too fast. If this is the case, the foundation floats briefly, and thereby the wave load will fully impact the foundation and thus also the crane. This is extremely dangerous, especially if using an onshore crane. This procedure must therefore also be planned in great detail prior to leaving port. Once placed on the seabed, the transport and lifting procedures are concluded, and the foundation can be driven into the seabed.

The preceding demonstration should not be taken as negative but should be viewed as various tips to be considered. Hopefully, they demonstrate the number of tradeoffs that are part of the decision-making process during the planning stages in this industry, which has not yet fully matured. Furthermore, there are several problems that should be taken into consideration, that were only briefly discussed here. Issues such as the onsite scour, which is related to currents whether tidal or another type, are also important.

Scour is the phenomenon that occurs when a solid body is submerged in the water and is standing on or penetrating the seabed. The water passing by will have to change direction and flow around the body. But since no holes arise, the water passing the solid body must speed up in order to meet on the other side of the body. This causes turbulence in the water passing and creates a scour effect that erodes the seabed behind the body.

This means that the water will erode the seabed behind the leg of the jack-up when it is fixed to the seabed. Over time this will cause the otherwise solid base the leg is standing on to disappear, and the leg will have no support any longer. This creates a dangerous situation where first the jack-up will start to lean to the side where the seabed is eroded, and if not detected, the jack-up may capsize.

This should, of course, be avoided at all costs, and the normal way to ensure that the scour is not affecting the jack-up when working is to penetrate the seabed deeply enough to secure to firm ground regardless of the scour effect. In short, the leg is normally flushed down through the top layers of the seabed to a depth that will be more than the calculated scour effect over the entire period when the jack-up is positioned and jacked on the position. This is a normal procedure to mitigate the scour effect on the vessel and should be observed always.

Related Images

Placing the bottom tower section on the foundation.

Lifting tower sections 2 and 3, which have been preassembled in port prior to load-out to save time offshore. Lillgrund Wind Power Plant, 2007-08-04. Installation of wind turbine from M/S *Sea Power*.

Lifting the nacelle in place after bolting the tower sections together. Lillgrund Wind Power Plant, 2007-08-04. Installation of wind turbine from M/S *Sea Power*.

Topping the process off by lifting and installing the rotor. Once this is done, the "mechanical installation" is finished. The turbine can be secured and left, then the crew moves on to the next one. Lillgrund Wind Power Plant, 2007-08-04. Installation of wind turbine from M/S *Sea Power*.

Monopile foundations installed and ready to receive turbines.

Lifting and placing a monopile in the pile gripper. This process is called "stabbing" and here the challenge is to get the pile completely vertical before driving it with the hydraulic hammer. The process may take a while if you don't get it right the first time. The pile will then be lifted slightly and placed again until it is perfectly in place on its final position.

Placing a 2-MW Vestas nacelle offshore. The Horns Rev 1 project was the first large-scale project carried out. While blessed with good weather and an early finish of the installation, the startup and learning process was difficult and cost the author significant hair loss.

Moving the installation vessel around on site may require a tug to assist; especially true for the less-maneuverable vessels, which require backup if anything goes wrong. Reaction time is extremely short when you are 20 to 30 m away from the foundation. With 50 to 80 turbines in your way, loosing steering power or an engine room blackout is a disaster. Therefore more and more vessels are now dynamically positioned, driving installation costs up.

Casting gravity-based foundations requires enormous amounts of onshore port space.

CHAPTER TWELVE

Vessels and Transport to Offshore Installations

The offshore wind industry developed very quickly from 2007 to 2010. This means that a number of new vessels to install the offshore wind turbines have been proposed and some are already under construction. This was not the case originally.

TYPES OF VESSELS

The first projects were installed using whatever was available at short notice and low cost. This was not really problematic as such since the projects were small in scale, comprising only a few turbines and in sheltered waters. The last project of this kind was Middelgrunden, which was mentioned at the beginning of this book.

This project became the defining point of the entry into large-scale wind farm installation. The project was installed using the jack-up barge JB-1 from Muhibbah Marine in Germany, and the solution was the smallest possible to install the turbines. The area where the Middelgrunden wind farm was installed is only about 2 miles from the old shipyard Burmeister & Wain in Copenhagen, so the foundations—fabricated and installed by Pihl & Søn—could be manufactured in the dry dock and lifted by an adapted lift barge from Eide Barge in Norway.

This system worked, with some small problems, such as running aground in the very shallow waters and some weather delays, but all 20 turbines were installed, and the project as such was a success. However, it led to the conclusion among other stakeholders in the industry that "real" offshore exposed sites could not be serviced by this type of makeshift equipment not only because of its limited capacity both in terms of lift and transport, but also the inability to negotiate the open sea during construction.

The research for new and better vessels had begun, and companies such as A2SEA and Mayflower Energy emerged with different philosophies about the installation of wind turbines offshore, one being a semijackable vessel (A2SEA) and the other a self-propelled jack-up vessel later known as a TIV (turbine installation vessel) from Mayflower Energy. The two companies would go on to dominate the installation of wind farms offshore over the next decade.

Basically there are three main trends in offshore vessels for the installation of foundations and wind turbines. The three types have been used widely and in many cases adapted to the specific task of installing a wind turbine offshore, since this is the most complex operation to carry out; they are described in the next sections.

Self-Propelled Jack-Up Vessels

This vessel is capable of loading foundations or wind turbines using its own crane and transporting them to the site under its own power and then jacking up with an airgap to the water. The foundation or wind turbine is then installed using the crane and various onboard rigging and/or lifting equipment.

Thus, this vessel's main feature is that it is completely self-contained. It does not rely on other vessels to assist in any of the operations. This type of vessel is the most expensive for installation, but for several reasons it is also the most cost-effective vessel of all. The self-propelled jack-up vessel is the preferred type of vessel for the future; a number of them will be built and start to come into operation.

There are, however, other types of vessels, also new builds, that are not self-propelled and that have great appeal to a number of people in the industry. These are self-positioned, but they have jack-ups and towed jack-up barges that cannot move to and from the site without the help of one or more tugs.

Tug-Assisted and Self-Positioning Jack-Up Barges

The main difference between self-positioning and self-propelled jack-ups is size and maneuvering capability. The self-positioning jack-ups are actually glorified jack-up barges with the capability to move around onsite within certain weather criteria. However, for transit to and from port, they need towing tugs to help them move.

The barges themselves are smaller than the self-propelled jack-up vessels because they are meant to stay onsite and install components, be it foundations or turbines. On the other hand, one or more towed jack-up barges will feed the installation jack-ups with components, thus making transit unnecessary.

The logic is, of course, striking, at least on the surface, and later in this chapter we will discuss the pros and cons of the approach and whether it actually makes sense to do so. The main reasons behind the strategy of the self-positioned jack-up, as well as nonpropelled jack-ups, are as follows:

1. The cost of building as a propulsion package for transit is significantly higher than the positioning package.
2. The jack-up can stay onsite and lift components while being fed from other barges.
3. The less costly transit barges can use weather windows better since the installation vessel is not wasting precious time going in and out of port.
4. The installation vessel, in general, can be built smaller because the loading capacity is surplus to requirements. The necessary capacity will be one or two units—normally focused on turbines—because the rest will be delivered by feeder barge.

That the system works is beyond doubt. A2SEA has tested it during installation of the turbines on the Robin Rigg project, and the installation concept performed well.

Floating Equipment

The third category is floating equipment, and this has been part of the industry from the early 2000s. This type has worked on the foundation installation side, although early in the mid-1990s, Vestas and Bugsier installed 10 smaller turbines offshore at the island of Tunø in Denmark.

For turbine installation in offshore and exposed waters, floating equipment has so far proved unsatisfactory. But for the installation of monopiles, the floating equipment has performed well, and various types of equipment have installed several wind farms in both the North Sea and the Irish Sea. There are, however, concerns that must be considered:

- The delivery of foundations from barge to barge is very cumbersome and takes a long time.
- The safety in this type of operation is not as high as working from the installation vessel only.

Therefore, the future will bring vessels that are self-propelled and capable of transporting the foundations themselves. For this type of heavy offshore work, crane vessels (e.g., *Magnificent* and *Stanislav Yudin*), those discussed in the next section, will become major contenders in the installation of the large and complex foundations in the offshore waters of the North Sea. The smaller jack-up vessels are not ideally suited to this type of work.

TRADEOFFS WHEN CHOOSING A PARTICULAR VESSEL

When making the choice of installation vessel, the wind farm owner or main contractor will have to weigh several criteria against each other. The cost of plant (COP) is significant to any project, and with daily rates of above 125,000 euros, the balance of people (BOP) must therefore be correct for the project. Nothing could be worse than having the wrong equipment for the work to be carried out onshore—and offshore this is ten times more costly—so logically the rental or tendering for the BOP must be performed with great care and the specifications must be delivered by skilled and experienced personnel. This is not always the case, mainly due to the low number of personnel in the offshore wind industry.

The COP is significant, and more often the owner has a low margin on the wind farm in operational mode. The offshore wind industry is not a spectacular market with high yields on turbines and large profit margins like those the offshore oil and gas industries experience. The revenue from an offshore wind farm is mostly known only for the lifetime of the farm. When this is given, it is easy to calculate what the owner can afford to pay to install and to operate the wind farm.

In general, the internal rate of return is around 7 to 10 percent, depending on the market and how well financed (and installed) the wind farm is. Therefore the wind farm owner will look for the most cost-effective solution to install and operate the farm, and this most often determines the BOP rather than the optimal solution, which looks expensive but is very likely much more efficient in the long run.

This favors the non-self-propelled jack-up and the less expensive floating equipment—which is not necessarily the case when all of the costs are calculated—and, as we said before, there is a dependency on more equipment when choosing the low-cost single piece of installation equipment.

Example Vessel

Let us use an example of 80 turbines to perform calculations. This also helps us to understand whether feeding or self-sustained installation vessels are the better solution. This is the going project rate at the moment, so it will serve well as an example. Further details are necessary:

- Distance to site is 100 nautical miles (NM).
- The vessels are towed at 6 knots if not self-propelled. This is the best towing window that can be achieved, and for this example, the speed is always 6 knots regardless of weather—for simplicity.

- Towing requires one tug and positioning onsite requires two tugs. With two feeder barges, this becomes three tugs.
- The self-propelled vessel will transit at 9 knots. These go at a speed between 9 and 12 knots, weather permitting, but for simplicity we choose 9 knots always.
- The feeder vessel will require two feeder barges. If the distance to the site is longer than it takes the feeder barge to do a complete cargo run (turnaround time), the requirement is two or more barges to keep the installation jack-up running. We have chosen two barges for this example.
- The self-propelled vessel will take 8 turbines per trip. This will be a reasonable average between the new vessels coming online over the next two years.
- The feeder barge will be able to take two turbines per trip. This is a reasonable number given that the size of the barge must reflect the cost of same.
- Cost of the individual pieces of equipment is based on today's average in the market:
  - Self-propelled vessel: 125,000 euros per day
  - Installation jack-up: 80,000 euros per day
  - Feeder jack-up: 75,000 euros per day
  - Tugs—40-foot Bollard pull: 8000 euros per day

This requires an explanation. The feeder barge must be jacked up in order for the installation jack-up to lift the components off the deck. The HSE rules and the Marine Warranty Specifications require this. In this case, the installation jack-up and the feeder jack-up will be identical, except for the crane. If not, the feeder barges will determine the pace of installation, since their operational envelope would be lower than the installation jack-up.

However, when purchasing a 500-foot offshore crane, you would pay around 8 million euros. Over a lifetime of 20 years for the jack-up, this corresponds to 1200 euros per day—nominally. At an interest rate of 8 percent, this will be a bit higher—say, around 1400. For ease, we have chosen round numbers and set the cost of the crane at 5000 euros per day. In any case, this is not a staggering difference.

We have excluded the cost of fuel, but it is clear that the tugs will use more fuel, and combined with the installation jack-up, overall consumption will be higher than for the self-propelled vessel. For this example we have chosen to regard the fuel consumption as equal. The installation of the turbines will take equally long, regardless of the method of transport.

Therefore, it is only—for this exercise—interesting to look at the BOP and number of days to install the turbines.

We have considered the cost per day for the feeder barge plus one tug for towing. This gives the following daily transit cost:

- One feeder barge: 75,000 euros per day
- One tug: 8000 euros per day

This is a total of 83,000 euros per day. This sum is marked with light gray in Figure 12.1.

Here we have considered the cost per day for one feeder barge, one installation jack-up, and two tugs onsite, giving the following anticipated daily cost:

- One feeder barge: 75,000 euros per day
- One installation platform: 80,000 euros per day
- Two tugs: 16,000 euros per day

All cost in euros		Self-propelled		Feeder barge concept	
	Unit		Cost/day/unit		Cost/day/unit
Distance to site	100 nm				
Transit time	kn	9		6	
Number of installation vessels	Nbr	1	125,000	1	80,000
Number of tugs	Nbr	0		3	8,000
Number of feeder barges	Nbr	0		2	75,000

Activity		Unit	Cost/day	Total cost	Unit	Cost/day	Total cost
Number of turbines per trip	Nbr	8.00			2.00		
Load outs	Nbr	10.00	125,000.00	1,250,000.00	40.00	75,000.00	3,000,000.00
Transit time	Days	0.50	125,000.00	62,500.00	0.75	83,000.00	62,250.00
Number of transits		10.00	62,500.00	625,000.00	40.00	68,250.00	2,730,000.00
Installation of turbines	Days	80.00	125,000.00	10,000,000.00	80.00	189,000.00	15,120,000.00
Total cost				11,875,000.00			20,850,000.00

Figure 12.1 Costs spreadsheet for vessel example.

This is a total of 181,000 euros per day. The program would then look like what is shown in spreadsheet format in Figure 12.1. This sum is marked with dark gray shading.

The conclusion is clear in this case. The self-propelled installation vessel can work just as fast—if not faster—and the cost comparison is such that the process of collecting turbines in port could be extended with the difference in cost of almost 9 million euros, even though we have allowed for double the loading time in port. Therefore, there is little doubt that the feeder system is more expensive and does not bring the perceived advantages.

So why should we use the feeder concept? The reason is simple: There are not enough self-propelled vessels available in the market that can handle the workload required. In addition, the vessels that are available are not capable of carrying large numbers of components—if 8 or 10 turbines are a large number, of course—and this makes it necessary to come up with a solution that will work on a project-by-project basis.

The solution is to use two or more vessels, of which one or more are not self-propelled, but have the ability to deliver a part of the complete scope. One vessel can transport and jack up but not lift the components, and another can lift the components into place onsite but is not very suitable for transporting.

The non-self-propelled equipment will be deployed in larger numbers to keep up the same pace as the self-propelled vessel. This is necessary in order to avoid waiting times offshore when installation is going on. By adding more pieces to the logistical puzzle, they can deliver the entire scope. This will not be less expensive for the customer, but in the end, it may be the only way to construct his wind farm offshore.

ASSESSING EQUIPMENT

Several factors must be considered when you wish to charter or buy installation equipment. The following are some of the main concerns to take into account:

- Impact on the logistical solution
- Intake versus distance
- Intake versus turnaround time
- To feed or not to feed

- Impact on project time scheduling
- Complete turbine installation—does it work?
- Specifications on different types of vessels and what their impact is on the project
- Cost and timing of projects based on the typical installation vessel's performance

The discussion is, of course, interesting, but we need to get some order into the process. It is important to understand what the various parameters will do for the process in order to assess which solutions are the best—maybe not in all cases but in the specific case we are working on.

For this we created what is shown in Figure 12.2, which provides the parameters that will influence our choices. Comments in the figure are elaborated on in the following sections. The figure can be read as a quick guide, where you fill in the vessel specifications and, if so desired, comment on what the requirements for each of the characteristics are to successfully install the project on which work is being done.

Basic Information

The basic information is what is obtained first for the process. Which vessel are we talking about? Is it available? Who owns the vessel? What is its main purpose—jack-up, transporter, cable-lay vessel, and so on? Once this has been established, we can allocate the vessel to the right individuals in the project organization.

Type of Ship

When contracting for the type of vessel needed, you must first determine which type to opt for. There are various options, as mentioned following, but they have different pros and cons that will influence your choice, such as price, capability, and, of course, suitability to the type of work to be done. This is, of course, important, and as we said before, the availability and specifics of the project may dictate which options to develop for the particular project scope.

Options

Would you prefer a jack-up, self-propelled, jack-up barge, self-propelled semi-jack-up, or floating installation vessel? It is important to start looking at the requirements for the sailing and jacking. What can the vessel do, in

	Category	Parameter		
	Basic information	Type of ship:		
		Shipowner:		
		Country:		
		Use:		
		Loa [m]:		
		Beam [m]		
		Height: [m]:		
		Max. draft [m]:		
	Operations/ bookings	Builder:		
		Charter cost [kEuro/day]		
		Charter minimun period [day]		
		Mobilization Cost [kEuro]		
		Purchase cost [mEuro]		
		Foreseen availability/ booking		
	Loading capacity	Max, Payload [tn]		
		Max, Deck Area [m2]		
		Max, load per m2 [tn/m2]		
		Deck Area shape/layout		
		Open/close deck availability		
	Lifting system (Jack-up)	System:		
		Maximum water depth [m]		
		Lifting speed [m/min]:		
		Carrying capacity (per leg) [tn]:		
		No. of legs:		
		Spudcan [m2]:		
		Max. penetration depth [m]		
	Propulsion and DP systems	Propulsion system:		
		Specification:		
		Manufacturer:		
		Performance [kw]:		
		Service speed [kn]:		
	Crane	Type:		
		Max. lifting capacity: [tn]:		
		At radius [m] (related to max lift capacity)		
		At height [m] (related to max lift capacity)		
		Hook height [m]:		
		Max. radius [m]:		
		With payload [tn]:		
		Lifting speed [m/min]:		
		Max. wind velocity [m/s]		
		Significant wave height [m]		
		Position in ship layout		
		Movement compensation system		
		Aux crane availability and characteristics		
OPERATION MODES	Transit mode	Significant wave height [m]		
		Max. wave peak period [s]		
		Wind velocity [m/s]		
		Needed visibility		
		Transit velocity [kn]		
	Jacking up/down	Significant wave height Hs [m]		
		Max. wave peak period [s]		
		Wind velocity [m/s]		
		Current velocity [kn]		
		Needed visibility		
		Lifting speed [m/min]:		
		Type of seabed		
	Waiting for weather	Significant wave height Hs [m]		
		Maximum wave peak period [s]		
		Wind velocity [m/s]		
		Current velocity [kn]		
		Needed visibility		
		Max. wave height for waiting on weather [m]:		
		Wind limit for waiting on weather [m/s]:		
	Jacked-Up / WTG installation mode	Significant wave height Hs [m]		
		Max. wave peak period [s]		
		Wind velocity [m/s]		
		Current velocity [kn]		
		Needed visibility		
	People transfer and accomodations	Vessel´s Wind turbine access system		
		Access to vessel		
		Movement-compensation system		
		Accommodations		
		Crew		
		Passengers		
		Helideck		

Figure 12.2 Parameters for a project can be displayed in a Vessel Assessment Sheet organized the same as this one or in another format that will provide the necessary information.

general terms? Is it a jack-up or a semi? This will be important for the process, and certainly the cost of the vessel is also determined by the capacities and specifics—that is, a self-propelled jack-up vessel is more expensive but also more effective.

The floating vessel or barge is less expensive but also less flexible and can only work in calmer waters with lower capacities of cargo, transiting, and jacking. In addition, as we already said, the availability of a specific vessel will perhaps even to some extent dictate the method of constructing the wind farm.

Ship Owner: Contract Partner, Vessel Operator

The ship owner is normally also the installation contractor. The main point of contact will be through this contractor who will offer the vessel or vessels based on the requirement of the client. If there are no specific preferences, the experienced installation contractor will normally be able to determine the scope of work to offer or that is appropriate for the project that is to be installed.

A general warning at this time is, however, necessary. Since the industry is still only about 10 years old, the number of really experienced companies is small. Therefore, you may receive quotes and scopes of work for offshore wind projects that are not feasible or will cost more and take longer than anticipated; this may be due to the inexperience of the contractor and/or the project owner.

The various wind farm owners require different scopes of work, and based on the contractor/vessel owners vessels and capacities in general, they can offer more or less of the installation scope. Furthermore, the various vessel owners will not necessarily be able to deliver the whole BOP or SOW but only parts thereof and only under certain contractual agreements. This is important to keep in mind when starting the process of acquiring the BOP.

Use

What is the vessel going to be used for? This is, of course, the most important question to ask before approaching the vessel owner/operator. There is little point in asking a company to quote for installing wind turbines if the vessel cannot jack up. The use of the vessel and its abilities will determine how large your BOP is in order to transport and install the turbines.

Loa: Meters in Length

This parameter is important for the seafaring capabilities of the vessel. The longer the vessel is, the better it will cope with the swell and waves coming from the bow direction.

The vessel length is important for seagoing behavior, and, of course, length is half the parameter in terms of deck space. But as mentioned before, the longer the vessel is, the better it can handle the larger waves.

Beam: Width of the Vessel

This is a very important parameter because the beam determines the ability to withstand overturning wave forces from side waves. The relationship between the hull beam and water depth is crucial. If the hull beam is narrow, the water depth in which the vessel can jack becomes smaller. If the beam is large, then the vessel can jack in deeper water. Therefore, the desire is to have a long and "beamy" vessel. The two parameters, length and beam, are the first criteria to be chosen in the design.

Height

The height of the vessel should be considered as two issues. Height, as in the height of the hull, is extremely important. Hull stiffness is given by the height of the hull plank. The stiffness of the hull plank determines the maximum distance you can achieve between the supporting legs. The shallower the hull plank, the shorter the distance. If you wish to have a shallow hull plank, you may opt for six instead of four legs on your vessel.

Furthermore, the possible freeboard of the vessel becomes an issue. The International Marine Organization rules state the freeboard height of the vessel when it is rolling to the side. You must avoid freeboard loss and thereby immersion of the weather deck. The freeboard must therefore be high enough to meet this criterion. Furthermore, the stiffness of the hull is important for the bending inertia of the hull and therefore the amount of load that can be put on the vessel.

Another issue regarding height is the air draft—the maximum height of the vessel when sailing—but this is less important. However, it should be considered when choosing ports and navigational routes. No overhanging cables, bridges, or power lines should be in the way.

Maximum Draft

In very shallow waters, this is critical since the vessel could run aground, or, more important, it will not be able to use the propellers for positioning. In deep water it also becomes important because a shallow draft and very beamy ship will have less directional stability if not vastly overpowered at low speed.

When positioning the vessel in a wind farm at low to zero speed, directional stability is crucial. A draft between 4.5 and 7 m is desirable for this type of work. Interestingly enough, it seems difficult for the new second-generation vessels to comply with this. The required number of turbines on board to make the vessel cost effective will dictate the size of the vessel.

The preceding statement regarding hull stiffness becomes valid since the steel required to maintain the stiffness will be significantly increased whenever the hull dimensions are changed to a larger capacity. This again brings the draft of the vessel to a higher level and the most important requirement for a large section modulus of the legs when jacking the increased weight out of the water.

So as you can see, design is an iterative process where a number of tradeoffs will be made to get to an acceptable design, where length, beam, draft, hull stiffness, and cargo capacity finally meet up. Once this is achieved, the cost to build the vessel must be within a reasonable budget that makes it attractive for the market to rent it.

Operations and Bookings

The sections that follow describe a number of items of the second major part of the spreadsheet in Figure 12.2.

Charter Costs

The charter cost per day is the price of renting the vessel for one day. There are several different ways of arranging the charter. They are normally agreed on to match the Baltic and International Maritime Council (BIMCO) box sheet terminology. BIMCO has developed a number of box sheets that determine the form of charter, the intended use of the vessel, the cost of the charter, and its duration.

Several different forms are used. For the wind industry offshore, the Time Charter Supplytime 2005 is normally used. The document is an agreed-on standard contract that everyone in the shipping industry can adhere to and is familiar with. However, the offshore wind industry has a number of requirements in addition to the BIMCO box sheets, such as weather downtime and delays where the charterer has special requirements it may try to impose on the owner of the vessel.

Therefore, the charter cost is the day rate, normally for a fully operational vessel, a time charter, where all costs of the vessel itself—insurance, operation and maintenance, and the crew and their daily provisions—are covered.

Added to the day rate is the cost of fuel to operate the vessel and the lube oil for the engines, which are normally invoiced at actual cost plus a 15 percent handling fee. Usually if the charterer and the personnel are onboard, they will have to be invoiced for their supplies' consumption and lodging costs, but this is normally individually agreed to outside the box sheet in a side letter or an actual installation contract.

Minimum Charter Period

Usually the charter period is determined when writing up the box sheet. This is important because the length of the charter will also influence the price, terms, and conditions on which the charter contract, or indeed the installation contract, is finally signed.

Most owners, if not all, will try to get a charter period for as long as possible. This is in order to secure the revenue that is required for the vessel. Therefore, the charter is normally fixed to a minimum 30 days or on a trip-by-trip basis. The owner will set a high price for these short-term charters, whereas 6-, 12-, 24-month, or even longer charters will be at a considerably lower cost per day, simply due to the fact that the vessel income is secured for a longer period.

Mobilization and Demobilization Costs

These are the costs for positioning and preparing the vessel for the installation work. Also, the costs of derigging and redelivering the vessel to the owner after completion of the work must be included here. This is also an important parameter regarding the actual cost of installing the wind farm. The vessel is usually occupied on another project when it is booked. The mobilization cost is therefore the cost of repositioning from the location where it currently is physically to the site harbor where the charterer wants to use it.

Furthermore, mobilization cost can cover items, such as surveying the vessel—the on-hire survey—and rigging all the specialized seafastening and grillages, that are necessary to load and transport the various components to the wind farm. Obviously this will take some time, and the cost of the vessel during this period must also be covered. However, for ease of understanding, the mobilization (mob) cost is the cost to position the vessel where it is needed.

Loading Capacity

The sections that follow describe a number of items in the third major part of the spreadsheet in Figure 12.2.

Maximum Payload

This is the actual weight of the turbine components you can load on the vessel, including seafastening. For the charterer, this is necessary information. Not only is it important to know what the cargo capacity is, but also it is necessary to break it down into the various components.

Often the vessel is specified as having a dead weight, a light weight, and a cargo capacity, but the relationship among the three is not always understood. Dead weight is the total weight of the vessel—the cargo, the ballast, fuel, lube, provisions, personnel, seafastening, and so on. The dead weight brings the vessel to the maximum draft it is allowed to have. Light weight is the vessel, only completely empty; no ballast or fuel is onboard. It is the weight of the totally empty vessel.

The cargo capacity is the amount of weight you can load onboard that can move around. However, this includes fuel, lube, provisions, and necessary ballast. Therefore, the cargo capacity will depend on the weight, volume, placing, and distribution of loads onboard the vessel.

It is important to have loads placed on board with a low center of gravity; (COG) then the balance of the vessel will be positive. The vessel has six free directions of movement—heave, pitch, roll, surge, sway, and yaw—but for this we will only consider the roll movement, since this is normally the most critical one. When the vessel rolls in the waves, the vertical center of gravity will move toward the side of the roll. This means that the weight distribution inside will shift—relative to vertical—and the vessel must in itself possess an equally large or larger moment of resistance that works in the opposite direction: the righting arm. The relationship between the two is called the metacenter height.

Normally the relationship between the COG of the vessel and the cargo and the righting arm is such that the metacenter height is reasonably high, which gives a vessel roll period of between 6 and 8 seconds. Anything shorter than this will be extremely uncomfortable, and anything much longer will cause the vessel to right itself very slowly; this may not be desirable in the open waters—especially in the weather we are sailing in offshore.

For a more in-depth explanation of vessel seafaring behavior, I recommend that you consult the vast knowledge base of marine engineering and ship design. This has opened up a completely new world of intriguing and problematic possibilities for me.

But the main issue of cargo capacity has another spectrum—namely, the execution of the installation program. The planning of the operation and execution program is partly determined by the number of components to transport and install per trip.

Having said that, it is obvious that the number of trips to be made depends, of course, on the number of turbines, or foundations for that matter, that the wind farm is to have. The tradeoff between vessel size and cost compared to the number of components to bring on board will also be impacted by wind farm size, the distance to the loading port, and the weather regime you may expect where the wind farm is located. So it is very important to understand the project's sequence when you look to place the order for the vessel(s) that will install the wind farm.

A low number of components (turbines or foundations) will mean a higher number of transit runs to and from port. If the number of turbines in the wind farm is small—say, 30—and you can transport 3 units, the number of days you will have to allocate to transit will be a minimum of 10. If the cost of the vessel is low, this is not a problem, but if the cost for transiting the vessel is high, project economics will suffer; given the duration of the installation of a turbine is 1 day, the transit time is 25 percent of your cost.

The problem will only grow bigger if the number of wind farm units increases. Thus, the 90-turbine wind farm will have the proportionate 30 days of transit, but the duration of the project over the calendar year will increase. The length of the project gives you the problem of more adverse weather, since the project stretches from midsummer to spring and autumn, so the availability of days on which the vessel can work will be less. This in turn means that the number of nonworking days will increase and thereby cost. Therefore, the number of units—foundations and/or turbines—that can be transported at any given time must be as high as possible.

Maximum Deck Area

The maximum deck area is the total available free loading space on deck. The loading space only counts the usable free deck area, so if there is an area behind a leg or between two obstacles onboard, they can only be counted—and used, by the way—if you can access them in one way or another. This is important in order to plan the layout of the components loaded on the vessel's deck. The reason is clear.

You can only utilize both to their maximum capacity if the area is free, accessible, and provided you can fit your components on the area. We discuss deck-bearing capacity more later, and this too is important because a deck area that is accessible but cannot bear the load is also useless. Therefore, the free deck area must be as large as possible—on the given vessel—and have a high carrying capacity.

Deck Area Shape and Layout

How is the open deck shaped? As just mentioned, the area should be accessible both for crew and, most important, for the cranes onboard. The deck needs to be able to receive the components in a way that it is possible to load and lift them offshore without having to move them around. The deck area can be large but poorly designed so that there are blind angles or obstacles that can prohibit access to lift a component and store it on the vessel.

Furthermore, some jack-up barges have a large deck compared to their size, but the installation crane takes up most of the space. For the untrained eye, the numbers seem great, but the actual usage is very poor. In some cases the components have been loaded onto the vessel by means of built-on seafastening *outside* the vessel's hull. This is not particularly desirable either from a practical or a stability point of view. Although it has worked in the past, it is not cost effective, and the safety issues of having crew climbing the components become complex.

Preassembly

Which type of preassembly of the turbine is considered? To what level of completion of the foundation and/or turbine have the components been processed? Which type of seafastening does this require? Which installation method is derived by the preassembly stage? How will transit to site be performed and under what circumstances? What will be the matching weather criteria?

If the components, such as blades, tower sections, nacelle, hub, and so forth, are loaded on the vessel as single items they will take up a certain amount of space and require a certain amount of handling both in port and offshore at the installation site. Preassembly addresses the number of components you can load—respecting the cargo capacity and the possible usage of available deck area. Furthermore, the number of offshore lifts that are necessary to install the turbine will be determined here. The less you preassemble, the more lifts you have offshore and thereby the longer it will take to install the turbine. The preassembly can also help to optimize the usage of a poorly designed deck area.

But the preassembly level is also dictated by turbine design in particular. If the turbine nacelle is not fitted with turning gear—a device that rotates the hub after having installed one blade horizontally, whereafter the hub is rotated 120 degrees to install the next blade, until all three are fitted—the only possible method of installing the rotor is to preassemble the hub and three blades onshore. But this will significantly reduce the number of rotors you can load, and it will handicap the transit mode, weather window, and passage of port and navigation channels due to the extreme width of the vessel and rotors.

Once again we have demonstrated that one tradeoff or another will seriously impact the manner, speed, and cost of installing an offshore wind farm.

Open and Closed Deck Availability

How is the vessel constructed? Are there a hold and hatches on the vessel so components can be stored on top of one another? This is one of my favorite arguments and the reason why the vessels I have designed and built look the way they do.

The majority of the installation vessels have open box-shaped decks, which means that you can load on deck and nowhere else. What you see is what you get. It is also very simple to perform the loading plan because once an area is used, you can only optimize it in height by means of very costly scaffolding or towers or another means of shelving, so to speak.

Therefore, the box-shaped open deck type of installation vessel—self-propelled or not—will either have a low intake (or loading capacity) or have to be built very big and voluminous to increase the intake. If a hold is covered by hatches, you can store components in layers on top of one another. This is an advantage because you can use the same area several times. But how does this work, since it is more costly to build a complex hull with hatches that can open and an empty cargo hold compared to the relatively flat box?

It is also an advantage to have the hold built for a number of reasons, as follows. There are more that could be discussed, but these are the major ones.

- The hull plank height. The higher the side of the vessel, the stiffer it is, and thereby you can increase the distance between the legs. Building a box that is empty inside is inexpensive, and the empty space between the structural members at the top and bottom of the hull plank will serve as a cargo hold—for free!
- The higher and stiffer the hull is, the less likely it is that you need six legs instead of four. But this also means a significant cost savings compared to the flat box hulls that are normally built with six legs, or of limited length.
- Placing the heaviest components—usually the nacelles—at the bottom of the hull will increase stability dramatically. If you can fit the towers to the tank top level (the floor inside the cargo hold is usually the top of the double bottom tanks and referred to as tank top), it may be possible to transport the towers fully assembled. This is a significant time saving when lifting offshore. If you only have to bolt the complete tower to the foundation as opposed to having to bolt all of the tower sections together, you can save time and money. Furthermore, the cables can be fully fitted inside the tower, which means a faster commissioning time after mechanical installation.

The Lifting (Jack-Up) System

The sections that follow describe a number of items of the fourth major part of the spreadsheet in Figure 12.2.

System or Type

Jacking systems come in several types. There is, of course, the expensive but fast rack-and-pinion system that is known from the oil and gas industry, where almost all drill rigs operate this type of system, or the less expensive hydraulic systems such as hand-over-hand from various manufacturers, which are slower but certainly can save money.

The choice of system is important because rack and pinion gives a high redundancy at a cost, but the vessels with hydraulic systems are much cheaper for the customer to rent. Until now, the majority of vessels operated hydraulic systems such as the hand-over-hand, with either a hydraulic pin connection to the leg or with a strong beam slotted into the toothing of the leg. New builds have been using the rack-and-pinion system, however, and it will be interesting to see which is used in the future.

Lifting Speed

Lifting speed is the speed at which the leg is lowered or raised. This information is given for two situations:

- Where the leg is lowered or raised, but with no load from the vessel
- Where the leg has engaged with the seabed and is lifting or lowering the vessel from or to the water

The rack-and-pinion system gives this value at full capacity. The lifting and lowering of the legs are usually twice as fast as raising or lowering the vessel—that is, under the load.

The hydraulic types are slower and have more complicated systems. But speed is very important, since the jacking up and down of the vessel is critical in the relatively high waves; therefore it is necessary to know the jacking speed.

There are tradeoffs if you choose the rack-and-pinion system; there are two options: electrical or hydraulic drive systems. The electrical is very environmentally friendly since no hydraulic oils or other fluids will escape by accident, but the installed power capacity on the vessel will have to increase more than with the hydraulic system. Increased power will also mean increased cost of the vessel.

Carrying Capacity

Capacity is measured in tons per leg—the highest load on one leg during the operation. The maximum active load you can apply to one leg during the operation is, of course, critical. Not only will you have the weight of the vessel to consider, but the larger the vessel, the larger the system has to be.

Earlier we determined that the larger vessel had more cargo capacity, and while you will not be lifting the legs—they go down to the ground, of course—you must be able to lift the vessel, including the cargo and ballast if required. So the system must be able to cope with the loads applied, and the system's fatigue life must be designed to perform the required maximum loadings of the leg and jacking system throughout the entire life span of the vessel.

Furthermore, it gives you the relevant ground pressure and jacking capacity of the vessel. You need to know exactly what ground pressure will be applied or else there is a significant risk that you will experience a punch-through—a situation where the ground below the leg gives and a hole is punched in the ground with the leg. This may very well cause the vessel to capsize, and the risk of losing vessel and crew is imminent.

This must be avoided and is dealt with by applying 50 percent more pressure on the seabed for a predetermined time. This preload is normally applied using passive pressure. This is in turn done by jacking to the height where the hull is no longer submerged or only just submerged in the water and blocking two diagonal legs and reducing the pressure on the other two—or four—legs.

Leg pressure is also relevant if you can experience uneven loads, such as when the main installation crane is fitted around the leg. If this is the case, then the crane and the leg can require a nominal seabed pressure 50 percent or higher than the other legs. This has to be dealt with during design; higher loads mean larger leg and spudcan dimensions. In operation, higher loads mean higher preload pressure.

Number of Legs

So what is the optimal number of legs? There can be three, four, six, or eight legs, but the most common setups have four or six on the jack-up vessel or barge. The number of legs gives the jacking capacity in that the total nominal capacity of the jacking system plus the number of legs equals the complete jackable weight. You need to subtract friction losses and so forth, but in general this is the way it is calculated.

Furthermore, the number of legs gives a certain level of security against a punch-through, since the loss of the sixth leg on a six-legged vessel can be considered less critical than losing the fourth leg on a four-legged jack-up. Losing the third leg on a three-legged jack-up. . . well, we can figure that out individually.

It should be repeated here though that the number of legs and their placement on the vessel can be in the way of the crane when installing components. Therefore, focus on the number of legs becomes relevant for a great number of reasons other than the obvious, which is jacking ability.

Spudcan

The spudcan is a box or octagonally shaped foot that will increase the footprint of the leg when engaged with the seabed. This means two things:

- Lower ground pressure, which gives a higher margin against a punch-through
- Less penetration into the seabed due to the large footprint

Some jack-up vessels work without a spudcan. The leg penetrates deeply into softer seabed sediments, thereby standing not on the ground but by friction of the leg against the surrounding soil. This is less desirable since the length of the leg must accommodate this, and the possibility that the leg gets stuck in the sediment is higher.

Furthermore, the spudcan is designed to penetrate into the seabed to some extent in order to avoid underscouring of the leg, thereby experiencing a punch-through. This is a very important feature of the spudcan; however, it poses another problem when retrieving the leg again. Therefore, spudcans should also be designed with flushing systems in order to flush it into the seabed and to flush out mud, sand, or silt when retrieving the leg.

Maximum Penetration Depth

The penetration of the leg—and spudcan, if any—is an important variable, but it also depends on the soil type. The penetration of the seabed should be limited to the largest amount of scour development while the vessel is standing on the position. However, in soft seabeds, the leg must penetrate enough to find firm standing ground. The parameter is, however, not fixed and therefore a project planning parameter.

Planning of the operation and calculation of the seabed penetration are crucial parts of the site-specific project manual that has to be produced before entering the offshore site. This must be verified by the classification society and approved by the insurance company of the client/charterer and the owner/operator of the vessel.

Propulsion and DP Systems

The sections that follow describe a number of items of the fifth major part of the spreadsheet in Figure 12.2.

Propulsion System

The topic of what is underneath the bonnet of the installation vessel is, of course, interesting, but it is less relevant than would be expected. Since some vessels are not self-propelled, the actual statement about which, if any system, is onboard becomes the act of picking whether there is one or not.

The vessel's type of propulsion, if any, should be examined to determine what the capabilities on the specific site while you are installing will be. Understanding this will also give you a good feel for the duration of the project—related to the metocean information given, compared to the vessels capabilities and the actual intended duration of the installation. Will they match or not?

If the vessel is self-propelled, a description of the propulsion system is relevant in order to understand and to determine the power and capabilities. For this, a capability plot of the dynamic positioning system, provided the vessel has one, will be required.

Dynamic positioning (DP) means that the propulsion and control systems of the vessel will respond to the environmental impact of wind, waves, and currents while on position and the system will detect, monitor, and adjust the vessel so that it will and can maintain the same position and heading for a very long time—or at least until the capability of the system is exceeded. This is why the DP plot, which shows the capability of the system, is important to have.

Specifications

Which type of engine and propulsion system is installed? The options are MDO or MGO Diesel engines tier 2. All of the engines onboard the vessel should be tier 2 due to pollution requirements for ships built after 2010. Tier 2 is the level of efficiency of the engines in the system and the amount of pollution emitted from the vessel while operating the engines.

Although older vessels do not require tier 2 engines, it should be noted that we are working in an industry that is perceived to be environmentally friendly, and we cannot afford to be seen as polluters compared to existing carbon hydrate fueled systems. Therefore, a requirement for all vessels to comply with tier 2 specifications should be implemented.

Performance

The performance—or how many kW the system has to work with—is, of course, interesting. Normally, this is designed into the vessel, and the only relevance it has is whether it can operate the vessel, jack it, and work in a sensible manner such that, depending on the size of the vessel, it has the ability to keep within the DP criteria set by the manufacturer of the vessel.

Service Speed

The speed of the vessel is, of course, relevant, especially for transit. However, it should be considered that the transit distances are reasonably short and therefore the tradeoff on speed versus cargo capacity is interesting. The more turbines you can load, the slower the vessel can go. The worst combination is, of course, a slow and small vessel with limited cargo capacity as discussed earlier. The transit speed is becoming more relevant, particularly for Round 3 sites in the United Kingdom and also for the German projects, which are located far off the German coast with considerable transit distances for the installation vessels.

Crane Types

The type of crane is probably the most important issue of all. The crane is the reason for supplying a jackable vessel—or platform in this respect. It is only in order to create a solid base for the crane to work from and thereby deliver the safety of the operation for the installation crew that is required. Therefore, type and size of crane are extremely important.

For a crane to be able to lift 500 tons or more at the height and radius required for installation of the nacelle, in particular, the telescopic types of cranes are not practical anymore. A telescopic crane is able to extend and retract the boom to the desired length. The cost is, however, that the weight of the boom itself and the size will become impractical at some point.

Furthermore, many mechanical parts work poorly with the saline environment of the offshore installation sites. Therefore, the choice should be for a lattice boom crane mounted on a pedestal or another type of lattice boom crane. The lattice boom luffing crane can be built to lift capacities of more than 7000 tons and work offshore on installation vessels. Therefore, it should be a luffing crane, pedestal mounted.

However, smaller installation vessels are often fitted with crawler cranes on tracks, which, of course, are also cranes with considerable lifting capacities—up to 1200 tons; but crawler cranes are not ideally mated with the vessel and originally were built for onshore work. Therefore, you should avoid crawler cranes free on deck.

Maximum Lifting Capacity

For the charterer, the lifting capacity of the crane is, of course, immensely important. The crane must be able to lift and handle all of the components loaded on the vessel and to do this with high accuracy and a high margin of safety. Therefore, prior to chartering the vessel, it is important to determine the heaviest component to be lifted to its required height at the required or desired radius from the crane.

The capacity for most nacelles should be around 350 tons at the required height and radius. Weight, height, and radius required must be decided during the design work.

Lifting Radius

The lifting radius in meters determines the distance to the foundation the vessel has to observe when jacking and lifting the components in place. The normal distance is around 25 m from the vessel to the foundation position for the crane to operate; this is also a reasonable distance between the crane and the turbine.

The crane operator must have the possibility to slew—swing—the crane around and work in front of it. If there is not enough clearance between the turbine and the crane, the crane will be working with the boom in the highest angle all the time. This will be a problem if a component doesn't fit properly. Then the crane operator must be able to move around to optimize the positioning and handling of the component.

Payload

The payload of the crane is defined in two ways:

1. The heaviest load the crane can lift at the minimum radius
2. The highest load moment the crane can achieve—that is, the load x distance to the slewing center of the crane

An example is the 600-ton foundation that is to be set on the seabed at a 25-m radius from the crane slewing center. This is the maximum load the crane can lift and the maximum distance the crane can lift it. Therefore, the crane is a 600-ton crane and $600 \times 25 = 15,000$-ton/m crane.

But a blade would weigh in at, say, 10 tons, so this load should be able to be lifted by the same crane at a radius of 1500 m; however, this is not technically possible. Furthermore, the weight of the boom itself will reduce the lifting capacity of the crane rapidly when increasing the boom length. However, lifting a blade at 70-m radius gives 700-ton/m load moment, and this is, of course, within the technical reach of the crane.

Basically the load moment of the crane is important, but the lifting capacity at the relatively short radii of this industry determines the capacity of the crane. Thus, a 600- to 1000-ton lifting capacity for the crane that would be used for lifting both foundations and turbines should be the target.

For lifting of turbines only, the capacity can be reduced to 400 tons at 25-m radius at a lifting height of 120 m; the lifting height is the final component. The length of the boom will, as already stated, be a reducing factor for the crane and, with the increase in boom length, the capacity drops rapidly.

Maximum Wind Velocity

The wind velocity is the maximum wind speed at which we can no longer operate the crane. This is a difficult area because the crane can possibly work at higher wind speeds than the turbine can be installed at. But, in general, if the crane can do 12 to 15 m/s, then this is enough, either due to crane limit or limit for installing the manufacturer's components. Waiting on weather is applicable to both situations

Transit Mode

Wave height limit is the point at which it is no longer possible to transit the vessel from one position to another. Usually this applies to the transit between the port and the installation site. The transit Hs criteria normally matches the criteria for jacking up or down. This means that we start transiting to the site when the Hs is slightly—say, 0.5 to 1 m—above the jacking criteria but on a downward trend.

In this way we can use the weather criteria for which the seafastening has been laid out to transit from the port to the installation site. This in effect creates higher usage of the weather windows onsite. However, the transit criteria is also limiting in that whenever the criteria is met or exceeded, waiting on weather will be applicable.

Maximum Wave Peak Period

How strong is the seafastening, and how is the seafaring behavior of the vessel? As already described, seafastening must be laid out to cope with the maximum operating transit criteria for Hs; this is important for the transit criteria. If this is low, the operational envelope is also low. Therefore, the designer—typically the naval architect—will look at the maximum G-force the turbine components will allow in any direction. This will be the governing parameter for the seafastening. The operational envelope and the maximum G-forces on the components are then determined by the vessel motion and the corresponding maximum seastate in which they occur. This will then be the operational envelope and thereby the design parameter for seafastening.

Needed Visibility

Even though there are radar and GPS, we still need visibility when sailing. However, in this case we probably need visibility for others. This is, of course, particularly the case when the components exceed vessel periphery. We sail out with rotors mounted on deck, which largely extend beyond the side of the vessel, and one can easily imagine the problem if an opposing ship passes too close. This is an accident just waiting to happen. Therefore, it is important to not only see others from our vessel but that they can see us.

If the components are wider than the vessel, it is also a safety issue when navigating narrow port entrances and so on. We may need to build up large units on deck for seafastening in order for the components to pass port entrance lighthouses, antennas, masts, and so on. This will be a planning issue right from the moment a method and vessel for installation are chosen.

The other issue is, of course, the ability to see through fog, rain, snow, and so on. For this we need to plan and implement a strategy for when we stop working and what the criteria are. As mentioned, we have radar and GPS, but in all instances the focus should be on avoiding situations that could potentially lead to accidents—if allowed to develop further.

Transit Velocity

How long is the vessel underway? This is perhaps less critical when you start thinking about it, since we do not travel for several days; if the distance is 50 NM, it doesn't matter too much whether you travel at 10 or 12 knots. But on the other hand, if you only travel at 6 knots, such as a towed barge would do, time for transit becomes an issue.

The speed in transit determines how long an applicable weather window must be. The slower the vessel, the longer the weather window has to be. This is fairly logical. As just mentioned, the transit criteria are a bit higher than the installation criteria; however, the transit window must be longer if the vessel is slower.

G-Forces Transmitted to Payload

How strong is the seafastening, and how is the seafaring behavior of the vessel? We already discussed this, but here the issue is turned around. We talked about how much G-force a load can take, but how much force the vessel puts on the load is different. If the load can take a 0.5 G-force as a maximum for up and down movement—heave and pitch—and the transport vessel is very stiff—beamy and full bodied—the acceleration forces will be high due to the relatively short distance the hull will travel up and down.

Thereby the acceleration force will be extreme, even at low wave heights, and this will determine the operational window, which could be considerably lower than the installation window. Therefore, we have to pay attention to the problem from both angles. But bottom line, this is important for the transit criteria. If this is low, the operational envelope is also low.

Jacking Up and Jacking Down

Jacking up and down is, of course, why we are there. Therefore, the criteria should match the environmental data for the best site possible. Following are descriptions of specific issues with regard to jacking.

Significant Wave Height

At what wave height limit can the installation vessel no longer jack up or down? This is one of the first questions the client will ask when approached by the vessel owner. The higher the criteria, the more usage of weather windows the client gets—and ultimately the security of being able to install the project in the available time frame. The significant wave height is an increasingly important issue since the offshore wind farms are being placed further and further out in the oceans. Thus, older and first-generation vessels that were built or applied will meet their ceiling for operation.

So whereas they were good for installing projects in near-shore waters, where you could mostly expect a decent if not good usability of the vessel with an Hs criteria between 1.2 and 1.5 m, the future sites require higher environmental criteria. This is simply because the Hs criteria of 1.5 will give an operational availability of less than 50 percent of the year.

This is not enough to secure the safe and timely installation of an 80- to 160-unit wind farm. If you increase the Hs to 2.0 m, you end up with 70 percent more or less of the year for installation, and this is what is required. However, this comes at a high cost, and the jacking ability is what hurts, since this is mainly brutal power and long fatigue life exposed to high loads and shocks from impact between legs and/or spudcans and the seabed.

Maximum Wave Peak Period

The peak period limit is important because in almost all cases, long, crested waves (swell) are limiting for the jacking operation. This is the case because the energy in the wave train that passes is higher than in the short-crested seas at the same height. The amount of water moving is higher, so the overturning moments are higher as well. Furthermore, the period corresponds to the vessel in that the impact, even on the jacked vessel, will exit the legs. This means that the legs will be agitated by passing waves, and if the wave period gets close to the eigenfrequency of the platform or vessel, it may cause damages to the legs or the entire structure.

Furthermore, for jacking up and down, long, crested waves will be able to lift or lower the entire vessel, particularly if the wave length matches or exceeds the vessel length. This means that the entire dead weight of the vessel will be placed on the leg stabbed against the seabed first. When this happens, the entire mass of the vessel is loaded on one leg. If it too moves— which is inherently what the wave tries to do—the mass and the directional movement of the vessel will load an enormous bending moment on the leg, which it also has to be able to withstand.

If the load of the wave is now increasing linearly from 1.5 to 2.0 Hs, the calculation is straightforward. But if this is not the case, the bending moment is exponential, and therefore the increase in the cross-section of the leg is dramatic in order to cope with the increased load and bending moment. Therefore, the cost of going from 1.5 m Hs to 2.0 m is drastic, and since the turbine will still only produce what the name plate says, the cost of installing only increases the project's capital expenditure without giving any significant gain in turnover, possibly only a shorter and safer installation period. This is one of the reasons why installation costs have risen dramatically over the years.

Wind Velocity

Until now, the normal limit for operating cranes and, more important, fitting components offshore has been around 10 to 12 m/s. However, the move further offshore and the cost of installation equipment drive the boundaries further upward; this presents a problem. For the crane and the vessel, the wind is not too much of a limiting factor. Only positioning the vessel before jacking and before leaving the jacking position becomes a real issue with regard to wind.

Very high winds can make positioning of the vessel difficult. But the new generation of installation vessels will be fitted with more propulsion power so that this problem can be mitigated. For installation of the turbines, however, the boundaries of operation are harder to move. The blades in particular are only good for one thing: catching the wind. When the winds increase, it becomes both a capacity problem in terms of crane operation and a safety problem in terms of holding and fitting the blade by the installation crew—it is just too dangerous. Therefore, the wind is a very important factor to observe when choosing a vessel—and a methodology.

Current Velocity

The current is important to observe when approaching and leaving the position where you decide to jack up the vessel. The maximum allowable current is normally 2 knots, and this seems fair in most cases. More power in the vessel will be very costly, and certainly the second and third generation of vessels will require an awful lot of power if this parameter must be increased.

If the vessel is 140 m × 40 m, length and beam, the environmental load from current and wind is gigantic, especially if it is fully laden with turbines. The wind surface is significant, and to hold this surface directionally stable toward wind, current, and waves, thruster capacity must be very high indeed. Therefore, a current of 2 knots is normally chosen, knowing that many sites have higher tidal currents. This introduces a time constraint because you may have to wait for the tidal current to slow down before jacking.

Needed Visibility

This is a safety issue, but why? When you have positioned yourself to jack, the visibility is not important any longer—or is it? Well, if there is a foundation where you wish to jack the vessel, it is important to be able to see it from a good distance while approaching. When landing the gangway, it would be nice to see where it goes. So, the first and foremost rule in this case is that you need visibility for all operations offshore.

Lifting (Jacking) Speed

A general rule of thumb is that the jacking speed should be higher than the tidal speed in order to safely jack up or down. There have been examples where the jack-up barge moved so slowly vertically that it was overtaken by the tide and lifted off the seabed again.

This is, of course, not desirable at all. Therefore, the jacking speed is one of the most critical issues and also determines the operational envelope. If the tidal variation in the area of operation is large—say, between 6 and 8 m—the equipment used must be significantly faster than this to safely carry out the jacking operation. Low jacking speed is not desirable because it limits the installation vessel's areas of operation.

Maximum Water Depth

The maximum water depth is the determination of the depth the vessel can safely jack up, according to its maximum environmental criteria. Operational parameter, the deep-water capacity, is important since this is the trend for future wind farms offshore. They will go farther out in rougher waters and be much deeper than seen today.

Type of Seabed

The type of seabed is very important to observe. The soft and silty seabeds will incur a lot of penetration from the leg and spudcan when pressure is applied. This means that the leg will sink deep into the seabed when jacking up. To control this, it is important to know the bearing capacity and type of seabed on which the wind farm will be standing.

The normal procedure is for the client to carry out seabed investigations to determine the soil characteristics, and by means of the seabed investigation report, the owner of the vessel will be able to predetermine penetration of the leg and spudcan and to develop a method for deploying and retrieving the legs. Specifically, the information will also be available with regard to retrieving the spudcan and the overburden of material that will hold it down.

Waiting on Weather

Waiting on weather (WOW) is the event that occurs when the weather limits of the vessel or the installation methodology for either the foundation or the turbine is exceeded. This normally happens when just one of the following events occur:

1. The wave height parameter is exceeded, so no work can be done—normally around 1.5 m Hs
2. The wave period parameter is exceeded, so no work can be done—normally around 12 m/s limit
3. When positioning the vessel prior to work or jacking up or down, the current exceeds 2 knots

It is critical to be able to see the top of the turbine and the entire deck area when working. Furthermore, work will be stopped when it is no longer possible to clearly see its top.

All of these events will trigger a WOW declaration, and when that happens, work is stopped after reaching and documenting that the weather limits have been exceeded. A statement of facts should be made to document all observations, loggings, and securing of every type of intermediate work that is carried out.

It is also very important to note that no discussions of whether it is appropriate or correct to stop work are carried out offshore on the vessel. All discussions occur onshore by representatives from all involved parties. In this way, arguments can be avoided. Stoppage of work, whether correct or not, should not be subject to a discussion about whether a delay is caused, or increased, or who is responsible for paying for the delays. This can cause people to exceed safety limits and jeopardize lives to achieve a milestone that could potentially be surpassed and thereby trigger a penalty. This should never take place in a forum that is offshore onboard the vessel.

People Transfer and Accommodations

The sections that follow describe a number of items of the final major part of the spreadsheet in Figure 12.2, in addition to some other issues related to the transfer of people.

Vessel Access to Wind Turbines

Access to the turbine is one of the biggest safety issues. The safe and easy access is paramount in order to achieve a fast, effective, and safe installation of the turbine in particular. The HSE requirements in all countries where turbines have been installed offshore so far have dictated the use of a fixed connection between the installation vessel and the turbine foundation.

The reason for this is that the offshore installation of wind farms is regarded as construction work and not a marine operation. This means that all onshore construction rules and regulations have to be adhered to. The gangway is expensive and necessary to access the turbine from the vessel.

The gangway must be able to stretch the inaccuracies of the positioning of the vessel onsite, which is done in harsh weather conditions and often at relatively low visibility and, of course, also at night. This means that the gangway should be extendable; it must be able to turn and pivot up and down in order to fit onto the foundation.

In case there is a large difference in height between the vessel where the gangway is fixed and the access point on the turbine foundation, an intermediate position of the access point, or a second and higher position on the gangway deck can be made. However, before the charter is agreed to, this is one of the more expensive points to clarify.

Furthermore, the gangway resembles the connection point of all electrical cables, hydraulics, and air hoses that are used during the installation of the turbines. If these are required, cable and hose trays must be available on the gangway without having to lay them out on the deck.

Installation Speed

In general, the time it takes to transport and install one turbine is referred to as the installation speed or the turnaround time. However, turnaround time consists of the preceding parameters, combined to form a program, or a program sequence, to give a detailed time schedule.

It should be noted that turnaround time can and will vary depending on the wind farm involved, even when using the same vessel. The reasons are a combination of distance, weather, and metocean conditions; turbine size and make; and so on. Actually, a combination of all of the preceding parameters applied to the vessel for the individual project comes into play.

Installation Envelope per Year

What is the operational envelope for the vessel? This in effect means which period of the year we can expect to use the vessel for installation and which period will have to be considered as downtime. In general we should go for as many days as possible, and again it depends on project location, size, and metocean data.

The same vessel will be good in one area—say, the Baltic Sea—for year-round installation, and it may be very cost effective. But if the same vessel goes to the Irish Sea to install, it may very well have a 40 to 50 percent annual downtime. Therefore, the old saying "horses for courses" is important. The right piece of equipment in the right place for the right job is necessary.

This is a major part of the technical due diligence that the project's wind farm owner, installation contractor, and financing institution *must* undertake in order to assess the viability of the method used for installation.

Summary

The time required to install one turbine, including loading and transporting to the site, is what is used for calculating duration of the project. This means that the net installation time for one turbine—or foundation for that matter—is calculated and put into the overall time schedule and sequenced so that the number of turbines—or foundations—loaded on the vessel is matched to the installation methodology.

We have discussed weather downtime for all kinds of operations off-shore, and the right time schedule will have weather downtime allocated at every appropriate stage. This is important because many time scheduling exercises simply add 30 or 40 percent for weather at the end of the form for the final time period to use. This is wrong, however, since allocating downtime in June is not the same as in December.

Furthermore, the drifting of the installation of, say, six units offshore could potentially mean that the completion slips into the following month, which will have a different downtime percentage. When allocating downtime where appropriate, you get a more correct schedule and end date than if it is tacked on at the end.

Finally, the normal speed for installing a wind turbine cold is 24 hours, and this is the time you should aim for achieving. This does become more and more difficult as the turbines become larger, but this is still the valid time frame to aim for per turbine.

●●●
Related Images

Monopiles loaded on a transport barge for travel to the site.

Test jacking offshore prior to start of installation work.

Leaving position after installing the monopile and transition piece.

Loading components and consumables in port prior to departure to offshore site.

Every inch of the installation vessel is used. Sometimes cargo hangs over in order to facilitate the best loading of it. This is not unusual and demonstrates the skill and ingenuity of the designers and planners to create an installation program that works for a project.

Steaming out to the site with only half of the turbines does happen but should be avoided.

CHAPTER THIRTEEN

Basic Information about Ports

The installation of an offshore wind farm depends, of course, on the proper setup in the port area. The port and logistics center must be developed in a way so that the various disciplines' tasks can be carried out quickly, safely, effectively and at the lowest cost that can be obtained when vetting all aspects against one another.

Since turbines, foundations, cables, and other main components cannot be manufactured "just in time," the port area must be extremely large. Normally the storage area requirement has been set at around 60,000 to 70,000 square meters in order to store and prepare all of the components for an 80-turbine wind farm.

This has worked out well for the first several projects installed since 2000. The projects—especially for the British Round 3—are substantially larger, and the German projects in general have larger turbines and certainly larger foundations. This adds more requirements to the dimensions of the pier area, and since the weight of the components are steadily increasing, the quality of the port and pier area, as well as the handling equipment, is growing. This obviously limits the number of ports that can be used for the purpose of installing an offshore wind farm.

CHARACTERISTICS OF THE IDEAL PORT

The port must be able to receive all of the components in a fast and effective manner so no more handling than necessary needs to be carried out. Therefore, the area available in the port must be paved and fenced in order to receive, store, and prepare the components for final unloading.

Project Ports and How They Perform

The project port is ideally located close to the production location of the turbine components. This has not been the case in the past. The reason is that turbine components historically have been transported by truck (i.e., the way onshore turbines must be able to transit from the production site to the installation site); therefore, the actual place of production has been the place where the turbines were manufactured anyway.

The offshore wind turbines are, however, much larger both in capacity and physical size. This effectively requires them to be manufactured on a pier area in a deep-water port. The 5-MW turbines and upwards are not transportable as one component on roads, and thus the turbine manufacturers are looking to position their production facilities close to a port area.

Therefore, the ideal port will have deep water with a large hinterland that can support the manufacture and storage of many turbine components. The components must, however, be ready for unloading so no onshore assembly is required after they have been manufactured and delivered from the factory. They should be stored at a point where they are easily rolled out to the pier and loaded onboard the installation vessel—or the transport vessel if the project is overseas.

Transport to and from the Port

As we just said, no transport on road or rail should take place when the components have been manufactured. The components, whether foundations or turbines, should only be rolled forward to the pier to be unloaded. This can be done by MAFI trailers (i.e., special port-handling ones for heavy goods that can be hauled by a so-called MAFI truck where the control cabin can turn 180 degrees so the driver has full control in all directions); SPMT trailers; or axle lines, which are also used for street hauling. But, again, the only transport is from the storage area to the pier front in order to be unloaded.

A PORT'S IMPACT ON THE PROJECT

The impact of a good or bad port on a project is significant. If the port layout and access lanes are ideal, the turnaround time for an installation cycle will be short. The opposite, however, will put constraints on all parts of the project. Here is an example: A small port with poor access (both on- and offshore) will mean that the supply of components will be slow and complicated. If the components have to leave port before others can enter in order to be prepared for final unloading, the logistical operation will be more complicated.

Imagine that the installation vessel will have to stay in port due to bad weather, but the supply of components is uninterrupted. This means the pier that must be used for loading and offloading is occupied by components that are stacking up. The hinterland transport of components for storage will collapse because the stored components are not unloaded due to the installation vessel not performing. A number of scenarios can be elaborated that are all valid problems for the PM during the project's execution.

Size and Layout of the Ideal Hub

So taking the preceding information into consideration, the ideal port will have easy access. This means that the entrance and navigational channels into the port must be wide, straight—considering the option of transporting fully assembled rotors—and not too crowded. In this way the components can enter and leave the port in all of the ways that are relevant for how the project can be executed.

Furthermore, the port area must be fitted with deep-water pier areas to accommodate the transport vessels—which are normally deep-draft vessels of up to 9 m when fully laden. The installation vessels are normally shallower draft, and therefore they are able to negotiate anything of 6 m and deeper.

The installation vessels, however, should have the benefit of a hard seabed surface in order to safely jack up at the pier. If this can be done, the loading can be carried out faster and safer as a result of the perfect stability of the vessel during the jacked-up mode. The pier area must be a hard surface for two reasons:

- The carrying capacity of the surface must be high enough to accommodate the loads of SPMT trailers, which can deliver an axle load of up to 34 tons. Therefore, regular asphalt surfacing is not enough. It must be much stronger so as not to crush under the weight of the components being moved around.
- A surface of either concrete or asphalt will prevent dust and nicks from pebbles from damaging the surface of the finished turbine components. This is important because all of the components of the turbine, in particular, are painted or fiberglass coated. The MWS and installation contractor will be very cautious about this type of damage because this could also pass as a scratch when loading or installing the component.

Finally, the port should have a good storage area in direct connection with the pier front area in order not to transport the components too far or for too long during storage, preassembly, or loading. As discussed earlier, the JIT principle does not work for this type of infrastructure project. The ideal area should be around 65,000 to 75,000 m^2 to store enough components before the project starts. So 50 to 70 percent of all of them should be delivered to the storage area in the staging port prior to project startup.

Combination of Vessel and Port

The preceding requirements are, of course, difficult to find, and when confronted with the ports available in Europe for installation projects, the conclusion so far has been that there is no ideal port. If it is close, it is small or badly laid out. If it is well laid out, it is so far away that it doesn't make economical sense to use it.

But in Germany, in particular, the infrastructure is being heavily built out, so in the future, German projects will benefit from new piers and infrastructure facilities. The Round 3 projects in Britain have also inspired turbine manufacturers to look for ports, and to establish new manufacturing facilities in close proximity to ports in order to be able to deliver the very large turbines that are expected for the far offshore projects in the British Round 3.

So once you have the ideal port, what does this mean as far as the installation vessels that are available or will become available in the next few years? The going rate these days is around six turbines loaded on the installation vessel, and this is feasible in at least Esbjerg, where Siemens and Vestas have been unloading projects for a number of years now. But in the British ports, this is still somewhat cumbersome, since the ports are not ideally suited to this type of business.

In the near future, however, installation vessels will be unloading 8 to 12 turbines per trip, and this will put a much higher strain on the staging port. The number of turbines already on the pier will have to increase by 30 to 50 percent or more to keep up with the installation vessel, and only a few ports have been considering this. So in order not to get a poor match between vessel and port, the ports in Europe will have to decide whether or not they want to be involved in this business.

If they want to be in the offshore wind industry, they will have to invest large sums in infrastructure to keep up with the development of the installation vessels for Round 3 and the German projects. So far, only the ports in Germany are showing some readiness to invest in this industry as a focused business segment.

OPERATION AND MAINTENANCE

The following are the areas of offshore operation and maintenance (O&M):

- Maintenance that is planned and carried out according to schedule.
- Maintenance that is unplanned—in other words, sending "the man with the van" to the turbine to restart it.
- Repair, which is when a main component breaks and needs to be replaced.

The maintenance is done by the service department. For more details about this, see Chapter 17.

Maintenance is—as stated—the planned annual checkup on the turbine to make sure it runs as contractually agreed. This is, of course, included in the warranty that the turbine supplier will offer as part of the availability of the turbines. The unplanned maintenance is when the turbine stops and a technician must go out to check and restart the turbine. In both cases, a service vessel is required to bring the person out to the turbine. This will be a small, high-speed vessel that can bring a limited number of people out to the wind farm to carry out routine work.

Repairing the main components requires a vessel that can dismantle the old part and replace it with a new one. If a main component, this would ideally be the installation vessel, but for repair work, the setup is slightly different in that only a part of the turbine would have to be replaced. This has led some people to think that the repair vessel should be much smaller with a smaller crane.

As we saw before, the crane is the smallest cost issue on the split between the jack-up and the crane. However, size is a determining factor when jacking up offshore. So if you want a high level of repair capacity, you need a large vessel with a strong jacking envelope. Plus, the crane is not the cost driver here—it is the jacking system.

Related Images

Towing jack-up barges in and out of port puts some severe requirements onto the staging port. First, the sway basin must be big enough to maneuver the barge, then in- and outgoing traffic should be ready to accept delays due to its movement at odd times. Finally, the length and beam of the tow is substantial, so self-propelled vessels are preferred because the time and effort to dock and undock them is much less.

Some problems related to jack-up vessels can be seen here. They need to jack substantially out of the water to properly operate the main crane. This makes the loading process a bit more complicated and crane positioning becomes an issue. Very large installation cranes move slowly and their luffing and turning in and out slows the operation down considerably. Thus, the latest vessels designed for use in wind farm installations, starting with the 2007 GAOH Offshore, fitted the main crane around one leg.

Monopiles loaded on a transport barge. The seafastening is considerable and, for the port to be able to handle this, extremely large cranes have to be present on the pier or brought in. This also limits the number of ports suitable for this kind of work.

Such cranes are present at the Harland and Wollf Yard in Belfast. The very large gantry cranes are perfect for loading this type of component.

Nacelles stored on the pier ready for preassembly. Note the massive use of space for one unit. When the blades are fitted, there is no possibility of using the space around the nacelle for other purposes until loadout has taken place. Therefore, space available along the pier is ranked highly in the vetting process for staging ports.

Siemens 3.6-MW turbines loaded on a transport barge before shipment out to a project. Note the two mobile cranes behind the barge; they require a massive bearing capacity on the pier and take up significant space. This is necessary because it would not be possible to load the barge to this extent using ordinary harbor cranes. They neither have the reach, height, nor the lifting capacity to carry out this type of operation.

CHAPTER FOURTEEN

Project Criteria

Several factors must be taken into consideration when planning an operation and maintenance (O&M) system. Some of these are wind, waves, and currents. The project must be examined to determine any possible bottlenecks, constraints on transfers, and bad weather that might impact accessibility to the turbines. The impact of possible downtime if the crew is unable to access the turbines should be considered.

It is very important that the operation and maintenance transport and transfer system that will be used accommodate the availability of the wind farm. This in effect means that the chosen solution must permit the crew and equipment to be transferred to its foundation during 95 percent of the turbine's availability. This percentage or more is the production period that the client has been using for his investment budget, so it is crucial that this is accomplished to avoid any penalty to the turbine manufacturer.

A high level of availability is necessary if the wind farm is to be economically viable, but the corresponding worst possible weather regime that the O&M solution must work in is calculated for most wind farms to be around 2 m significant wave height (Hs). This in effect means that waves up to and above 3.5 m from trough to top will be encountered, and based on the wave period (the time it takes the full length of one wave to pass a fixed point), the waves can be extremely long—possibly up to 150 m.

For ships and other vessels to be able work in such a harsh environment, size becomes extremely important, and corresponding to size is, of course, the cost of the vessel. This can pose a problem since it is important that the overall O&M cost of the wind farm be kept as low as possible. Subsequently, the transport and transfer system becomes the single largest cost in the entire O&M plan.

OFFSHORE ACCESS SYSTEMS

The first factor that must be determined is what you want to access; getting access to the vessel and to the turbine are two different issues. Why is this true? The reason is that if you use a crew boat—assuming you are not using the installation vessel for O&M—you can access either the turbine with the crew or access the repair vessel from the crew vessel, again to deploy the crew.

When the preceding situation is the case, the system requirements differ depending on whether it is the vessel or the turbine. One involves vertical access (in case of accessing the turbine and the repair vessel from the crew boat), and the other involves horizontal access (if accessing the turbine foundation from the repair or installation vessel, which are both normally jacked up). Further, the jacked-up installation vessel gives steady access from the vessel to the turbine, whereas in the case of the crew boat, it depends on whether the vessel or the turbine is afloat. In the latter case, the requirement is higher.

For our purposes, we will only look at the parameters for accessing the turbine from a crew vessel. The crew boat will approach the access ladder on the foundation, and by means of it, a Browing system, or other device, the crew members are deployed onto the foundation and can then use the ladder to climb onto the platform.

Certain regulations govern the ladders used, and these come from the HSE regulations in the specific country of deployment. The access tower or ladder must have resting platforms above 6 m in height and for every 6 m thereafter. This is important because crew members must be able to rest before climbing to possibly 25 m above sea level. Doing this in one go in a survival suit is difficult and certainly a very uncomfortable experience.

Normally, Hs is between 0 and 2. The criteria are set by HSE authorities and should be adhered to so as not to inflict injury on the crew or the vessel. The maximum required access criteria is 2.0 m. However, this is difficult to achieve, mainly due to the movements of the crew boat. The 2.0 m Hs criteria is desired by wind farm owners, since it will make it possible for them—in most cases anyway—to keep the turbine's availability at around 95 to 97 percent of the possible production time. This is also required in order to make the wind farm cost effective.

Generally speaking, the access system must be safe, easy to operate, and able to fit on the crew vessel on which it is deployed. Thus, no system should be deployed that is not failsafe and too hard to operate in these conditions. Even though it is commonly said that weight and size do not matter offshore, it should be noted that the offshore vessels for crew transfer are usually under 24 m.

So using the large, complicated offshore systems for access, which alone weigh enough to sink the vessel, is probably not the right way to go. Why do we say that? It's very simply because the numbers just don't add up! If the crew vessel needs to be 50 m to carry the access system, the cost is going to be north of 10,000 euro per day. This will impact the financial viability of the turbine because this vessel has to operate year round for 20 years in the wind farm. The cost is just too high.

Furthermore, the vessel can possibly carry 30 or 40 technicians, but it would take too long to deploy them in sets of 2 or 3 people onto the turbines. This is actually counterproductive to what we want to achieve—namely, a high availability percentage for the wind farm.

WAVES

Transport to and from offshore turbines requires a transport vessel of some kind. If the weather was calm at all times, the requirement for the vessel would only be the capacity to transport 12 to 16 crew members per trip. However, the wind and the tidal streams offshore create both waves and swell, and this must be taken into consideration when deciding on a suitable means of transport because the vessel will almost always be subjected to waves and currents during transport and, most important, during the transfer of personnel. Waves, in this respect, will induce significant forces on the vessel; particularly during station-keeping in front of the turbine foundation, this becomes apparent.

The following briefly presents some of the practical problems that have been encountered when transferring personnel to and from offshore wind turbines. As mentioned previously, waves are just one more problem when comparing offshore and onshore wind farms. The waves are, of course, a recurring problem related to bringing the turbines offshore, and the waves encountered in the area in question are extremely large—up to 10 m high. Therefore, both the foundation and the access design must take into account that extreme forces will have to be dealt with.

First of all, the definitions of waves and swells should be clarified. *Waves* are short, crested seas with a wave period of less than 4.5 seconds. Waves with a period above 4.5 seconds are called *swells* because the seas tend to have a longer and more unidirectional passage by the foundation.

The short, crested waves often will pass in many different directions relative to wind and wind shear. Therefore, they are more unpredictable, and for a short vessel, this can be a serious problem. Even though the short waves will not have the same energy potential as the longer swells, they will create unpredictable movements in the vessel, making transfer difficult.

This problem can to some extent be dealt with by using a larger vessel. However, this will significantly increase both CAPEX and OPEX costs, since the larger vessel is more expensive to purchase, as well as to operate.

The larger swells will, however, have a significant impact on vessel movement relative to the foundation. As a wave train passes, the energy it releases will be enough to move the entire vessel a large distance. Therefore, the transfer method, whether a ramp, gangway, or something else, must be capable of a large stroke length, as well as a very fast adjustment system, in order to maintain the desired position relative to the foundation.

The vessel will have six directions of free movement when in the exposed water:

- Heave
- Pitch
- Roll
- Sway
- Surge
- Yaw

It is therefore crucial that the method of transfer be able to counteract these movements. It is particularly important that a vertical stroke length in combination with the horizontal movement of the vessel in two directions can be absorbed by the system.

A vessel moving in 2 m Hs waves will be able to move a large distance within one or two seconds; the system must be able to cope with this movement. Therefore, it is also important to look at the transfer systems available in connection with the vessels available in the market today.

Offshore on the site, the wind farm owner has usually placed a met mast, and some measurements have been made from the mast and the data have been evaluated. However, the met mast should also include a wave buoy, which would give the characteristic wave data for the site. The buoy will record waves over the entire lifetime of the wind farm.

The wave data should provide the average of the times when the waves are not higher than 2 m Hs. Normally, the annual service check of the entire wind farm is conducted during the summer. The wave climate should therefore be favorable for this purpose.

WIND

When working offshore, it very quickly becomes apparent that wind is an extremely important factor and one that will also determine whether or not it is possible to transport and/or transfer personnel to the wind turbines.

Obviously, the turbines are positioned in areas where wind is constant and has a reasonably high mean speed. Therefore, working with repair or retrofit outside the turbine is defined largely by wind speed, as long as the vessel carrying out the repair is stable on the seabed.

But the wind will also generate waves. The time delay between an increase in wind speed and an increase in wave height is approximately 1 hour under normal circumstances. This in effect means that the systems used to carry out the installation—repair and O&M work—must be able to handle the wind and wave forces and be flexible at the same time.

When one or more turbines are manned and there is a deterioration in the weather conditions, the crews must be able to safely reenter the transport vessels. As the wind picks up speed, the crew onboard the vessel must be able to forecast the time left for recovery and the wave height to be expected in order to plan and execute the safe movement of the crew and equipment.

Wind will thus be a determining factor, not only because it will influence the workability of repair vessels, in particular, but also because the wind-generated waves are the most frequent on the open ocean. Swell is almost always caused by storms that pass in the open waters of the world. Thus, the swells generated in the North Sea often are caused by storms in the Atlantic and so forth.

CURRENTS

Currents also have a large influence on the transport and particularly the transfer method because the vessel must be on a fixed location while transferring the crew. This means that the vessel must be able to counteract the currents while working around the turbine. These currents can become extremely high, and ones up to 5 knots during tidal variation are not unusual.

It must be possible to fit the vessel with a thruster capacity that is high enough to stay in position even in the strongest current on the site. This is an important point when deciding what type of vessel will be used, since both the catamaran and the SWATH types struggle with the ability to fit bow thruster capacity. However, they can, to a large extent, work by counterrunning the propellers in the torpedoes, which will then make the ship rotate around a center axis.

The best possible station-keeping will be achieved, however, by employing both bow, stern, and main thrusters in a system referred to as dynamic positioning. This system, however, requires an enormous installed kW capacity, which is not economically possible for this type of operation.

The currents measured in the North Sea range from 0.5 m/s up to about 2.5 m/s, and this will have an impact on the way the turbines are approached and what vessels are or will be used. Furthermore, it very much depends on the area in the North Sea (or elsewhere) where the wind farm is located.

Therefore, less emphasis is put on the station-keeping capabilities of the vessel as a result of currents influencing the vessel in the eastern part of the North Sea, where currents are low but swell is higher, compared to the west side along the coast of the United Kingdom, where the currents are high but the swells are predominantly smaller.

However, one must always consider whether the approach to the foundation is safe since the combination of wind, waves, and currents can make for unfavorable conditions, thereby creating a higher risk when trying to approach the foundations.

Related Images

Lifting and installing the "bunny ear" configured nacelle.

Removing the lifting equipment from the installed nacelle. You would have to look very hard for a job with a better view.

Every inch of space on the installation vessel is used.

The final result. A wind farm ready for operation.

CHAPTER FIFTEEN

Transporting Wind Turbines

This chapter provides an analysis of the different methods of transporting and transferring personnel and wind turbines to offshore locations and highlights the most successful operations. Based on previous experience, we have listed several criteria that should be taken into consideration before deciding on the optimum solution for the operation and maintenance (O&M) strategy for a project.

Since the first commercial turbines were installed offshore in 2001 on Middelgrunden, a number of variations on the method used at that time have been put into operation. The overall approach has been to sail either a small RIB, which is an inflatable boat that can carry up to four people, or a small vessel that can hold up to 12 people to the landing arrangement of the turbine foundation. The personnel then climb up the ladder onto the platform. This method has generally been successful, and so far no problems or accidents have been reported. Some problems have occurred, however, on several other fronts:

- The landing arrangements on Horns Rev were not strong enough to support the force of the larger 12-passenger (pax) vessels and had to be replaced in 2004.
- The method is not approved by the Danish authorities to more than 1.1 m Hs and therefore does not solve the related availability issue.
- The distance to shore that the small vessels can travel cost effectively is rather short. And, since the transport time counts as working hours, the number of actual "in turbine" hours will be reduced by a factor of two for each hour of transport (one hour each way means two hours less work in the turbine).
- The solutions found that are capable of staying offshore with larger numbers of pax are very expensive due to the Special Purpose Ship (SPS) code applied to vessels with a pax capacity of more than 12. This means that the overall O&M cost will in fact exceed the budget.

Therefore, the three main problems related to the O&M strategy are:

1. More flexibility in terms of boarding the foundations in adverse weather is required.
2. More suitable offshore vessels at lower costs are needed.
3. The O&M crew must be placed as close as possible to the wind farm.

For an offshore wind farm, many factors must be taken into consideration to determine the optimum place for an O&M base. The following are some of them:

- The O&M base should be close to the wind farm in order to cut transit times from the shore to the location as much as possible. In this way, the crew can be moved on short notice, taking advantage of shorter weather windows.
- The shore base should have several good harbor facilities to support the O&M vessels.
- It may be useful to look at the possibilities for EU regional support and funding of jobs in an underdeveloped zone or area.

If the shore base is far away from the offshore site, the distance would make prediction of weather windows very inaccurate, and that could result in a substantial amount of excess downtime.

If we look at some of the German offshore wind farms, it may be sensible to use Helgoland as a base for the overall O&M solution, since the benefits are substantial. Plus, it would be a clear bonus for the island to attract jobs in sectors other than tourism (Figure 15.1).

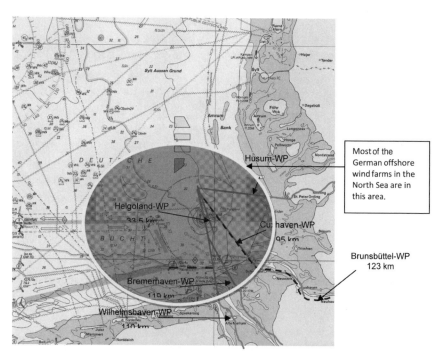

Figure 15.1 Distances to harbors suitable as O&M bases. *Courtesy of Winkra.*

TYPES OF TRANSPORT VESSELS

Many different types of transport vessels are available for use in the offshore wind industry today. The one common denominator is that they all promise to be the best and the most stable platform from which to work. However, the methods employed to become the best and the most stable platform are very different and to a large extent show the degree of ingenuity that is applied by naval architects, engineers, and contractors.

The problem of overcoming the physical challenge for the vessel to stay stable in this environment is further complicated by the fact that the statistical methods for determining vessel behavior, such as the strip method, are not valid for vessels that are smaller than 40 or 50 m in length. Therefore all design must be based on previous experience, guessing, or scale model testing.

The testing scenario is, however, very costly and does not appeal to the predominantly small companies that offer this type of service vessel to the market. And certainly investing some 50,000 euros in a scale-model test seems unappealing in a situation where no apparent contract is in sight. Therefore, the following information on types of vessels and their characteristics is based solely on previous experience and not actually on scientific research. However, the vessels do work, and that is why they are mentioned here.

This, in effect, means that before entering into any form of relationship with a supplier of service vessels, the wind farm owner should outline a number of requirements for the vessel in question and create a test series for all bidders who can supply this type of vessel and have done so in the past. The only valid alternative for new suppliers will unfortunately be to build on speculation or to perform a scale-model test.

In general, vessels come in four types, as described in the next sections.

Monohull Vessels

These are the typical kind of oceangoing vessels. Their advantage is that they are stable and can handle severe weather conditions. Even a relatively small vessel will have very good seafaring characteristics from all directions, and although they are not as comfortable in head seas as the catamaran or SWATH, they are superior to them in beam seas. Furthermore, they can take on big cargo because they are often fitted with a large hold with waterproof hatches.

Several suppliers in Europe build this type of vessel, and many specialize in vessels that can handle high wave forces. To name them all would be too extensive. Furthermore a number of used vessels are available.

Catamarans

Characterized by the two parallel hulls, catamarans have a deck suspended between them. The deck contains the bridge and accommodation module. They are characterized by good seafaring capabilities until the weather becomes extremely severe. They travel at high speeds and have a small water plane area, which can be a huge advantage when traveling. They are more "steady" in the head waves than the monohull, which gives a comfortable heave and pitch with very short periods. However, this means that acceleration will be greater when the waves increase in size.

Their disadvantage is that they become uncomfortable in beam seas when the weather deteriorates. Furthermore, catamarans cannot carry any significant payload unless the vessel is rather large. They are often made from GRP or aluminum to keep their weight down, which makes them vulnerable to the rugged type of work they should perform, particularly when docking onto the turbine foundation. Furthermore, they are rather expensive both to buy and to operate.

There are a number of suppliers of this type of vessel, and to name two: Baltec in Germany and Alnmaritec in the United Kingdom. Several more can be found by searching the Internet.

SWATH Vessels

The Small Waterplane Area Twin Hull (SWATH) vessel is a derivative of the catamaran. The main difference between the two is that the SWATH has two torpedo-type buoyancy hulls that are placed approximately 2 to 2.5 m below the surface. This makes the vessel more stable because the waves will not have a large surface in the splash zone to assault. Thus, the majority of the wave forces do not have an influence on the vessel, which makes it more comfortable to work from.

This type of vessel also has the ability to remain steady in the water when waves and swells pass in the vessels' longitudinal direction. However, this type of vessel suffers even more from low cargo capacity than the catamaran and is even more expensive to build. It also becomes difficult to install bow thrusters because of the poor possibilities of redistributing weight in the torpedo hulls.

The two main suppliers are Abeking & Rasmussen of Germany and Lockheed Martin Littoral Combat Ship Department in the United States.

Liftboats

Originally, liftboats were developed in the United States, where they serve as workhorses in the near shore oil and gas industry. They are three-legged jack-up vessels that are self-propelled, and operated by a crew of 3 to 5 people. They have a capacity of up to 32 people on board, but that can be somewhat uncomfortable because it means 8 people are in each cabin.

The vessels are designed to work in depths of up to 200 feet, or some 65 m of water depth. It is not certain how the European naval authorities consider them in terms of SPS codes and other safety regulations for seagoing vessels. They are, however, a very good and rugged type of vessel that can take care of the problem of deploying people to a turbine in bad weather. All that is required is that the front of the turbine be jacked up.

Once jacked, however, it may be necessary to abandon the platform if the jacking criteria are exceeded for any reason. Furthermore it is limited to deploying people on the turbine that is in front of the jacked area. If people are deployed to other turbines and the weather deteriorates, they cannot be rescued by the jack-up.

Following are some examples of all four types of vessels.

Monohull Cargo Vessel

This small cargo vessel is equipped with a boom for loading (Figure 15.2). It has a foldable hatch suitable for carrying spares and so forth. The boom can be exchanged with a large knuckleboom crane for offshore work (Figure 15.3). Crew accommodations in this type of vessel are fairly basic (Figure 15.4).

Figure 15.2 Monohull cargo vessel.

Figure 15.3 Monohull cargo vessels have booms for loading and hatches that fold.

Figure 15.4 Crew accommodations.

Note the beam of the vessel, which is much wider compared to the length. This gives stability when operating offshore. The high fo'castle gives security and distance to waves and seawater spray. The photo of the beam shows another monohull in the background that also has high-speed performance but not the same rugged seafaring characteristics as the cargo vessel (Figure 15.5).

Figure 15.5 Beam on monohull cargo vessel.

Catamaran

Note the wide beam as a result of the two hulls on the catamaran (Figure 15.6). This helps to keep the catamaran steady in head seas, but in beam seas the roll will be very severe due to the light weight and the low draft.

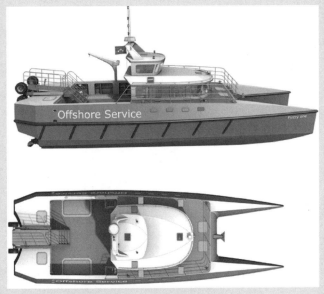

Figure 15.6 Two images of a catamaran. *Courtesy of Fintry 1.*

Figure 15.7 Photograph of a common catamaran used in the United Kingdom.

Another type of catamaran is aluminum; it is very popular in the United Kingdom (Figure 15.7). They are good for this type of transport, and in some cases, the work space is better than with the other type of catamaran. It depends on the site-specific criterion.

SWATH Vessel

The Small Waterplane Area Twin Hull type of vessel is designed specifically for calm conditions on the high seas in order to keep the vessel steady and cost effective in terms of sailing and fuel consumption (Figure 15.8). The draft on the vessel tends to be higher because of the deep placement of the torpedo hull, so a 2.5- to 3.5-m draft is not unusual.

Figure 15.8 A SWATH vessel has a very slim protrusion of the water line.

Liftboat

Liftboats have a very basic design, but they have the necessary features such as a large crane, a roomy cargo capacity, and comfortable accommodations. They are essential for may types of activities carried out on the ocean (Figure 15.9).

Figure 15.9 Wind farm platforms need liftboats available at all times.

Table 15.1 provides a summary of each type of vessel's features.

TRANSFER SYSTEMS

The vessel itself is, of course, a major part of transporting O&M crews to and from the offshore turbines. However, the transfer system in combination with the vessel will determine the workability of the entire system. It is therefore crucial not only to look at the best vessel for the specific site, but it is also of paramount importance to combine the vessel with a suitable means of transfer.

An assessment of the different transfer solutions must be made to determine which system is the most suitable. This is done by creating a matrix in which the different systems are compared to one another. This makes it easy to determine which system is most appropriate for deployment for the wind farm.

Available Systems

One would think that the many years of offshore mineral exploitation had generated a great number and variety of different solutions for access to structures and that these methods of access would be applicable to the

Table 15.1 Vessel Types

Type	Monohull Pilot Type for Crew and Cargo Transfer	Monohull Cargo Type	Catamaran for Crew and Cargo Transfer	SWATH for Crew and Cargo Transfer	Liftboat Data from 245- Class Liftboat
Manufacturer	Several	Several	Several	A&R, FBM	Several in the United States
Speed	15+ knots	15 knots maximum	20+ knots	15+ knots	8 knots maximum
Max Distance to Shore	20 NM	Various but can be unrestricted	Various depending on class	Various depending on class	Various depending on class
Max Hs in m Operational	1.1 m	1.1–1.5 m depending on size	Expected 1.0–1.25 m	3 m	1.1–1.5 m for jacking
Number of Crew	2–3 for 14 hrs	4–9 for unrestricted service	2–9 depending on class	3–6 depending on class	4–6 depending on class
Max Number of Passengers	12	12	12	12	12

Size	10–20 m	30–70 m for this industry	10–70 m depending on use and material	Min 20 m, max approx. 100 m	From approx. 33–100 m depending on water depth and class
Max Number of Cargo	Approx. 500 kg	Up to 200 tons depending on size	1–20 tons depending on size	1–20 tons depending on size	Up to 334 tons for 245 class
Cost	1–3 million euro estimated	1–5 million euro estimated	A200 (pictured) 2 million euro	FBM 4.5 million euro	42 × 29 m liftboat US$18 million
Seafaring 1–5	3	1	2	2	3
Rating 1–5	2	1	2	2	3
OPEX/24 Hours	Approx. 1750 euro	Approx. 1210 euro (DIS)	Approx. 2700 euro	Approx. 4000 euro	Approx. 5000 euro

offshore wind industry as well. Unfortunately, this is not the case. Since the offshore oil and gas industry rely on dynamically positioned, large, rugged vessels designed for deep-sea operation, and subsequently form very stable platforms where even large helicopters can land, both the size and cost of these vessels would exceed the O&M budgets of the offshore wind industry.

First of all, the price, which normally runs up to 100,000 euros or more per day, makes this type of equipment unattractive. The sums involved are negligible for the oil and gas industry, but they are significant for the wind industry. It is simply not affordable.

Even though the stability is desirable, the size of the equipment is not practical for the access to wind turbines offshore for two reasons. First, the turbine would suffer damage if it came into contact with the vessel. Second, the water depths in which the shallowest turbines in offshore wind farms are positioned make it impossible to approach them because of the more than 6 m draft of the vessel. However, this is not the case for this project.

So only a few solutions are available today, and except for the first two of them, they are only experimental:

- Offshore Access System
- The Ampelmann
- Browing system, Lockheed Martin Littoral Combat Ship program
- Boat landings
- Viking Selstair
- Transfer crane solutions: PTS, Grumsen, and others

The common denominator for these systems is that they must be fitted on a vessel or on each foundation in order to transfer personnel to and from the offshore turbines. Therefore, their success largely depends on the seafaring characteristics of the vessel to which they are fitted. Furthermore, except for the boat landings, the Browing system, and Selstair prototypes, they are not in use at the moment. This is, of course, important because the owner of the wind farm probably has no interest in applying a prototype solution to a large-scale wind farm. The access system must be one that is reliable, tested, and certainly legal.

As mentioned earlier, the seafaring characteristics are determined by the wave height and the length of time the vessel can withstand. In addition, the transfer system must be fitted to the vessel in a secure manner in order to safely transfer personnel. This means that the system itself must be very rigid and able to cope with high forces, accelerations, and bending moments. This is often not the case. Other systems, such as the Waterbridge, have experienced this problem.

Furthermore, crane solutions for lifting people are actually illegal. This is true in the Danish sector, where health and safety regulations are governed by the health and safety authorities and not by the marine, energy, or offshore ministries. In Denmark, for instance, it is prohibited to lift people with cranes, regardless of size and make. The only exception is occasional light work on a single position where scaffolding is not viable but certainly not as means of regular access to a structure. Suspending a person from a steel rope on a remotely operated crane offshore can be treacherous and is not recommended as an access method. The number of things that can go wrong is infinite. Finally, it is not economical to install such a device on every platform; this is expensive and will require huge amounts of unnecessary maintenance.

Another system is the Lockheed Martin Browing system for offshore transfer of marine personnel that has been used for many years by the Britain's Royal Navy. Basically the system consists of a gangway that can be extended and retracted with the same speed as the wave-induced movements of the two vessels it is extended between. However, to function against a fixed offshore structure, the system must be reinforced to take the higher forces from the vessel moving toward and away from the foundation.

Combined with a SWATH vessel, the system could be a very good solution when reinforced as just described. When fitted to the SWATH, the Browing system is approved for crew transfer of up to 2 m Hs and should therefore be an adequate solution when operated properly.

The access system has been tested and operated on a daily basis in offshore waters for more than 10 years. The system, however, is designed for two vessels in motion. This will be different from the fixed structure against which the system would be used if it is applied to the O&M solution for an offshore wind farm where the foundation, of course, is fixed.

Therefore, careful design and alterations must be carried out if this is the case. The fact that the system is working against a fixed structure will increase forces and bending moments on all the movable parts, so it is important that they be reinforced. If this is done, the system will be smaller, lighter, and faster than all the other solutions.

The boat landing was developed for the offshore oil and gas industry years ago and has as such been in operation for the longest period of all the solutions. This method has also been chosen for the previously installed wind farms and has been successfully used for deployment of personnel without any problems.

However, the system has only been approved for operation in up to 1.1 m Hs and is therefore, in the current configuration, not suitable for access to the turbines by itself. This does not give the accessibility that is desired, since the demand is 2 m Hs in order to give 95 percent accessibility. This can to some extent be counterbalanced by using a vessel type, such as the SWATH, that has benign movements in the higher-wave climates.

Selstair is marketed by Viking as a means of lifesaving, and as such the system works well. However, there is a problem with the system that makes it unattractive to this industry. As presented in other studies, the Selstair is fitted onto the transfer vessel and a line is run up to the turbine platform. However, no instructions are provided for how to do it and how it is supposed to work. This means that the only other possibility is to fit a Selstair to each turbine, which is certainly unattractive. Also, the fact that this system would have to be remotely operated means that higher maintenance costs would be incurred.

Finally, the Offshore Access System (OAS) and the Ampelmann are two systems that attempt to mitigate the effects of waves. The idea is good, but the experiences so far have not been impressive. Both systems are very expensive and require large craft in order to be stable. The fact that a gangway of more than 10 m is suspended from the stern or loin of a vessel is in itself a heavy burden, and the smaller vessels would easily capsize in this event.

This means that the cost is significant and, in the case of the Offshore Access System, the fact that the gangway is fixed to the turbine foundation means its rigidity—and for the foundation, for that matter—must be very high. This leads to another increase in cost, which is undesirable.

Offshore Access System

The OAS is an extendable gangway suspended in two hydraulic cylinders and motion stabilized in order to pick up the vessel motion relative to the foundation (Figure 15.10).

The Ampelmann

The Ampelmann is the same type of solution (Figure 15.11). In effect it is a set of hydraulic cylinders that are fitted to the deck of the vessel and hold a suspended platform in order to keep it steady for the gangway to be placed

Figure 15.10 OAS heavy-duty version. *Courtesy of Offshore Solutions.*

Figure 15.11 The Ampelmann. *Courtesy of Ampelmann Operations.*

onto the foundation. This way, vessel motion is taken up by the Ampelmann, and the gangway and platform are kept steady for ease of access to the foundation.

Table 15.2 lists the features of each system, so that you can have a readily available overview of the various types of information that can be obtained. For further details, the websites of several of the companies offer lots of data and some have pictures from ongoing projects.

Table 15.2 Transfer Solutions

Name	Ampelmann	Boat Landing	Browing System	Offshore Access System	Selstair	Transfer Crane
Manufacturer	TU Delft	Various	FBM Babcock	Fabricom	Viking	Various
Description	Gyrocompensated platform	Fixed structure	Gyrocompensated gangway	Hydraulically operated gangway	Flexible vertical safety staircase	Remotely operated cranes for lifting of crew
Max Hs	2 expected	1.1 actual	3 max, 1.5 practical	1.5 expected	2 expected	2 expected
Size of Vessel Required	30 m+ due to weight of system	15 m+ for safe seafaring	20 m+; see SWATH description	40 m+ due to weight of system	Unknown	15 m+ for safe seafaring
Maintenance	High	Low	Medium	High	Low	Medium
Estimated cost 1–5	5	5 due to number required	3	5	5 due to number required	5 due to number required
Rating 1–6	4	2	1	5	3	6

Related Images

Loaded and ready for departure with the first batch of turbines for the Lillgrund wind farm. It is extraordinarily simple from a technical point of view to carry out this work, provided you know what you are doing and have planned correctly for the project. Once started, you just repeat the process until the last turbine is installed.

Off to a good start placing the bottom section on the foundation.

Horns Rev 2002. Weather: sun, sun, and sun. Wind 3 to 4 m/s, waves 0.5 to 1.0 m Hs. So unusual that people in the Esbjerg port talked about us being "extremely lucky." But this is what it is all about—good planning blessed with some luck, in this case a lot of luck.

Sailing out to the site with only towers onboard. Sometimes onshore production cannot keep up, so to be sure the vessel is occupied and offshore production is running, this step is taken. It's also a graphic reminder of how important planning of all operations from the factory onshore to the foundation offshore is.

CHAPTER SIXTEEN

Deployment Strategies

When reviewing the data presented in Chapter 15, it becomes apparent that the resource allocation for an operation and maintenance (O&M) department, and certainly the hardware involved, is considerable. Furthermore, the entire spread of surveyors, classification intervals, personnel expenses, and administration is a significant proportion of the OPerational EXpenses (OPEX) costs. This is due to the fact that the naval authorities and the regulatory requirements they implement do not discriminate between one small organization with one vessel for offshore transfer use and a full-scale shipping company.

The safety, personnel, and technical issues are the same for sending people out to sea whether it is only a few individuals or many. This in effect makes sense in order to safely navigate the oceans, but it puts extra costs on the single vessel that the owner requires for the wind farm, so other solutions rather than investing in a full-blown offshore operation should be considered.

The desire of any offshore wind farm owner is, however, to have access to equipment and personnel whenever it is required and preferably without delay. Therefore, the goal is to own equipment and also to employ the crew and repair technicians needed to be able to carry out O&M and also repair whenever it is required. However, previous calculations made in cooperation with the repair department from Vestas show that the amount of money spent per day adds little cost efficiency to the project.

Based on the figures obtained from Horns Rev, the conclusion is that it is possible to spend around 3500 to 4000 euro per day for a vessel equipped with 3 crew members and 12 technicians. This is, however, very close to the limit of what is possible to operate when the weather criteria are also taken into consideration. Weather downtime will be significant, and therefore the number of technicians should be higher.

Unfortunately, a number of 12 pax on board is the maximum before you must enter a higher SPS code of compliance. SPS is the safety code of practice for cargo and passenger vessels. This regulatory code is international and consists of increments of 12, 50, 200, and 1000. When the passenger level is higher, a number of criteria must be raised dramatically, and subsequently the cost efficiency of the vessel becomes very poor for this type of work.

Therefore, the desire to have 18 or more technicians onboard results in demands for the vessel to carry rescue equipment for 50 people, which will dramatically increase a number of technical features such as damage stability capability. Other issues, such as hull thickness, may make it impossible to use an existing vessel for conversion, and today the cost of a new vessel is very high due to the worldwide market situation.

SHARED ACCESS TO OFFSHORE EQUIPMENT

The alternative solution to owning and operating an entire O&M spread with vessels, crew, and technicians is to share the resources with the wind farms that are located close to the project. The advantage is, of course, that it is possible to significantly reduce the cost of crew, technicians, and plant. The disadvantage is that you may not have instant access to everything when you need it. Therefore, a method of sharing the resources must be developed in order for all wind turbines to be serviced as quickly and efficiently as possible.

In the following, we discuss the repair of offshore wind turbines. It is clear that the frequency of main component breakdowns in an offshore wind turbine does not justify investing in a very expensive repair vessel. Therefore, a framework agreement has been developed by ship owners to facilitate swift repair of offshore wind turbines at a reasonable cost.

In our opinion the only sensible way to address the operation and maintenance of offshore wind turbines is to do exactly the same: design a framework in which two or more wind farms share the crew transfer vessel, thus significantly reducing the cost of plant per kWh. This also makes sense because the number of times at which even a minor spare for the turbine is needed is relatively small. The lead time for larger components will be several days, if not weeks.

Therefore, the task of planning maintenance is not as difficult as it might seem, simply because it is not often the case that you can just sail out and restart the turbine. The parts needed for repairs, whether small or large, are not at hand, and it is not economically viable to have one or two turbines in stock as spares. Furthermore, maintenance schedules are typically organized around the best access opportunities during the year—that is, during the summer.

So the key to the optimum O&M system is to calculate the number and type of breakdowns experienced in the desired turbine so far, compare this with the lead time for spares, and then calculate the necessary response time for the single repair. It is also apparent that the number of kWh generated in one single turbine has a relationship to the cost of O&M plant. If the turbine generates 2000 euro per day when operating at full load, it is allowable to have it standing idle for two days before the cost of plant is less than the production loss if the choice is a SWATH vessel.

Within a two-day period it is very likely that a vessel can be sent to carry out either the small repair or the restart needed. So the recommendation is to search the market for possible relationships with adjacent wind farm owners to reach an agreement on sharing the O&M spread. The cooperation does not necessarily have to start when it is installed, but it would be advantageous if the neighboring wind farm (or farms) agrees to buy into the spread that the wind farm owner operates whenever they are built. In this way the cost and degree of idle time for the equipment can be minimized, and subsequently the cost per kWh can be reduced.

A further advantage is that the technicians will become more accustomed to being transported and working offshore, which will further decrease the number of idle hours because rough weather has a large impact on the level of performance in the turbine. The technicians will suffer less from seasickness and other transport-related fatigue factors. The final task is to define the means of transport and the actual setup on site for the O&M solution.

Using Helicopters

It has been debated whether the added security of using a helicopter for transport of personnel when the weather is too rough for the transfer boats is worthwhile. The previous experience is, of course, limited in the way that Horns Rev and Alpha Ventus are the only offshore wind farms where such a backup solution has been applied. Therefore, the logical step is to ask the wind farm owners who have experience in this field for their feedback and then report their responses.

Basically, the contract with the helicopter supplier is that he will have a standby service agreement, where the client could ask for assistance within a certain time frame from the blackout of a turbine until the point in time when the helicopter must deploy a technician on the turbine in question. Flight time for Horns Rev would be 0.5 hours as opposed to the 1.5 hours transport time with the crew boat. Distance is, of course, a factor that must be considered carefully.

The fact that Esbjerg Airport has a heliport for the Danish offshore sector only a couple of kilometers in the right direction of the flight path to Horns Rev makes it a reasonable consideration. Furthermore, the helicopters are in other long-term service agreements that have made the price more attractive for the wind farm owner when the contract was negotiated.

So compared with the daily production at full load of approximately 2000 euro, the cost of the helicopter should be fairly low for it to be attractive. However, this service will cost the owner much more than the production of one turbine per month, and compared with the total production per 2 MW Vestas machine on Horns Rev, this would in effect mean that you need to keep two additional turbines running for the entire year in order to make up for the cost.

Therefore, the solution only makes sense in the months in which you could expect near full production—for example, January and February. However, in such cases, you may be prohibited from reaching the turbines by boats for days. For the remainder of the year, it is not economically viable at all.

For the Alpha Ventus project, the information received is that the helicopter service was of paramount importance and that the wind farm would not be efficiently serviced unless such a means of transport was available. If and when the helicopter backup solution is transferred onto the wind farm, it must therefore be looked at very carefully. Although the turbine at full load has more than double the daily production, the distance to shore, as well as a suitable location where a helicopter is sitting idle most of the time, must be assessed.

Since the Horns Rev figures do not add up with only 30 minutes of transport, the distance to shore and pricing structure must be radically different for the wind farm in order to work. This is doubtful because the pricing structures for the helicopters do not vary greatly in the North Sea region due to the amount of work and the actual capacity present. We cannot at present recommend the deployment of a helicopter service unless the owner can negotiate dramatically different rates than have been achieved in the past on the offshore wind farms that have previously been installed.

This point can be debated at great length. If the helicopter service was so good and important, one would think that this would be a standard requirement for all offshore wind farms, but it isn't. Various project developers are all following different approaches to this problem, and it will possibly take a bit of time before anything significant can be deduced about the viability of helicopter use.

Additional Safety

This gives rise to the question: Is there any additional safety that can be deployed that will increase accessibility to the turbines? From here, the point of view is that there is not. You can double the spread that is originally intended for the O&M solution, or you can design two spreads to perform different tasks altogether. The wind farm owner could, for example, benefit from the use of a SWATH or monohull and a jack-up for a larger repair.

This would not provide a significantly better operational envelope in the wind farm, but it would increase the number of tasks that can be performed by the owner before requiring bigger and more expensive vessels. Unfortunately, this also significantly increases the transport and maintenance costs and therefore subsequently lowers the profit per kWh.

Finally, it should be noted that the deployment of equipment as mentioned earlier in this chapter would also apply for the shared O&M solution between two or more adjacent wind farms. This would also have the benefit of increasing the availability of craft for deployment of technicians, since there would now be two spreads to choose from.

Cost Model

As we saw in the table in Chapter 15, which compared the different types of service vessels, there is a great variation in the cost of the plant. Unfortunately, there is also a great variation in cost related to the operation of the same. As an example, the same SWATH from Lockheed Martin will come out in several OPEX cost figures on fuel and lubricant consumptions, depending on the speed at which the vessel needs to sail. An increase in cost of more than 30 to 40 percent is not unusual, and comparison of the vessels based on this is virtually impossible unless a tender document is issued stating the specific criteria the vessel must meet.

As can be seen in Table 15.1, however, the probable purchase price and OPEX cost are listed where it has been possible to obtain one. If the wind farm owner wants to issue such a tender, Advanced Offshore Solutions would most certainly offer their services to achieve the best cost and quality of vessels.

FUTURE TRENDS IN THE SERVICE VESSEL INDUSTRY

Even though great efforts have just been made to describe what the issues are regarding the offshore service vessel industry and its specifics, it should be noted that the trend is driven forward by market development in various ministries of energy. Why is this so?

Well, it is simply because Round 2 and Round 3 in Britain have moved the wind farms further offshore. Also, in Germany, the initial frame of mind was to claim they didn't want to see them from the shoreline, so almost all offshore German wind farms are more than 20 km from the nearest beach.

This means that the daily deployment of technicians for maintenance becomes unrealistic in terms of the small service crew boats that sail at high speed to and from port. The sailing time to the nearest suitable port will often be more than 2 to 3 hours away, and transport time does form part of the working time for the technicians. Therefore, the transit time of 3 hours in each direction will mean that 6 hours of a maximum 12-hour working day per technician will be lost in unproductive transport time.

A new business is therefore emerging, where the technicians are kept offshore on hotel vessels in the near vicinity of the wind farm. In this way it is possible to use the small crew vessels as deployment vessels in the wind farm and host the technicians for a full week onboard. This significantly increases productivity, and the trend is to also use this type of vessel for the installation.

This raises the bar for cost, however, and therefore small vessel operation is threatened in the coming years as bigger companies with the balance sheet to invest in and operate this type of vessel are entering the market. As always, only older, less expensive vessels are available in the beginning, but as a consequence of the development of the industry, in the coming years there will be a new generation of offshore wind farm hotel vessels with high standards and abilities.

CREW VESSEL SELECTION CRITERIA

For the crew vessels, however, a process similar to the installation vessels can be employed. A number of criteria for the proper use and proper characteristics can be developed in a schematic form (Table 16.1). The table provides an explanation of various details so the reader can understand why the specific item is important.

Table 16.1 Crew Vessel Specification Sheet for a Catamaran

Basic information	Type of OAS	Crew boat double hull, up to 12 pax
	OAS owner	Various: Denmark, Fintry Marine; U.K, Alnmaritec; WindCat Workboats, The Netherlands
	Country	DK, UK, NL, Germany
	Trading distance from nearest safe port: 50 NM	Vessel will have a maximum trading distance from port
	Fuel consumption: 250 l/h	Fuel consumption high due to excessive speed
	Use or purpose	Bring service and installation crew from port to turbines; access via normal gangway
General parameters	Width: 10 m	The dual hulls are wide apart in order to achieve stability and low draft
	Length: 20 m	As small as 15 m and as large as 25 m; typically, the catamaran is smaller than the monohull in order to keep costs down
	Transit speed: 25 kn	Very fast craft
	Wind speed (limit): 12 m/s	Vessel will not cope well with high wind and seas
	Significant wave height: 1.3 m	Unless the vessel is significantly larger than 25 m, the Ampelmann or OAS will not fit. Then the only access is via the ladder, and this cannot be done in very high seas by the catamaran unless in head seas, which is not always possible. However, the catamaran is better in performance in general than the monohull.
	Weather condition limits	As per above
	Anchoring/ mooring system	None; vessel will push against the access ladder

(Continued)

Table 16.1 Crew Vessel Specification Sheet for a Catamaran—Cont'd

Loading capacity	Max load: 4 ton	Could be a little higher, but in general 12 pax is the load it will take, along with fuel and water; the catamaran will not transport a large amount of supplies due to the hull form
Operations/ bookings	Supplier	Numerous: Alnmaritec, WindCat Workboats, Fintry Marine, etc.
	Builder	Various yards: Alnmaritec, Hvide Sande Skibssmedie, Damen Shipyards, etc.
	Charter cost: 2500 euro/day	Vessels are generally inexpensive to build, own, and operate
	Minimum charter period: 30 days	However, the owners will seek long-term employment if possible
	Mobilization cost: 50,000 euros	Easy to move and plenty in supply
	Purchase cost: 2.5 m euros	In this range, could however be more
	Foreseen availability/ booking	Plenty of suppliers all over Europe
Waiting for weather	Significant wave height: 1.3 m	
	Maximum wave peak period: 6 s	
	Wind velocity: 10 m/s	Due to installation capacity of component suppliers
	Current velocity: 2 knots	For positioning/leaving the site
	Needed visibility: 200 m	Worse weather calls for more visibility
	Maximum wave height for WOW: 1.5 m	
	Wind limit for WOW: 12 m/s	

Table 16.1—Cont'd

People transfer and accommodations	Vessel's wind turbine access system	Mooring arrangement at bow for climb access
	Access to vessel	Mooring arrangement at bow for climb access
	Heave compensation system	None
	Accommodates: 12	The normal standard as per SPS rules is 12/60/250
	Crew needed: 3	Depending on number of pax

The crew vessel specification sheet seen from the charterer's side could look something like Table 15.1, although there are more details that the individual charterer will probably want to know.

BASIC INFORMATION RELATING TO CREW VESSELS

Which type of vessel should be rented? Is it a monohull, a catamaran, or perhaps a SWATH that is considered best for the project?

Type of OAS

The type of vessel will be determined by the actual use in the project and location of the same. If the project is near shore and in sheltered waters, the criteria do not have to be so high, but if the vessel is for open seas and heavy weather, small catamarans and monohulls will not be able to cope.

OAS Owner

Who is the owner? This may not be as relevant, but the contract has to be with a serious and respected partner; otherwise you run the risk of chartering the wrong piece of equipment or, even worse, that neither the equipment nor the owner will perform. The contract for a unit, such as a crew vessel, will be for a minimum of five years—normally—and therefore it is a long period to be committed to the wrong equipment and supplier.

Country

The vessel should not be chartered from a country or company too far away from the job site. The reason for this is to be able to secure smooth and fast communication lines to conduct business. If the owner is too far away, you will be working through agents or people contracted in; this will make things slower and more difficult. The optimum is to have a supplier right in front of the wind farm. This, however, is not always the case, and therefore the closest acceptable suppliers should be chosen.

Trading Distance

The trading distance from the nearest safe port is an extremely important feature for the crew vessel. Since they are small craft, normally with no specific accommodations for the marine crew, they cannot stay out longer than 16 to 24 hours. Further, life rafts and other safety equipment must comply with more stringent rules and regulations in order for the vessel to trade farther offshore. For most crew vessels this is not the case.

Fuel Consumption

Fuel is expensive, and since we need to travel at high speeds to reach the offshore site quickly, the vessel consumes a lot of fuel. The numbers in Table 16.1 are arbitrary, yet quite accurate for the smaller craft. If a 21 to 25 m catamaran is transiting at 25 knots, the fuel consumption could easily hit 350 to 400 liters per hour; thus, it becomes a significant cost parameter for the charterer.

Given that the vessel trades 360 days for 5 years and the traveling distance is 3 hours each way to and from the offshore site, the cost would be 5 years × 360 days × 6 hours of steaming × 350 liters = 3.8 million liters of diesel, or close to it. At 1.40 euro per liter, this comes to around 5.3 million euros, so possibly the boat will be completely paid for. This is a significant cost.

Use

Why is the vessel being chartered? Is it for crew transport only? Will equipment be transported as well? Is the wind farm close to the shore? Are the waters and the site exposed or sheltered? Does the crew have to stay offshore overnight? These questions are crucial to the charterer and, of course, also to the owner in order to find the right vessel. The costing of the vessel will be very much dependent on these issues, and therefore the specifications for the actual job to be carried out must be determined in rather great detail. Otherwise there is a good chance that the wrong vessel will end up on your site; that is not what you want—from neither a cost nor a safety and operations point of view.

General Parameters

For the vessel charterer—the wind farm owner or operator—the length of the beam and draft, as well as the speed and consumption, are important parameters to determine whether the vessel is suitable or not for the project. Therefore, a number of parameters are given as a minimum list as in the following.

Beam or Width

The beam of the vessel (m) will tell you something about the seafaring qualifications, insofar as whether the vessel is slim and fast, if it is a monohull, or whether it is beamy and deep in the water, thereby often slower but more stable to work from.

For a multihull, the speed is less dependent on the beam, but the stability in transverse waves is much more important in this case. The reason is that while the multihull takes head seas well, it can have difficulty lying in transverse waves. This should be considered when opting for a catamaran, and often the extra cost for the SWATH is worth considering.

Length

The length of the vessel (m) is a very important parameter. The longer the vessel, the better it normally behaves in the waves—to a certain extent, at least. Length and beam go together, and a long and decently beamed vessel will behave nicely in rough waters offshore as long as the speed is kept right. A high-speed vessel in bad weather is a sure recipe for disaster, and the possibility of losing the vessel to a large crossing wave or sailing it into an oncoming wave is high. Therefore, care must be taken to equip the vessel with the right amount of power and to adjust this to prevailing weather when steaming out of or into port. Again, a full-bodied vessel with a reasonably deep draft will make the passage slower but safer in bad weather.

Transit Speed

The transit speed (kn) is almost always referred to as the most important figure of all. The faster we can be onsite, the more work we can do. But speed is dependent on the weather, as we said before. So the speed of the vessel can be really high, but at high speed in marginal weather, the ride can be horrific because seafaring behavior is poor and the trip may become agonizingly long even though the distance is only 20 or 30 NM. Therefore, speed and seafaring behavior always go hand in hand.

For several offshore wind farms, however, the distance is longer, and two to three hours of steaming can be experienced. Since the technicians are under the 11-hour rule (there must be 11 hours of rest between two working periods), the time spent traveling to and from site must be deducted from the actual working time. So for 2 × 2 hours traveling to and from the site, that leaves a maximum of 8 possible working hours per day. But the 4 hours per day traveling must be paid for, and thus the cost looks like the following with 12 technicians on board:

$$5 \text{ years} \times 360 \text{ days} \times 4 \text{ hours travel} \times 12 \text{ technicians}$$
$$= 86,400 \text{ hours lost over 5 years of service}$$

Add to this the cost of fuel, and you would get the results in Table 16.2, including, of course, the average day rate for the vessel plus marine crew.

So, as can be seen, the cost of bringing personnel to and from the wind farm for service purposes is equally spent on the vessel, consumables, and technicians. This table is relatively crude and should be refined. But it serves as an example that this industry and the cost of services must be taken seriously, especially if we calculate this cost over 20 years where it would be around 85 million euros to bring 12 technicians out on a daily basis.

Table 16.2 Service Vessel and Crew Cost

	Number Years in Service	Number Effective Days	Cost/ Day (€)	Cost/ Hour (€)	Consumption/ Hour	Total Cost (€)
Vessel, including crew	5	360	4500			8,100,000
Fuel consumption	5	360	2940	490	350 liters	5,292,000
12 technicians	5	360	4200			7,560,000
Grand total						**20,952,000**

Wind Speed (Limit)

The maximum wind speed (m/s) or velocity the vessel can and will operate at is, of course, relevant for the charterer. The wind speed at which the vessel can transit is extremely important. Even more critical is the ability to keep on station in front of the foundations and transfer personnel to and from the turbine. For this purpose, the wave climate is determined, and as previously mentioned, waves are first and foremost wind-generated.

Significant Wave Height

Once again, the waves are the prominent factor for the vessel and absolutely all work offshore. The waves and their significant height (m) have been discussed at length previously, but this is important for the crew vessel as well. This is especially true because wave height, in this case, makes it either possible or impossible to service or repair the turbine. For the installation, it was a question of whether the installation vessel could jack up or not. For the service and repair vessel, it is a question of when and how it can deploy personnel on the turbine. It may seem insignificant compared to the cost of the BOP for installing the turbine in the first place, but the difference between 1.8 and 2.0 m Hs for the crew vessel can cost millions of euros over a 20-year lifespan.

Weather Limits

The weather limits are consolidated in the documentation of the preceding criterion. Thus, wind, waves, and currents are significant factors in determining the vessel operational parameters. However, the normal procedure is to deliver the entire set of specifications to the charterer, including the weather limitations of the vessel. Weather limits are not necessarily the maximum limits for the vessel to operate in but for the safe transfer of passengers offshore.

This is important because some vessels have a maximum criteria of 1.2 to 1.5 m Hs, but this is for transfer of passengers, not transiting. Imagine if the transit capability was this low! In that case, the wind farm would never be cost effective if the transfer of personnel, and thereby the service, would only be possible in this weather climate. However, transit could be in worse weather conditions, which means that the vessel can move personnel to and from the turbines in a downward and rising wave situation, respectively. In other words, this would mean transitting out when the waves are from, say, 2.0 m Hs and descending and transiting back to shore when the waves are 1.5 m and rising.

Double fenders to fit the bow of the crew vessel.

Note the propellers' forward thrust to push the vessel against the fenders to dampen movement from the waves.

Figure 16.1 Photograph of a vessel anchoring or mooring system to access a turbine.

Anchoring or Mooring System

Anchoring in the wind farm is normally not possible, so the anchoring or mooring system against the turbine foundation has to be developed specifically for the wind industry because the foundation has specific requirements. The access to the foundation is normally via an access ladder fitted with two or four vertical tubular steel fenders wherein the bow of the crew vessel can push against it—or moor afloat (Figure 16.1).

Loading Capacity

When dealing with the crew vessels, it is necessary to understand that their sizes are usually small, and therefore they are first and foremost intended for the transport of personnel to and from turbines. However, a small amount of cargo is normally expected, and therefore any cargo capacity in the 1- to 5-ton region is acceptable. The cargo size and nature are also limited to the fact that the vessels are often aluminum and therefore not ideally designed to carry heavy point loads. Thus, cargo such as grease, small tools, and so forth are often all the vessel can handle. Large components are not transported on crew vessels.

Operations/Bookings

As said before, the booking of a crew vessel is generally long term. Usually it is a 5-year service contract that the wind farm owner must give out to a vessel provider so the turbine supplier can perform service and maintenance during the warranty period—normally 5 years. After this period, the wind farm owner takes over the service obligation, and the vessel provider, assuming he has done the job well, will be in a very favorable position to continue chartering the vessel for an additional 15 years.

Waiting on Weather

Weather downtime is calculated in the same manner as for the installation vessel. It is important to understand that the crew transfer vessel will predominantly be delivering personnel and small items to the site, but this will normally only be done in daylight hours. The criterion for doing this is, however, the same as for installing the turbines. Thus, the weather window—when deducting the good weather periods at night—will be significantly smaller than for installing turbines. So when calculating weather downtime, this part of the puzzle should be considered very carefully.

Personnel Transfer and Accommodations Offshore

The main role of the crew vessels is, of course, to deliver personnel on the turbine. This is why we hire them at a very high daily rate for a 20-year period: the lifespan of the wind farm. It is therefore equally important that the vessels can deliver the job at a high standard with excellent safety and do this consistently day after day.

So, according to the previous general description, where the service and crew boat market is presented, it should be made clear that not only the vessel must perform but also the method of crew transfer from the vessel to the foundation, and vice versa. In general, access must be safe to use even at larger wave heights such as 1.5 to 2.0 m Hs. This is the case with some of the newest types of access: the Ampelmann and the OAS. But this ability comes at a price and weight, and the question is whether it is economically viable—yet—to use these types.

The reason for this uncertainty is that the access system must fit the vessel. Therefore, large systems require a significantly larger vessel than the normal crew boat. That is why access systems have mainly been deployed in the oil and gas industry until now, although the first wind farm installations have used these means of access with good results. But as always, cost is the issue.

Future projects, such as Round 3 in the United Kingdom and several of the German projects, will require offshore accommodations for the installation and maintenance crews. The reason for this is that transport will take up almost all of the available working time for the employees. Therefore, the only viable solution is to use an accommodation vessel of a decent size.

Until now, the vessels have been old, converted night ferries taken out to the offshore wind farm, and it is from such "mother ships" that the technicians are deployed. The future in the installation and O&M industry, as far as crew deployment goes, will be from these types of vessels. However, transport from the mother ship to the individual turbine still has to be carried out using runabouts. Whether monohulls, catamarans, or SWATHs will be used is, in this respect, up to the individual wind farm owner.

Related Images

Waiting is a great part of the time spent offshore. Waiting on weather, waiting for components to be rigged, waiting to move location, and so on. It takes a good deal of patience to accept that the time is regulated for all activities when working offshore.

Maintenance and surveillance of all operations are crucial to a seamless installation project.

CHAPTER SEVENTEEN

Repairing Offshore Wind Farms

The operation and maintenance (O&M) strategy is normally not combined with the repair plans for offshore wind farms. This is because of the fact that the approach to the two types of work is different. The O&M of offshore wind farms generally requires a small- to medium-sized vessel without large lifting and loading capacities, since it is a matter of bringing crews out to the turbines and safely on and off again when servicing has been completed.

Repairing offshore turbines is considered to be the case when a major component—a blade, generator, gearbox, and so forth—must be replaced or repaired. For this type of work, you need a large vessel with the capability to stand firmly on the seabed for a long duration—12 to 24 hours uninterrupted—because the sheer weight and size of the components require cranes and other lifting equipment to be replaced and refitted.

Therefore, companies like A2SEA A/S in Denmark have developed models for cooperation between wind farm owners and themselves to achieve two goals:

1. To bring down the price on the large and very expensive equipment
2. To give customers the maximum satisfaction in terms of cost, planning, manning, and operation

The idea is that the owners and operators of offshore wind farms join one another in an effort to reduce the cost of a repair. The problem is that the crane vessels or jack-ups that must be deployed to carry out any component replacements are very expensive.

Furthermore, the equipment is not available in large numbers, nor does it lie idle in harbors close to wind farms. Therefore, the day rates for this type of equipment, regardless of which type is chosen, are extremely high. Normally a day rate of 30,000 to 40,000 euros is charged, making any type of repair costly.

There is only one way to bring this price down and that is to use the vessel as much as possible to keep the number of idle days low. This can only be done if owners, suppliers, and operators of the offshore wind farms share the cost for one single vessel among themselves.

HOW DOES IT WORK?

One crane vessel can service (carry out repair parts) around 250 to 300 turbines during a one-year season. No wind farm is that big as of yet, and furthermore the turbines do not break down in those numbers. Actually, the average number of turbine failures during one year is estimated to be around 4 to 5 percent.

So if the repair vessel only has, for example, four turbine repairs on Nysted or Horns Rev per year, which ship owners must be able to carry out at any given time, in order for Vattenfall or DONG to keep the availability of the wind farm above 95 percent, the cost per repair would be astronomical because the vessel or jack-up would have to be kept within reach of the wind farm and thus unable to take on other work. Therefore, the idea is to have several wind farm owners share both the availability of the necessary repair vessel and also the cost of keeping it available 24/7/365.

Other derivatives of this have been discussed, such as the wind farm owners and/or operators accepting a grace period of anything between 1 and 14 days, which would allow the repair vessel to do other work and thereby enable owners to lower the vessels' day rates. This makes sense because the spare components for the turbines are not stockpiled.

It would be very expensive for the single wind farm owner and not effective, since the lead time for a spare component is generally around 14 days or even more. So a system where the offshore wind farm owners and/or operators and suppliers join in a framework agreement to obtain repair capacity at a reasonably low cost, combined with the gracing strategy just mentioned, would be a very sensible solution.

To clarify even further, forecasting the availability of the offshore wind farm and the related availability of repair and maintenance vessels, makes little sense because the spare components can never be readily obtained anyway. This means that within the 1 to 14 days when the component may or may not show up, there may or may not be a weather window for the repair to be carried out.

So determining the availability of the wind farm based solely on the weather, and certainly on wave forecasts, makes no sense. The only practical way to address repairs offshore is to be as well prepared as possible and to carry out repairs at the first opportunity. A system where wind farm owners in the same region join in a framework agreement with an equipment supplier is the best and most cost-effective solution of all.

Related Images

Checking the monopile while driving it into the seabed. It is crucial that it goes into the ground exactly vertical; otherwise, it may not be possible to fit the turbine on afterwards. Therefore the monopile is checked often and particularly in the beginning of the pile-driving process to see whether it needs adjustment.

Accessing the turbine using RIB is no longer a common sight. Second- and third-generation wind farms are all in much more exposed waters and require a more rigid approach.

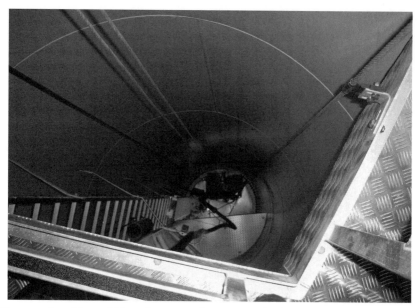

No room for old men! There is not much space inside the turbine. Notice the steel ropes on the side. They are for the elevator, which can hoist two men—of limited size—or tools to the top of the tower. To climb the tower, you need to be in good shape.

A transfer vessel is moored against the turbine foundation. This way it is safe for the crew to enter and exit the turbine without the risk of falling into the water.

CHAPTER EIGHTEEN

Environmental and Other Issues

The offshore wind industry is a green industry—as is, of course, the rest of
the wind industry. The jobs that are created are green jobs, and the industry
is trying very hard to keep a green label attached to all of the disciplines of the
work carried out.

PROTECTING THE ENVIRONMENT

The companies in the industry are determined not to cause any damage to
the environment by manufacturing or installing their products. This is, of
course, an important issue, since the industry's reputation is based on the
perception of causing no harm to the environment while harnessing natural,
inexhaustible resources such as wind and solar power.

Therefore, all of the projects planned, and when building permits are applied for, must demonstrate that the environment will not suffer any damage
or harm from the installation, operation, and dismantling of the wind farm.
This part of the permitting process is fairly rigorous and in the past (and present) has created a number of challenges that have not been easily dealt with,
since the authorities in many cases have raised the bar for what should be
achieved in this specific area.

One example, of course, is that waste management is specified in great
detail. The reason is very simple: We—society—do not want any pollution
of the marine environment during or after installing the offshore wind farm.
In addition, we want to know how to remove the wind farm after it has
spent its productive life offshore. We wish to have a clear plan for the management of the process from the beginning to the completion of the project.
This is until now fairly unique for the wind industry offshore.

The concept that we need to know where the wind farm is to go after
decommissioning is both sensible and logical. When I buy a couch from
IKEA, I may not have given a great deal of thought to how to get rid of
the packaging—and there is a lot of cardboard to get rid of—but I have
to deal with it after the couch is in my living room. Obviously, I can just
throw it in the trash bin, but then I will have used up my available volume

for my garbage in general. I can burn it, but if you live in an apartment on the seventh floor, it's not a great idea. So what does one do?

Often you end up taking it to the recycling center, but recycling rules vary among states and countries. So you can see that even when it's just purchasing furniture, the process raises questions. The argument that it is not really my problem doesn't work, whether it's a store or the manufacturer. The fact remains that the garbage is there, and we have to figure out a way to get rid of it.

The same goes for the waste produced offshore. We can't really store it on the blue shelf (drop it in the ocean) because this would inevitably create another problem, and solving one problem by creating another is not really practical. So what do we do? Most of the authorities that approve wind farms have solved this problem: They ask the developer to define a waste management plan and, in addition, a decommissioning plan for the entire wind farm.

The decommissioning plan is fairly simple. It is the installation process in reverse, so to speak. But the stakeholders—contractors, decommissioning facilities, and so on—may not look the same in 20 years, so the plan is normally reasonably basic. However, this will be seen in real life roughly 9 years from now when the first offshore wind farms are scheduled to go offline.

This is also interesting considering that market forecasts for installation of wind farms and that the inherent shortage of equipment is forecast to continue over the next 10 years. It seems logical that the vessels that will decommission the wind farms will be the same ones that installed them to the extent that they are still around.

As a parallel, the oil and gas industry is facing this problem, particularly in the North Sea, where some 800-plus installations are scheduled for commissioning—although it is not happening yet. One could argue that oil prices will dictate when it happens, so the problem remains largely unsolved. For offshore wind farms this is not the case, for the following reasons:

- The fatigue life of structures is used up after 20 to 25 years, so revamping them is not an option.
- The leases for sites generally run for 20 years; this could be renegotiated, but then the preceding point takes precedence again.
- The possibility of putting larger nacelles and rotors on existing towers and foundations doesn't work due to material strength—and fatigue life again.

So all in all, one should accept that the forecast is right, they *must* come down after 20 years. This does not mean, however, that they can't be replaced. They can. But that just adds to the number of vessels, contractors, and general stakeholders in the industry. This is a fascinating positive upside in terms of jobs and opportunity.

WASTE MANAGEMENT

As with the decommissioning plan, the waste management plan is part of the scope of work (SOW) for the contractor—and the owner. But whereas the decommissioning plan is more a statement of intentions sometime in the future, the waste management plan has to be real and delivered prior to starting work both on- and offshore.

The waste management plan basically describes the flow of materials to and from sites: who is responsible for cleaning up, informing others, collecting data, and so on. Figure 18.1 shows this process, which is necessary for the flawless monitoring and disposal of waste materials, whether they be liquid, gas, or solid. The plan gives an overview of all the main stakeholders in the project execution. It defines their role and what is expected to flow to them and what they deliver back on that basis. Thereby four types of actions can be derived:

1. Information, which is basically telling the appropriate stakeholders what has happened.
2. Data collection, which describes the action of counting and controlling all of the waste and its proper processing procedure.
3. The flow of materials, demonstrated by the passing from one stakeholder to the next until finished on the offshore site—the same, of course, goes into the onshore part of the project.
4. The flow of waste, demonstrating the route from the place of building the materials and the way back to the waste processing plant.

In this way, all process and information flows have been covered, and the chance of materials lying around somewhere without a responsible owner is reduced. It is not gone because we have made a plan and are monitoring it. The waste produced must be collected—all of it—so it is important to develop and maintain systems to make this happen.

This is effectively done by measuring the weight of the waste—the difference between the component and the container it is delivered in (whether

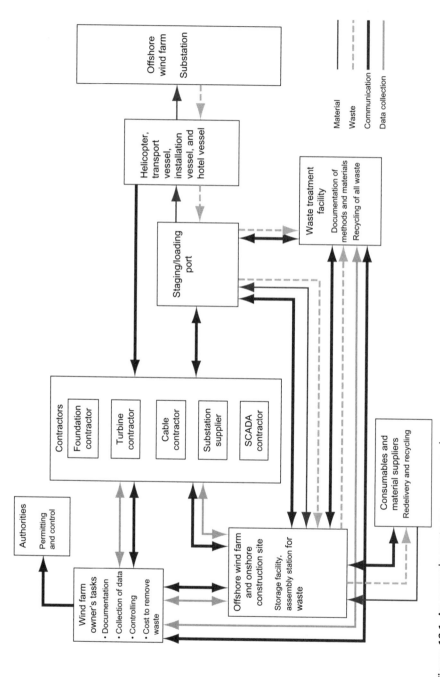

Figure 18.1 An example waste management plan.

cardboard, wood, or steel—it doesn't matter), and this difference is the one that must be delivered to the waste processing plant and accounted for. For fluids it is also reasonably simple to log the number of liters, gallons, or tons delivered and the remainder to be delivered back to the receiving waste processing plant. It is, of course, more difficult for gases.

The gases or fumes discharged into the air, whether on purpose or not, are normally assessed by estimate. You can, of course, calculate the NOx and SOx, the CO_2 emissions, and so on. But emissions' measurements for unplanned hot work with a welding torch, or the test running of an engine in the offshore substation, is more difficult.

Sometimes, for example, the turbines have to be installed without the power cable connection; it is not desirable, but it happens. If this is the case, a generator must be placed on the foundation in order for the turbine to be supplied with power. This is necessary because the turbine must be supplied with a slightly higher air pressure inside to keep moist, salty air out and thereby prevent corrosion of the materials inside the tower. Furthermore, lighting and some heating are needed to maintain the integrity of switchboards and all the electronic hardware inside the turbine. If this cannot be delivered otherwise, the generator must be installed for an interim period until the cable connection is carried out.

This is, however, an unplanned action with a number of waste management concerns, such as fuel consumption, refueling of the generator (this actually requires a specific procedure since it falls under offshore pollution laws such as MARPOL), emissions from the generator, and bringing the generator out and back again when it is no longer needed. Further, emissions must be accounted for, and this is a bit more complicated since we do not exactly know how much fuel we put on and how much is actually used during the period it is deployed. But we must certainly do our best to determine this.

So we have developed the plan and we know how to measure it, and now we need to monitor and collect the data in a format that is acceptable to the receivers, in the end usually the authorities; the waste processing plant; the wind farm owner; and the personnel who are supposed to carry out the monitoring of the process.

A specific template for this type of documentation should be developed, and this should be done to the standards of local authorities in the country where the waste ends up. The waste documentation can prove to be more complicated since, for example, the turbine and foundation components do not necessarily originate from the country where they are installed. A Spanish turbine can end up in the United Kingdom, and a German foundation can end up in Holland.

This, of course, raises some unique challenges in the documentation process since the amount of waste—say, packaging in Spain—ends up as waste in the United Kingdom. In neither of the countries is the amount in and out the same. How this is handled depends on the individual project, but it should be noted that this is a problem.

POLLUTION ISSUES

Offshore wind farm installation environmental issues are, however, more than just waste management. We wish to establish the offshore wind industry as one where we discharge nothing harmful into the surrounding environment. This goes for all aspects of the work carried out. We make sure that oil spills are not happening, and we take the MARPOL convention very seriously. We make sure that all components are fabricated in the most effective and environmentally friendly way. But, in one instance, this seems not quite to be the case. When we construct the installation vessels, we tend to seek out the least expensive supplier possible, sometimes sacrificing the environmental respect that we hold high elsewhere.

It should be clear that nothing is wrong with the work carried out in Third World countries; the workmanship is fine and adequate to the point where we can be satisfied. But in some cases, the HSE standards applied are not so spectacular. The issue is, of course, that when we solve the environmental problem in Europe by means of installing offshore—and onshore wind—in a safe, efficient, and environmentally friendly manner, we participate in an ongoing environmental disaster in the yards in Asia. Note, in Figure 18.2, the lack of any kind of pollution barrier to the ground. Also notice the very unsafe access ways—provided this can be called an access way. This is where the standards for HS and certainly E are much lower, if they exist at all. This cannot be right.

What use are health and safety regulations in Europe if people risk their lives to build the jack-ups (see Figure 18.3) we use for the installation of offshore wind farms? This, of course, saves money, but European clients should not accept this type of procedure to cut costs. If we want to hold the HSE banner high, we should do so all the way through. We cannot solve our problems by neglecting the same problems where we buy products and services.

The result you gain from rigging the exorbitantly expensive rope in dirt and rain water is substandard. The risk of quickly damaging rope, drum, and sheaves is extremely high. Dragging the rope over a wet dirty concrete

Figure 18.2 Jack-up barge construction in Asia.

Figure 18.3 This rigging of the A-frame crane for the European fleet of jack-up barges demonstrates that people are working without any regard for safety.

Figure 18.4 Although this is an inexpensive way to rig the crane, it contradicts everything the author has learned in more than 27 years of crane rigging and operation.

yard will cause rust in the strings immediately and the rope on which the life of the crane operator and surrounding personnel depends is no longer correctly functioning (see Figure 18.4).

As an industry, it is vitally important that we apply the same rules and regulations throughout the entire supply chain. Otherwise, we become exactly what we accuse less respected markets and suppliers of being. For example, are they wrecking the vessel in Figure 18.5 or building it? The photo of course begs the question as to whether the wind farm developer has done any subcontractor evaluation in the yard. Probably not. If this yard had been evaluated by a European developer, the contract would never have been signed.

As the preceding figures show, the supply chain stops in Europe—for the moment. We would not be able to turn the clock back to the days when the construction sites looked like this in Europe, and therefore, the only way to prevent competing against manufacturers who have low HSE standards is to turn their clock forward and ask the same obligations of them. This is not unreasonable. At some point, pollution and lack of respect for the environment and health and safety will come back to haunt us as will the inability to deliver a cleaner, safer place.

Figure 18.5 This photo shows the construction of a jack-up for the European wind industry.

THE WORKING ENVIRONMENT

This leads us to the next issue of environmental concern or care: Offshore work is a hazard to marine life and to the sea. As we said, the risk of discharging waste, fluids, or other materials in the water must be mitigated. For example, driving foundations creates earsplitting noise and plowing sub-sea cables disrupts the seabed. Therefore, authorities worldwide have developed regulations for both.

Seabed disruption is not in itself desired, so the permit to dredge—which is what plowing cables involves—is scrutinized vigorously to make sure no more harm than necessary is done. You have to appreciate that the cables will be buried into the seabed using a plow that will dredge a 1- to 1.5-m deep trench into the seabed. This has serious effects to the benthic life on the site where the cables are buried. Therefore, utmost care must be taken to ensures that fish, and especially shellfish, are not harmed by suspended ma-terial in the water column.

The method of cable burial is determined ahead of time in the design phase in order to address any concerns that may arise. Furthermore, to determine the short- and long-term effects of any dredging or plowing, test fishing is often done in order to define the before and after situation for marine life. This, however, also leads to discussion about what the marine environment is actually like in the proposed site area.

A good example of this was a Danish offshore wind farm where marine biologists had to determine the fauna and to what extent the local fishermen were suffering losses—mainly because the dispute always arises that the proposed site is, per incident, the best fishing bank. This, of course, gives grounds for claims of damages to the industry—or, in this particular case, fishing.

The local fishermen claimed that the fishing grounds were on the site. The biologists therefore test fished the area for quite a while and didn't catch anything significant. On presenting the findings, the fishermen then claimed that the biologists were using the wrong equipment. The biologists responded that they didn't quite understand this argument since they had only used the equipment confiscated from the local fishermen. No claims were subsequently filed and no damages were awarded.

PILING NOISE

Piling noise is the final great challenge the industry must face. When a pile is driven—whether a monopile or an anchor pile—into the seabed, the normal procedure is to use a hydraulical hammer. This hammer pounds on the top of the pile with a force of 200 to 400 tons when it drops down on the pile. The vibration is transponded into the water column in the way that the pile vibrates due to the blow from the hammer. The noise intensity is above 200 dB in some cases. For comparison, a jet plane is around 130 dB, and whenever noise increases with a value of 3 dB, the noise level is doubled. This is because the scale for measuring noise is logarithmic.

So what does that mean? Well, above 160 dB the noise can burst a marine mammal's eardrums. Higher noise levels can mean permanent injury to the mammals, and if they do not leave the area, they may suffer serious damage. Therefore, normally the procedure is to put pingers in the water prior to driving the piles. Pingers are specialized underwater noisemakers that send out a loud ping in order to deter the mammals from entering the area. This works well and, under normal circumstances, takes care of the problem.

However, as a conscientious industry we must consider another problem: The installation process of the offshore wind farm must not cause any changes in the behavior of the marine life in the area. Furthermore, it is logical that the mammals should not be physically harmed either, since this would lead to a change in the local biotope as well. In theory we are talking about two different types of changes: changes in behavior and changes in typology; neither is acceptable. This has been agreed to all over the European Union, but only Germany has fully implemented rules for offshore piling work.

Unfortunately, the nature of installing an offshore wind farm is such that we almost certainly need to pile drive foundations regardless of where the site is located. Therefore, consider the following scenario. If installing 80 tripod foundations, we must drive 240 anchor piles first. This is critical because the duration of this process may cause marine mammals to leave the area for good. This is not the intention, and therefore the rules in Europe have been developed such that a noise level of 160 dB must not be exceeded when measuring the noise from driving 750 m away from the pile being installed.

This is clear and simple. Put a hydrophone in the water 750 m away from the pile being driven and measure the noise. The peak levels of the hammering should not be more than 160 dB. The bad news is, of course, that water is not a good isolator for sound. On the contrary, sound moves well through water. Therefore, the question is how can we dampen the sound or completely isolate the pile driving noise from the water column?

To add insult to injury, sound travels into the water from pile driving in three ways. Figure 18.6 shows that the sound passes into the water via the air,

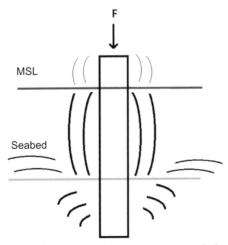

Figure 18.6 Sound leaving the pile upon hammering once with the force F on the top of the pile.

which is insignificant; the ground, which is significant since the noise intensity increases with the necessary force required to drive the pile once almost full penetration has been reached; and the water column, where the noise passes unfiltered through the pile wall. The noise levels can only really be adjusted when we focus on the part that passes through the water column.

There are several effective methods of damping the sound level. The only problem is that the majority of methods today do not give the necessary amount of damping to get below the 160 dB. The damping you can achieve is in most cases between 5 and 15 dB. Even though this may sound like a little, it is actually a significant amount of damping, but it is not enough. Therefore a very focused effort to be able to reduce sound further is going on right now.

Today four main routes of noise mitigation are followed to reduce the overall discomfort to marine life. One is a bubble curtain, which is basically a plastic tube—very large—that is perforated to allow pressurized air to flow into the water to create bubbles that will lower the noise level. This system has been tested on a number of occasions, and like all the concepts mentioned here, it works. The remaining problem, however, is that it is difficult to make the bubble curtain effective enough to bring the noise down sufficiently.

First of all, the bubbles expand as they rise to the surface; at some point they break up into two or more bubbles, and thereby the efficiency is reduced. Second—and this is the worst part—the bubbles flow with the water current. This means that the bubbles start around the pile, but as they ascend to the surface, they drift away, sometimes exposing the pile to the water. Several methods have been suggested and tested to improve the overall efficiency of the bubble curtain.

The latest development is to make a large bubble curtain around the entire installation vessel and thereby make it more effective. However, the air pressure required to create the bubbles is immense and cumbersome to achieve. To lay out the tubes requires an extra vessel. Then you need to apply enormous pressure to reach around the entire vessel and pile. Thus, the problem remains that you cannot reach the 160 dB level.

So, in essence, the method is labor- and equipment-intensive; it is slow; and the air pressure needed for a long period of time requires a great deal of fuel to drive the compressors and, of course, adds to the overall pollution of the environment. Plus, what you wanted to do still hasn't been achieved.

It should be mentioned that tests are ongoing at the moment to evaluate the actual efficiency of bubble curtain and several other methods of noise mitigation. Hydrosound dampers are another method of reducing noise off-shore while driving piles. Since we have established that the air bubbles will reduce the noise, it is tempting to look into why. The theory is fascinating: For every frequency there is an ideal bubble size that will negate—under optimum conditions—the frequency completely. Keeping in mind that the bubble will increase in size when surfacing, the hydrosound damper is therefore made of rubber balls that are placed in a net to keep them free floating and always in the same position.

Deployment and maintenance of the system have not yet been tested, and in the real world there is no such thing as optimum conditions. Theoretically, though, the noise damping could be significant, and if combined with other methods, this could be an interesting way of achieving higher sound damping than what has been seen so far. The inventors of the system do, however, also accept that more research into the area is needed.

Using double-insulated pipes around the pile is also currently being fully tested offshore. The idea is that a large tube fitted with an insulating system will dampen the sound. This is also the case, and with a bubble curtain con-tained inside, the combination is very good. Handling of the system, however, seems to be very difficult, so it is not yet considered an implementable system; therefore the first installation projects to use it are yet to come.

The dewatered cofferdam is the most radical of all. This system allows the pile to be contained in a tube and is quite similar to the double-insulated pipe system. But the cofferdam is watertight, and therefore the water is pumped out completely. Now this is, of course, more radical than the other systems, and the open air connection allows for one big bubble—the surrounding atmosphere—to dampen the noise.

A more detailed description of it is presented in the following section. It should be noted that the cofferdam method is a patent that the author has obtained, so no evaluations or comparisons with other systems are made to avoid appearing biased.

COFFERDAMS

A cofferdam is an empty space created in the sea in order to be able to work in an air-filled environment. A cofferdam is therefore a box fitted to the seafloor, and under normal circumstances the water is pumped out and the seafloor is exposed to the open air. It is then possible to work without

Figure 18.7 A cofferdam created from sheet piling to recover a historic shipwreck.

using breathing apparatus. The cofferdam type shown in Figure 18.7 is often seen at bridge and port construction sites, where the necessity to work on a dry seabed requires the water to be pumped away.

In the case of driving monopiles, the cofferdam must be constructed out of a single tube. This is necessary to move it from one position to the next in a fast, efficient, and safe manner. The cofferdam must also be reusable during the entire project, and since the installation process of an offshore wind farm will be continuous for 80 or 100 monopiles or more, the process of moving, installing, and retrieving the cofferdam must be effective. Therefore, a sheet-piled cofferdam is not an option. The single large–diameter tube is the right choice. This is also the theory behind the pile in the previous pile system, but whereas the water stays inside the noise mitigation system, the cofferdam is pumped dry.

The cofferdam in Figure 18.7 is, however, driven into the seabed. This is done to keep the water from penetrating through the seabed once pumped out. For short working times of 12 to 24 hours for the wind farm foundation installation, the sheet piling method is not possible because it would produce even more noise than driving the piles, it would take far too long, and retrieving the foundation would not be effective.

Furthermore, the sound would partially pass through the seabed into the cofferdam and thereby out into the water column again. This is actually counterproductive. So, to solve this problem, the cofferdam must be a tu-bular design with a bottom lid through which the monopile is deployed into the seabed. The sealing at the bottom of the monopile is important.

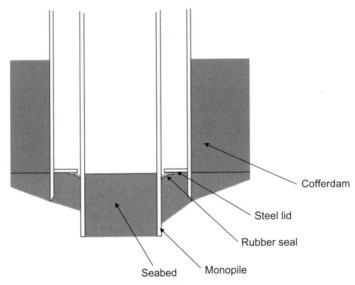

Figure 18.8 Principal design of the cofferdam.

When the water is pumped away from the cofferdam, the pressure on the seabed is 1 atmosphere, or 1 bar, but the surrounding water pressure at 35 m is 3.5 bar. This difference would cause the surrounding water to press the seabed backward up the cofferdam. If this happens, the ground would literally explode upward and ruin the top part of the seabed. The consequence of this is that the pile will have to be driven even farther into the ground to achieve the sleeve friction needed to be firm enough for the turbine to stand on it. Therefore the bottom lid has been developed (Figure 18.8).

The gap between the monopile and the bottom lid is sealed off with a diaphragm type of seal, which will allow the pile to pass through during the driving operation, but prevent the water from entering the cofferdam. This is shown in Figure 18.9. The bottom diaphragm seal is known from vacuum cleaner bags and diving suits where a flexible but tight-fitting seal is required. This is necessary because the sound will travel unhindered through a solid seal and thus the sound mitigation effort would be in vain. The diaphragm seal is shown in principle in Figure 18.10.

Finally, the cofferdam is designed as a telescopic body (Figure 18.11), which means it can be deployed at both shallow and deep positions without having to change the length of the tube. The cofferdam is simply extended or retracted to fit the distance between the installation vessel and the seabed, regardless of depth—until maximum extension, of course.

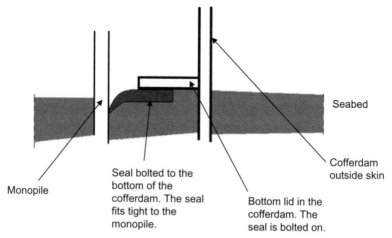

Seabed

Cofferdam
outside skin

Seal bolted to the
bottom of the
cofferdam. The seal
fits tight to the
monopile.

Monopile

Bottom lid in the
cofferdam. The
seal is bolted on.

Figure 18.9 Sealing of the pile and cofferdam.

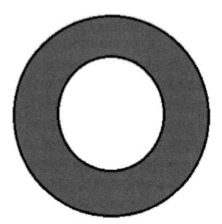

Figure 18.10 Diaphragm seal seen from above.

In this way, the method becomes feasible to use in any water depth as long as it is within the boundaries of the cofferdam's length. This is unique compared to other systems, although in all fairness it should be noted that a bubble curtain or hydrosound dampers can be deployed in several levels of water depth.

Proposed Method of Installing Cofferdam Foundations

So what is the actual method of deploying the cofferdam? Well, the idea is to make the system cost-effective and fast to use. This should be the number one feature of any system deployed—it should be cost-effective and not slow

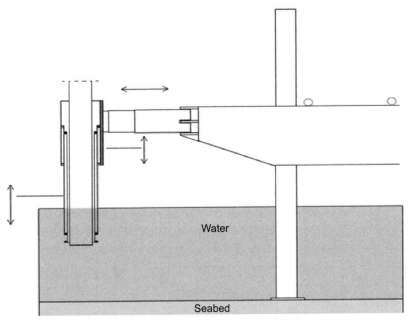

Figure 18.11 The cofferdam shown being extended toward the seabed.

down the process of installing a monopile or an anchor pile. When installing monopiles, they must be used as either:

1. Monopile foundations, where the monopile itself is large diameter and forms the foundation. This is, until now, the most common way to deploy piles.
2. Anchor piles, which are smaller diameter (2–2.5 m), and are used to anchor the tripod or jacket foundation to the seabed.

Installing Monopiles Using a Cofferdam

The first instance where the monopile is the entire foundation is when the pile is high and wide but the distance to the installation vessel is short. In this case, the monopile is loaded into the cofferdam on deck—horizontally—and tilted over a cantilever to the upright position. The method is known from the MPI vessel *Resolution*, where an upending tool is placed on the stern of the vessel.

This tool makes it possible to tilt a much heavier pile than the crane can lift over the stern. This is a good method of deploying piles—whether monopiles, or in the other case, the anchor piles. Regardless, the system is developed in order to both upend the pile and handle it when moving it across the weather deck of the installation vessel. This is demonstrated in Figure 18.12.

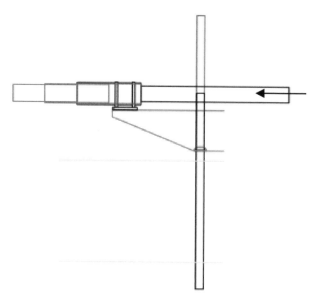

Figure 18.12 Monopile loaded into the cofferdam.

Once the pile is loaded into the cofferdam, it is upended and positioned over the stern of the installation vessel. Then the cofferdam, including the pile, is lowered to the seabed, where the pile is stabbed and checked to be sure that it is straight vertically. Just the fact that the cofferdam is holding the pile in the upright position and pinning it to the seabed pretty much tells you it is straight. Furthermore, the cofferdam can hold and adjust the pile using very small movements, so the adjustment process to get the pile vertical will be quick and simple, which is the main focus.

In Figure 18.13 note how the cofferdam is empty when the pile is stabbed into the water. This is because the pile is fitted through the diaphragm seal on deck before upending it. This means that no water is in the cofferdam when lowering it to the seabed (the pile is, of course, still filled with water as it is opened). The end effect is that the process of lowering the pile is not extended because water has to be pumped out, or a bubble curtain has to be established.

Once the pile is stabbed, the hammer can be fitted and the piling can commence. It is important to notice that the cofferdam, with the pile, is empty when it is deployed due to the seal between the pile and the bottom lid of the cofferdam. Therefore, only the water that enters during the piling operation will be pumped out by three submersible pumps fitted permanently to the cofferdam. In this way, the procedure is not delayed by having to empty the cofferdam prior to pile driving.

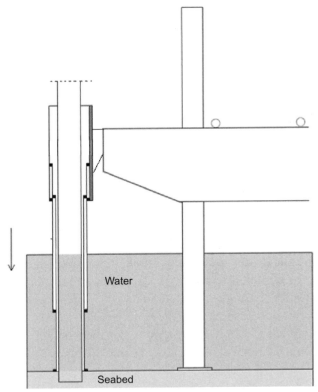

Figure 18.13 The cofferdam is telescoped down to the seabed, and the pile is stabbed.

So all in all the method for installing a monopile should not be an added feature that will lengthen the installation process but a feature that, in the end, will help with the installation from another point of view—namely, efficiency.

Installing Anchor Piles Using a Cofferdam

The second method is to drive three or four anchor piles to secure the tripod (three piles) or jacket (four piles) to the seabed as just described. In this instance, the actual method is the same. Piles are fed into the cofferdam, and then upended and positioned. However, the positioning is slightly different, of course, because the geometry of the tripod is triangular at the base, and the jacket is square. This creates a unique problem.

The tripod and jacket geometries are specific. We cannot accept that the anchor piles are slightly out of position. If one anchor pile is out of its geometrically correct position, the jacket or tripod cannot be installed. This makes it necessary to have the cofferdam adjusted precisely for each pile.

The cofferdam also requires traveling at much larger distances to reach the appropriate positions. The cofferdam must be able to cover a geometrical figure of 25 × 25 m area in order to place the piles. Therefore, the smaller cofferdam has to be able to accommodate the geometry, and therefore it must be able to swing to meet the correct positions of the anchor piles, as shown in Figure 18.14.

As can be appreciated, the cofferdam must be able to move in and out according to the geometry of the foundation. Thus, the jacket will require four positions to be reached. Furthermore, it is understood that the side length can be substantial, and therefore the construction of the cofferdam and the arm on which it sits must be very large. However, the pile diameter is smaller, so the weights in general can be kept reasonably low.

Once the pile is fitted to its final position, the cofferdam can be retrieved again using the winching device by which it is lowered to the seabed; this is a significant feature. It is, of course, logical that the cofferdam must be able to release the pile—whether anchor pile or monopile—completely.

If this is not the case or if the crew misjudges the required jacking height, the cofferdam will not be able to release the pile. In this case the jack–up will

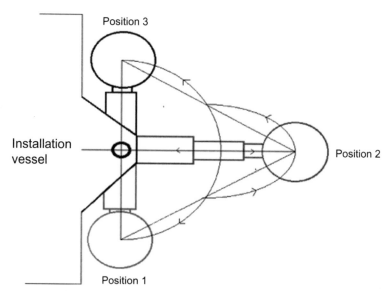

Figure 18.14 The various positions of the cofferdam when driving the three anchor piles for a tripod-type foundation.

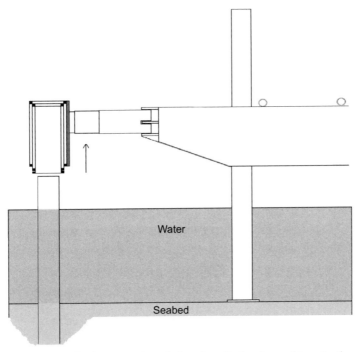

Figure 18.15 The cofferdam is retrieved after the pile has been driven to the correct position and depth.

be stuck in position, and this is, of course, not desirable. Therefore, an adjustment feature must be added where the cofferdam will travel above the baseline of the vessel to always be able to release the pile. In this way, the cofferdam can be removed easily and quickly in order to move the installation vessel swiftly to the next position (Figure 18.15).

As can be seen, the cofferdam is a viable solution that can be deployed for both monopiles and anchor piles. But in all fairness, it is a concept that has to be proven, the same as the other concepts discussed in this chapter.

Hopefully, what's here will encourage you to use your imagination to form opinions about the subject and to study the area in depth. The problem is relevant and present today, and if no solution is found, the industry will have an issue that will prevail for a long time and put restraints on the installation of driven piles offshore. This is, of course, not ideal for an industry that wishes to be seen as both a growth area and a forerunner in the environmental preservation area.

●●●

Related Images

Offshore turbine installation. Although we frequently talk about WOW downtime, from time to time we enjoy nice weather conditions.

A jack-up barge in port receiving supplies by means of an onshore mobile crane.

CONCLUSIONS ABOUT O&M TRANSPORT SYSTEMS

The task of designing an operation and maintenance (O&M) system that will deliver the highest possible availability of offshore wind turbines is difficult. This is mainly because the object of the matter is to join together two different interests into one solution. The desire is to have the wind turbines always running and most particularly when the wind is blowing to the maximum mean speed on the site. But, contrary to onshore wind turbines, the possibility of reaching the turbine during bad weather rapidly decreases. Therefore, the consequences of an unplanned production stoppage are difficult to handle, simply because the turbine is inaccessible when access is needed the most.

VESSELS

The solution to this catch is to design the transport and transfer system in a manner that will make accessibility as high as possible. This can be done by combining different systems. The vessel will be one half, if not more, of the solution, but the size of the vessel in question does not fit the modeling software available today. Thus, there are only two ways to determine the capacity of the vessel:

1. A scale model test, which is costly
2. Full-scale, experience-based knowledge of the vessel's behavior, which is what is available today

Therefore, the only valid way to determine the best vessel for the purpose is to look at the experiences so far with the different types.

It is the opinion of most in the field that the best vessels have been the monohull type because they have benign movements in all weather conditions. However, the vessel must be larger than the 15 to 20 m that is the standard size for crew vessels. This makes the market very small because 30- and 60-m vessels are almost extinct as a result of poor economical viability under normal circumstances. In addition, the offshore supply-type vessels that are the right size are far more expensive to own and operate because of the propulsion and positioning systems they have.

The SWATH types are also very benign but only in longitudinal wave headings. But the effect of passing waves has been dramatically reduced due to the specific design in the water plane area. Therefore, the vessel will perform well when a good transfer system is fitted to it. Furthermore, using the SWATH design is only possible when the vessel exceeds the 20-m length, which speaks in favor of this type over the catamaran. So it is best to consider the following two types:

- A large monohull cargo vessel for conversion, as discussed earlier. Again, it should be fitted with a good transfer system in order for the crew to pass safely in poor weather conditions.
- A new build SWATH-type vessel that has benign movements in bad seas except when transverse to the hull, where the capabilities are reduced. Again, it should be fitted with a transfer system that allows the crew to pass safely in poor weather conditions.

TRANSFER SYSTEMS

The difficulty with the transfer systems is first and foremost that they are designed to counteract wave motion. The problem (in our opinion), however, is more complex because wave motion is induced to the vessel, and thereby forces and bending moments are increased dramatically. Plus, wave motions are not confined to the transfer system but mainly to the vessel.

It is extremely important that the transfer system doesn't fail when a wave moves the vessel 2 m or more in as many seconds. The only two systems the author has reviewed, and that are currently available, that will counteract this effect are:

1. The boat landing, wherein the vessel will keep a forward thrust larger than the wave and current effect.
2. The Browing system, which is designed to react to wave motion and is not physically attached to the foundation and thereby does not suffer damage by loss of contact. Furthermore, the Browing system is designed to cope with the possible event of loss of contact with the foundation, which is a major safety advantage.

Finally, it is reasonable to state that so far no human lives have been lost as a result of the use of them. The other transfer systems mentioned in this book are either prototypes or are necessary to install on each foundation, or both.

The number of units is unappealing, of course, to an owner, and the proto-types have not yet proved to be workable. Therefore, wind farm owners should not put themselves in a position where they must choose between untested solutions and those that are certainly not solutions that need to be installed and maintained in large numbers.

RECOMMENDATIONS

In lieu of the findings I made during the writing of this book, these are some recommendations for an O&M transport system. They should not, however, to be regarded as the only possible solution to the problems encountered during offshore wind farm deployment. In addition, the recommendations here do not take into account that new systems are currently under develop-ment and systems, such as the Ampelmann and the Offshore Access System, are being used today in more and more projects. Therefore they can be regarded as viable solutions for safe transfer to the turbines.

The systems, however, come at a price; that is, the transfer vessel can no longer be just a small runabout that would be used in the more near-shore projects. For Round 3 projects in the United Kingdom and the far offshore projects in Germany, crew vessels must be of a substantially larger type than the small catamarans and monohulls used so far. Thus, the size of the access system can increase and generally this may make bigger, heavier systems more attractive.

I suggest that the wind farm owner pursue the option of entering into a framework agreement with a supplier of the following equipment:

- A monohull vessel of a length that is more than 30 m, fitted with a Browing system
- A SWATH-type new build fitted with a Browing system

The recommendation is that the owner enter into a relationship with a supplier because my belief is that the economics of scale are relevant. The possibility for the wind farm owner to attain reasonable financial viability based on one vessel is minimal.

The supplier should be one with a number of vessels under its manage-ment. Only in this manner will OPEX cost be optimal. Crew administration costs, insurance, and vessel running costs will be divided among more vessels and thereby be as low as possible for the client.

There are several possible suppliers for this setup. We suggest that the wind farm owner issue a tender for design, build (conversion), own, and operation of a service vessel in connection with the tender for installation of the wind farm. In this way, the time for the build and/or conversion of the service vessel will be adequate, and the owner will not be subjected to excess financial exposure.

When suggesting a monohull, it is because the vessel has benign movements in bad seas all around. Furthermore, a vessel of this size will have a mass inertia high enough to remain stable in front of the turbine foundation in poor weather, and this is a crucial help to the transfer system.

When also suggesting a SWATH vessel, it is because this is the second best type to operate in poor weather conditions. Furthermore, the Browing system has been mounted on a SWATH for the British navy, and the combination has a good track record. When suggesting the Browing system, it is because it has been operational for more than a decade in the British navy.

It is, however, crucial that the system be recalculated and reinforced to cope with the larger forces and bending moments experienced when docking next to a fixed structure offshore. But when this has been carried out, it is in my opinion that it is the most flexible and advanced system of the six mentioned in this book.

Related Images

A wind farm standing idle is not only serene, but also a scaring sight. This means that either the wind is not blowing, or the turbines need immediate attention.

Sundown at Horns Rev. Beautiful weather during the installation was part of the program. A great photo opportunity but not good for business.

FINAL THOUGHTS

The industry is clearly still in its infancy. With 12 years of commercial installation of offshore wind farms behind us, we are still far from having a set of established standards and procedures, but the work continues—on all levels. The HSE professionals are trying their best to detect all the potential and latent dangers that personnel working offshore, as well as onshore, could possibly encounter. The devoted effort to make the industry safe and reliable—as reliable as wind energy itself—is of paramount importance to everyone.

The vessel designers and operators are looking into a very hazy crystal ball to determine what the ideal installation vessel should look like, and their efforts are not in vain. However, they must see a healthy market with some clear benchmarks for what vessels must be able to deliver to design and build the necessary ones for the installation, operation, and maintenance of offshore wind farms.

So far, there are as many solutions as there are designers, operators, ideas, and concepts in the market. The turbine manufacturers are forthcoming in the way they are open to any good ideas that may be presented to them. But the willingness to listen to ideas may be counterproductive because it can sometimes make things go off course and detract from the main point: installing turbines offshore.

Therefore, the industry needs to mature enough to set a framework—or a standard—for how a wind farm is to be installed. In this way the designers, the project managers, the owners, the operators, and the turbine and foundation suppliers can all define what they want, and operators can then build it. If not, we will battle for years over which of the following to do:

- A complete turbine should be preassembled, transported, and installed because this is optimum.
- A preassembled turbine and foundation should be transported and installed.
- The foundation's and the turbine's transport and installation should be completely separate.

The author believes in the last choice, but the reader must form his or her own opinion to validate what will work for a particular situation. It would be a pleasure to engage in a discussion on the subject with those people who have given this entire topic some thorough consideration.

INDEX

Note: Page numbers followed by *b* indicate boxes, *f* indicate figures and *t* indicate tables.

SENIOR GOLF

SENIOR GOLF

Robert O'Byrne

Foreword by Julius Boros

WINCHESTER PRESS

Library of Congress Cataloging in Publication Data

O'Byrne, Robert.
   Senior golf.

   Bibliography:
   Includes index.
     1.   Golf.   2.   Aged—Recreation.   I.   Title.
GV965.029   796.352'3   77-21853
ISBN 0-87691-231-5

WINCHESTER is a Trademark of Olin Corporation
used by Winchester Press, Inc. under authority
and control of the Trademark Proprietor

Printed in the United States of America

Published by Winchester Press
205 East 42nd Street
New York, N.Y. 10017

Designed by Mel Brofman

TO THE MEMBERS OF THE DEAL GOLF CLUB,
WITHOUT WHOSE HELP THIS BOOK MIGHT
HAVE BEEN FINISHED IN HALF THE TIME

Contents

Foreword

A book on senior golf has long been needed. This comprehensive one by Bob O'Byrne completely fills the bill: It's the right book at the right time.

Who is a senior? The USGA defines a senior golfer as a man who has reached the age of fifty-five. For women, the age is fifty. Current estimates put the number of golfers in this category at close to three million. As the birthrate decreases and the life span increases, seniors will soon be golf's dominating age group.

This book by Bob O'Byrne has something for all seniors: Whether for the beginner or the seasoned veteran, it's all here. Designed as a complete handbook, it covers just about everything the reader wants to know about the world of senior golf. There are chapters on getting into condition, joining a senior organization, retiring to a golfing community, plus golf vacations, history of senior golf, results of major competitions, as well as the men and women who have won USGA senior championships.

Although not basically an instruction book, there is an excellent section that reviews the fundamentals of playing each club. A chapter which particularly interested me was "Tips That Work." I thought I knew them all, but even a senior pro can pick up a couple of tricks here.

Bob has been a fisherman as well as a golfer most of his life. In fact, his job as travel editor for *Sports Afield* requires him to go fishing—which is nice work if you can get it. I am also one of those folks whose golf keeps interfering with my fishing. In the chapter on golf resorts, the author has indicated a number of places which have good fishing either on the course or immediately adjacent. To me, this is a most welcome feature.

I recommend this book without reservation, even for those who are just thinking of becoming seniors.

Julius Boros

Introduction

I don't suppose anyone ever looks forward to the day he or she becomes a senior golfer. A man at fifty-five is almost past the dangerous age and few women like to admit being fifty. In my own case, I hailed the day as not only freeing me from the tyranny of flippity-wristed youngsters, but providing me with the opportunity to "clean up" in minor seniors' tournaments. In the three years since I attained my seniority, I have competed in some fifteen senior events at the club, state, invitational, and USGA qualifying level, as well as one on the international level. In none of these did I really distinguish myself, but something very important began happening to me. I was enjoying golf more than I had in the previous forty-five years I had played the game. The reasons are hard to pinpoint, but chief among them were the chances to play new courses, to develop new friendships, and to renew old acquaintances, and—above all—to talk over my golf problems with other seniors plagued with the same difficulties.

As my enthusiasm for senior golf grew and as I talked to more and more seniors, I was surprised at the number who asked me how to get started in senior golf, who had no concept of the senior golf picture. This book developed out of a desire not only to look at senior golf as it is today, but to share this wonderful world with new seniors whose enjoyment, I hope, will prove as great as mine.

Robert O'Byrne
March 1977

SENIOR
GOLF

1

The World
of Senior Golf

Senior golf is a relatively recent phenomenon. Despite the long history of golf in Scotland, there seem to be no records of competitions where age was a condition of entry. There are records of golfers who continued to win amateur and even open championships while in their fifties, but no tournaments specifically for seniors. Part of this is because in those days there just weren't that many people around over the age of fifty. Now, although longer life expectancy is worldwide, senior golf—except for a few competitions—centers in the United States.

The tradition of senior golf—its origins, its tournaments, its players—is an American innovation. It had its beginnings in 1905 when the Apawamis Club of Rye, New York, held its first Senior Invitational. It was won by James D. Foot of the home club, who was good enough to win it four more times in the next six years. His winning score for two rounds was 179. (In recent years the winning score has been as low as 141, which is some kind of a commentary on the improvement of senior golf through the years.)

Credit for organizing this first event goes to Horace L. Hotchkiss, who singlehandedly turned out forty-five golfers from nineteen clubs. He established the age at fifty-five and over, and this has held ever since for all male senior amateur events. By using the term "senior," he made sure for all time that no one would ever call it a

competition for "old men." The Apawamis Senior Invitational continued through 1916, when it became the Senior Golf Association
for three years, and finally in 1920 and ever since, the United States
Seniors' Golf Association. Today the Association has about 900
members from forty states and Washington, D.C. Its annual championship is held on three courses concurrently: Apawamis, Round
Hill, and Blind Brook.

Membership in the Association is by invitation, and the waiting list is long. Traditionally, being a low handicapper was no
guarantee of membership. Because of this, the Association has been
criticized in the past for having a weak field for their championship.
However, this was certainly not true in 1975 and 1976 when Dale
Morey, a new member, won the championship. It looks as if we will
see many more of our very best seniors competing in the future.
This association, which has done so much for senior golf, certainly
deserves the best field.

In the years since senior golf got started at Apawamis, events
for seniors have proliferated to an astonishing degree. The Appendix contains a fairly complete list of these, with addresses, to write
for further information.

A quick look at some of the major events for seniors will give
some idea of how well this sport is organized today. The most important championship for seniors and the most difficult to win is that
of the United States Golf Association. The qualifications for entry
are uncomplicated. You must be a member of a USGA Club. You
must be fifty-five or older. You must carry a handicap of ten or less.
For an entry fee of $25, you will be put in a foursome and sent out to
qualify on a designated course in your region. Quotas are assigned
to each regional course based on the number attempting to qualify.
One hundred and fifty seniors who make the grade then proceed to
the venue selected for the championship that year. After two rounds
of medal play, the low sixty-four compete at match play for the
championship. There is a great deal of opinion, pro and con, as to
whether the match-play format should be abandoned in favor of 100
percent medal play. I, for one, think the format is just right. The two
rounds of medal play in the regional qualifying plus the two rounds

of medal at the championship site ensure a top field for the final sixty-four at match play. If you are a senior and your handicap is ten or under, I urge you to attempt to qualify at the local level. It is one way of meeting many friendly seniors and getting to know the various organizations they belong to. Usually, the qualifications are held on some of the finest courses in each region and you will enjoy playing them. A great many seniors play in these qualifying rounds who never score low enough to make the final tournament, since only about 10 percent of the field make it to the finals. If you are interested in playing, write the United States Golf Association for an application; the address is listed in the Appendix.

If you have not been a regular competitor in senior golf events, one of the best ways to get started is to play in your state or regional golf association annual tournament for seniors. If your club is a member of the association and you meet the handicap requirements, you are eligible to compete. Handicap requirements vary from state to state, but usually something in the neighborhood of fifteen is the maximum allowed. Applications should be available through your club pro well in advance of the tournament date. Once the application is completed, send it with your entry fee to the State Golf Association and you will be notified as to playing partners and starting time. These tournaments are usually a one- or two-day affair with a prize for the low gross as well as one for the low net. In most cases, competition is followed by a happy hour and then a banquet. Since the company is invariably pleasant, this is an enjoyable way to get started in senior golf.

The first place to begin senior golf is, of course, right in your own club. Your club may already have a competition, and in many instances, the age requirement is fifty rather than fifty-five, the accepted age in most senior events.

I have some reservations about the manner in which many club senior tournaments are conducted. In many cases they are one-day medal play for qualifying and then the champion is determined by match play at handicap. This seems to me a downgrading of the senior championship. It would appear that just as the club championship is played with a medal qualification and then flights, the senior championship would be considerably more important if it too were played in flights, with the championship flight match played without handicap. If your club does not already have a senior championship annually, why not see if you can get one started? Most clubs have plenty of seniors to make up a field and I am sure that once begun, it will be an annual event bringing a good deal of pleasure to all the competitors, whatever flight they may wind up in. I would suggest that since most of the major competitions for seniors are at fifty-five or over, the senior championship within the club should set the same age limit, if there are enough golfers. A further refinement, which might prove of interest in clubs that have a good number of seniors, is the presentation of prizes for seniors competing in various age groups. For instance, the championship flight might well be fifty-five to sixty-five; another flight for those sixty-five to seventy; another flight for those seventy to seventy-five; and if you have enough seniors, another award for those over seventy-five. For most of the senior tournaments discussed so far, there is a counterpart for the women seniors. The only place one doesn't seem to find tournaments for senior women is at the club level. It's odd that more women don't initiate seniors' tournaments on the local level even though the field is probably limited. Perhaps the necessity for a woman to declare her age is a factor.

In addition to the senior competitions just mentioned, there are a number of senior associations throughout the country as well as many senior invitational events. Some of the better-known ones would include the Bellaire Senior Invitational, the American Senior Golf Association Championship, the Francis H. I. Brown International Team Matches, the World Senior Championship, the North and South Senior Women's Championship, the North and South Senior Men's Championship, the Southern Seniors' Golf Association, and the Western Seniors.

So far we have considered senior golf mostly from the standpoint of the private club golfer, but even more senior golf is being played on public courses as more and more communities are beginning to recognize golf as the ideal sport for their senior citizens, providing, as it does, mild exercise in pleasant surroundings.

Many seniors, some with the aid of a cart, play golf into their nineties.

Seniors are among the fastest growing of our age groups. As of 1975, there were more than twenty-two million seniors in our country. With the growing concern for healthful activity for older people, cities, counties, and municipalities are beginning to offer discounts at publicly owned courses to seniors. Some privately owned public courses offer special rates to seniors who play on weekdays. Certain private clubs have discounts for longtime members.

Harry G. Eckhoff, National Golf Foundation's Senior Consultant, took a look at some of the things that are happening in this sector. Here are some of his findings: "The Bethlehem, Pennsylvania, municipal golf course permits seniors to play anytime Tuesday through Friday until 4:00 p.m. for $1. At the Virginia Beach, Virginia, municipal courses seniors may play any weekday for $2. The regular rate for eighteen holes weekdays is $4.50. The Greensboro, North Carolina, and the Northern Virginia Regional Park Authority (greater Washington, D.C., area) offer half-price rates on weekdays, which reduces the eighteen-hole price to $2.50 for seniors. At the Milwaukee, Wisconsin, County Park Commission's fourteen golf courses, seniors sixty-two and over may play eighteen holes of golf any weekday for $1.10; the nine-hole rate is 55¢. The regular adult rates are $3.50 and $1.80 respectively. In 1966, the first year that Milwaukee County offered a separate senior golf permit, senior play accounted for 29,983 rounds, or 4.7 percent of the 644,514 rounds played on all the county's golf facilities. By the end of 1974, senior play had soared to 124,062 rounds, or 18.1 percent of the total 684,633 rounds played."

The story of Tucson's senior golf program is of interest to anyone hoping to start a senior program in his community. "The city of Tucson has a popular senior citizen golf program in operation at its three eighteen-hole municipal golf courses. Tucson's program came about largely through the efforts of the local Senior Citizen Golf Association, which succeeded, in late 1971, in getting a Tucson City Ordinance passed that authorized reduced green fees along with special rates at the city's golf ranges. A senior citizen golf pass entitles the holder to play nine holes for $1; eighteen for $2 Monday through Friday excluding any holidays which fall on a weekday.

"Tucson senior citizens worked long and hard to get the city ordinance authorizing reduced rates for golfing activities. When they first approached the city officials for special privileges, they were told reduced rates could not be considered; that reduced green fees for senior citizens were not popular in the United States, and that if granted they would establish a precedent.

"Senior golfers were determined to prove that a reduced senior rate program would be beneficial for Tucson. They solicited the aid of the local Chamber of Commerce and formed a Senior Citizen Golf Association, which now has over 700 members. A survey of forty-five cities was initiated in October 1970, requesting data on their golf programs; forty-four replied. Thirty-one, or 67 percent, had senior citizens' programs with reduced fees. It was apparent that granting senior citizens special golf play privileges at municipal golf operations is almost a national practice.

"When Tucson city officials were apprised of the survey findings, they were still unimpressed. They then contended they could not consider this type of program, as their golf courses were already playing to capacity and could not accommodate increased play. But Tucson's senior citizens were not about to give up! They initiated a survey of the Tucson municipal courses to obtain definite playing figures. Sixteen senior citizens worked in shifts from 7:00 A.M. to 5:00 P.M. for almost two months to obtain the necessary data. In early June 1971 the golf association issued a notarized 'To Whom It May Concern' letter outlining their detailed results, which in conclusion stated city golf courses were playing to only 66 percent of capacity or saturation.

"The survey findings were a shock to the city officials. The mayor and councilmen began to realize some positive action should be taken. After fifteen months of hard work, including 360 pages of correspondence to eighteen city officials, thirty-two published news items, and the accomplishment of two important surveys by the seniors, the city council in October 1971 unanimously approved the

Tucson Resident Senior Citizen Golf Program. Today the program plays a very important role in the overall municipal golfing picture."

During fiscal year 1973, the first full year the senior special-rate program was in effect, senior citizen play accounted for 13.6 percent of the year's total play. Says James A. Marr, Tucson's Golf Course Operations Coordinator, "For the past year and a half there have been twenty-two to twenty-seven senior citizens per month signing up for the golf program. Over 1,600 senior citizens have applied for their senior golf card. It is estimated that at least 1,000 seniors play golf regularly from one to five days a week. In the summer months seniors make up approximately 60 percent of the early morning weekday play. So you can see the program plays an important part in the overall municipal golfing picture."

To me, the heartwarming thing is that in a number of cities seniors are not waiting to be helped but are going out and getting things done themselves. It should be an inspiration to seniors everywhere.

Let's look briefly at the accomplishments of a senior who set some standards on what can be done after age fifty-five. The late John Ellis Knowles won the United States Seniors Golf Association's Championship six times, the last time at age sixty-seven. When he was seventy-three, on the two days of the competition, he bettered and equaled his age with a 72–73. His 145 would have been good enough to win the championship most years. When he was eighty-three he beat his age twice in the tournament when he posted an 82 and an 80.

Shooting one's age or better is no mean accomplishment but thousands of seniors do it many times in the course of a year's play. Not all of us will do it but it is a goal within reach of many. In fact many golfers have done it who didn't take up golf until they were in their sixties.

If you're fifty-five or over and have some leisure time, now is the time to take up the game in a serious way. It is astonishing how you can improve if you practice a couple of times a week. Take a good look at your whole game and try to evaluate your weaknesses. Especially work on your short game. Good putting and approaching are hallmarks of every winning senior's game. If you have lost a little distance over the years you can make up for it around the greens. The important thing is to realize it's not necessary to have your game get a little worse each year. It's really an attitude of mind, and now that you have more time to play the game, you may soon find that you are playing the best golf of your life. I guarantee if you play it with seniors you will enjoy it more. Life doesn't have to begin at forty; it can begin, at least for senior golfers, at fifty or fifty-five.

Golf Equipment and the Senior Player

We are fortunate to be playing golf at a time when technology within the equipment industry is advancing at an unprecedented rate. Seniors, especially, are lucky to be heirs to the developments in lightweight shafts, new systems of matching clubs, and improved designs in both woods and irons. As we have grown older, clubs have become better, and that progress has helped many of us to maintain our handicaps.

Today there are more different models and brands of clubs on the market than ever before—over fifty companies at last count, not including the dozens of foreign makes. So if you can't find the driver or 5-iron or sand wedge you want, don't blame the manufacturers. On the other hand, selecting the right set of clubs is not an easy task. Several areas must be considered, particularly by senior golfers. This chapter will attempt to help you sort fact from fiction in the club business and find the sticks that fit you best.

The Grip

Since the golfer and his club come together at the grip, that's a logical place to begin our discussion. For most of us, the standard size of grip that comes on all stock sets of club is suitable. The average hand fits comfortably on these grips. For a person with small hands a thinner grip might be of some use. For a person with larger hands or with long fingers an oversize grip is advisable. Ladies should take special note here, because long fingernails often indicate a need for oversize grips. Also, female fingers are generally longer and more slender than male fingers, and often the standard ladies' grip is too small.

There is a relatively simple test you can make to determine whether the grip on your present set of clubs fits you properly. Just grip the club normally with your left hand, and then check the position of your fingertips. If they are pressing into the base of your thumb, the grip is too small for you. If your fingertips fall well short of the base of your thumb, the grip is too big. The proper-size grip will allow you to just barely touch the base of your thumb with your fingertips.

If you need a smaller or larger grip you should probably not ask for a change of more than $1/32$ of an inch. In extreme cases you can increase or decrease the size $1/16$, but be careful not to overdo it. The best procedure is to test clubs with varying grip sizes before you go to the trouble and expense of regripping your clubs.

You'll often hear that the size of your grip can have a direct effect on the flight of your golf shots, that an undersized grip will lead to a slice and an oversized grip to a hook—or vice versa. Well, the truth is that it can work either way, so don't go changing your grip size in the hope of curing your duck hook or banana ball—at least not until you've first had a talk with your pro.

The material of your grip is another important consideration. Leather is a great sentimental favorite among seniors, but it may not be the most practical alternative. It absorbs perspiration, becomes slippery in wet or cold weather, and after long use it is relatively difficult and expensive to replace. Nonetheless, many golfers claim that leather has unequaled "feel," and who can dispute questions of feel?

If you don't have a personal hang-up on leather, then the all-weather rubber grips made by Golf Pride, Master Grip, and others are a very good choice. While they don't have the natural tackiness of leather, they do stay fairly dry under all but monsoon conditions, and they're extremely durable.

In recent years a company called Tacki-Mac has come up with

the best of both worlds, a rubber grip that has the look and feel of leather. The Tacki-Mac grip is injection-molded from a blend of rubber and Kraton, a petroleum by-product from the Shell Oil Company. Over one hundred pros presently are using this grip, and its comparative lightness has made it attractive to a number of club manufacturers, so you can expect to see it as standard equipment on sets in the near future.

The Shaft

Without question, the most important part of a golf club is the shaft. Although it never touches the ball, the shaft is directly responsible for the way you bring the clubhead into impact. For that reason alone, the flex, length, material, and weight of the shaft you swing are of paramount importance.

We seniors all remember the days of the hickory shaft and, of course, the introduction of the first relatively heavy steel shafts. Since that time the quest for the ideal golf shaft has led us through a series of marked changes. Through these developments researchers and manufacturers have strived for one major design goal—lightness.

By reducing the weight of the shaft more weight may be added to the clubhead without increasing the total weight of the club. This means that the heavier the clubhead and the faster you can swing, the harder you'll hit the ball. With a lighter shaft a golfer can either swing the same weight clubhead faster or swing a heavier clubhead at the same speed, thereby increasing his distance. Armed with that formula the industry has pressed for lighter and lighter shafts.

The first of the light shafts was aluminum, which was introduced several years ago amid much fanfare. Any golfer who read the thousands of words on the subject became convinced that he would get more distance out of aluminum shafts. But in the end aluminum did not prove to be much lighter than the steel shafts of the day, and in addition many golfers felt that it lacked the feel and responsiveness of steel. A final point against it was its appearance; although lighter in weight, it was bulky.

The roar over aluminum subsided after a couple of years and

the search and research went on. Next came a series of lightweight steel shafts. Manufacturers found a way of introducing special agents and alloys that aided in lightening the steel and permitted the shaft to have a thinner wall without danger of breakage or wear. Among the first of these shafts was the Apex, developed by the Ben Hogan Company. Today there are as many lightweight steel shafts as there are manufacturers, each with a slightly different set of properties. The new Superlite Steel Driver Shaft, by True Temper (supplier of 90 percent of the industry's shafts), and the UCV-340 chrome vanadium steel shaft, by Union Hardware, are examples of the lightweight steel shaft at its most streamlined stage.

There are other lightweight shaft materials as well. In 1975 the Zirtech-Titanium shaft hit the market with claims that it was as light as any shaft ever produced and stronger than any lightweight steel shaft. The jury is still out on these shafts, which are quite expensive. Metal alloy shafts are also available, the best known being the Titanalloy shaft, which blends six different metals and is presently about half the price of the pure titanium shaft, although it's a little bit heavier.

Last but not least there's graphite. A regular graphite men's driver weighs about 1.5 ounces less than the regular steel shaft. Frank Thomas, presently the technical director for the United States Golf Association, developed the first graphite shaft, in 1972, when he was working with Shakespeare & Co., and it is still the lightest shaft material on the market. Until recently it was also the most expensive, but graphite prices are at last beginning to go down, and good pure graphite drivers are now available for $55, down from $155 and more. I wouldn't be surprised if competition forces the prices even lower, which will make these beauties very attractive indeed.

What does all this talk about lightweight shafts mean to you as a golfer? The answer is complex. The senior golfer who will get help from the lightweight shaft is the person who has played golf most of his or her life with a good swing but probably not a powerful one. Perhaps you hit your best drive somewhere in the 220- to 225-yard range when you were much younger and today find yourself somewhere in the area of 200 yards off the tee. With your smooth swing you should be able to adjust to the lighter shaft very quickly, and with your driver you could well expect to pick up an additional 15 to 20 yards, or, in other words, go back to your old distance. This is certainly a boon to seniors. It is a thrill to enjoy the pleasure of hitting them way out there once more.

Another advantage of the graphite shaft is that the golfer can go to a stiffer flex without difficulty. Frank Thomas feels that regardless of the material of shaft being used you are well advised to play the stiffest shaft you can handle comfortably, that is the stiffest shaft that you can bring into the ball with consistency and accuracy. Most experts and clubfitters agree with that viewpoint. Stiffer shafts are more "forgiving," meaning they give you reasonably solid shots even on off-center hits. So before you move from steel to graphite ask your pro whether you should also move up one step in the stiffness scale, either from an A flex (flexible) to an R (regular), from an R to an S (stiff), or from an S to an X (extra stiff), though that last jump is reserved for the really long hitters. Ideally your pro will have a couple of different flexes for you to test. In the past such trial and error has been an expensive service for the club pro to offer, since he did not want to be stuck with unused graphite demonstrator shafts, but now that the price of graphite seems to be lowering, graphite test drivers should start to become more common.

One more note on graphite drivers. You may be able to increase the length of the club a half inch without losing any feel in the clubhead. Ask your pro about that possibility, too. All things (and swings) being equal, a longer driver will give you a longer drive.

Swingweight

Swingweight is used primarily by equipment manufacturers to guarantee that as each club gets shorter it swings with the same kind of feel. For our purposes there are three ranges of swingweights — light, medium, and heavy. The lightest swingweights would be from C-1 through C-8; these are used almost exclusively by women and are designated as extra-light. Male golfers normally play with clubs swingweighted in the next three ranges: a light range (from

C-8 to D-0), a medium range (from D-1 to D-3), and a heavy range (from D-4 to D-6). Your primary concern should not be to get a precise swingweight, say D-1, but rather to select a range that feels best when swinging. Most pros have scales that accurately determine the swingweight of any club as well as the club's total weight in ounces. If you have a club that has good feel, which you have been using with good results for some time, it would be worthwhile to ask your pro to check out the swingweight.

Let us assume that you do have your favorite club checked and that it turns out to be D-2 on the scale. Now, if you were to weigh any other club in the set from which your favorite club came, the swingweight of that club should also be D-2.

It is quite easy to change the swingweight of a given club, so easy in fact that you can do it unknowingly. The weight of your golf glove, for instance, which might be as much as an ounce when you're holding the club, can lower the swingweight. If you let mud accumulate on the clubface it can raise the swingweight as much as two points. Even the wearing of a wristwatch or ring can alter swingweight. The important thing is that you find a swingweight range that is comfortable, and then stay with it.

If you would like to find out if a heavier swingweight would be more satisfactory, you may do so with the aid of lead weighting tape available at many pro shops. This special tape can be stuck onto the back of the club in layers sufficient to change the swingweight in an upward direction two or three points. The next time you watch a pro tournament take a close look at the clubs the pros use. You'll see that more than half of them are fiddling with this tape to give them an ultimate exactness in swingweight.

Of course it is also possible to take weight out of the head of a club to bring it to a satisfactory swingweight, but this is a job for a pro or clubfitter.

Senior golfers are best advised to stay with the light to middle ranges of swingweight (C-8 to D-0 and D-1 to D-3). If you find, however, that you are losing distance as the years go by, try dropping down in swingweight. You'll be able to swing the lighter swingweighted club a little bit faster, and the increased clubhead speed you generate will probably restore your old distance. However, if you have experienced no distance loss, it's unwise to tamper with the swingweighting on your clubs.

New Ways of Matching Clubs

Although most clubs today are matched according to the swingweight method we have just discussed, there are several new methods gaining attention within the golf industry. Frank Thomas has said that the matching of sets of clubs is far behind the other areas of club manufacture in the extent to which clubmakers have experimented and used available technological capabilities.

Several manufacturers are now searching for the ultimate matched set of clubs, a set that will allow you to make the same swing with every club, instead of having to make the minute adjustments (both conscious and subconscious) you now must make between the driver and the 9-iron. To achieve such uniformity, it seems that clubs must be matched dynamically, meaning they must be matched while in motion, rather than just statically or at rest, which is the only measurement swingweight provides.

Several companies are trying to perfect these dynamic matching methods. The criteria for matching the sets include moment of inertia, frequency of vibration, and center of gravity. Some companies are now matching clubs exclusively according to these methods, others combine one or two dynamic methods with the static swingweight measurement, yet most companies still adhere to the swingweight scale as the final standard. The reason for that adherence has as much to do with the golf market's familiarity with swingweight as it does anything else. As time goes by, it will be interesting to see whether any of these new dynamic methods will catch on. Keep your eyes on such companies as Sounder, which matches according to moment of inertia, Console, which uses frequency of vibration, and Square Two, which matches according to center of gravity.

The Clubhead Features

There are a number of features you should consider with regard to the bottom part of your golf club, the clubhead. By these features I am not referring to the age-old debates between forging and investment cast irons and between persimmon and laminated woods. It has been all but proved that none of us can tell the difference between a forged iron and an investment cast iron, and the same is true between persimmon woods and laminated woods; not even the pros can tell them apart.

But there are certain design factors that every senior golfer should be aware of and many senior golfers should take advantage of. When choosing your irons, for instance, pay attention to where the weight has been distributed on the clubface. If you miss the "sweetspot" a lot and hit shots out of the heel and toe of your irons, you will probably benefit from a club that is heel-and-toe weighted, or, as it's also known, perimeter-weighted. These irons take a little weight away from the center of the clubface and put it on the outside, for those off-center hits.

Likewise, if you have trouble getting your iron shots up in the air, you should seriously consider getting a set of sole-weighted irons, where most of the club's weight is near the sole of the club, so that you can get more power down under the ball. Irons with an offset clubface can also help you get the ball up more easily.

Slicers might consider buying woods with a hook face. Many companies make their stock sets with a hook face, and you should look into that possibility.

Set Makeup

Not enough golfers give serious thought to the numbers on the clubs they carry. Many of us still carry a 2-iron, even though we have trouble hitting it. That's unwise, especially for senior golfers. If you haven't tried one of the more lofted woods, such as the 5, 6, or 7, do so. Not only are they easier to hit than the long irons, but they can go a lot of places the irons never will—into the roughs, traps, and other poor lies. You should also consider carrying some sort of utility club, such as an extra wedge (in addition to the sand wedge and pitching wedge) or a chipper. As we get older it pays to improve our finesse around the greens. It's easy to drop one of the longer clubs and improve your short game with one of these handy auxiliary clubs.

Picking a Putter

Selecting a proper putter is one of the most important things a senior can do to improve his or her game. Some golfers are such fine putters that they need no help in this quarter, but for most of us putting is a subject of constant worry. I personally believe in the old adage, "You can't have too many putters." Of course, I don't mean this literally, but it is reasonable to have as many as three or four putters that you can rotate to suit different putting conditions, or just so that you can change when your stroke seems to go sour.

Basically, putterheads come in three designs: mallet, blade, and center-shafted. Some golfers, having selected one of these styles, will cling to it for the rest of their days, while others will experiment with all three styles, changing from one to the other as the mood suits them or as their putting woes afflict them. Anyone who follows the televised golf tour is aware that several of the best pros change putters constantly. Regardless of what category you fall into or what putter you choose, there are certain considerations you must confront. How long should the shaft be, and what lie should it have? What stiffness should the shaft be, and how much loft should the face have? And what kind of grip will be best?

The average putter is about 35 inches long, and unless you either stand up very very straight or bend way over to putt, this length should be suitable. As far as shaft stiffness is concerned, a few players like a shaft that is fairly limber so that they can feel some sway in the head as they waggle the putter. Others like a stiff-shafted putter.

Getting the right lie may well be the most important of all considerations. If, in your putting stance, you stand well away from the ball, you probably require a flat-lied putter. If you stand quite close you will need a more upright lie. The object, of course, is to keep the sole of the putter flat at all times, except on severe uphill or downhill lies where there has to be some slight change in the angle. It would seem that a good many golfers prefer the upright lie, as they stand relatively close to the ball, usually playing it inside their left heel with their left eye directly over the ball. However, there are certain golfers who like to keep the ball some distance from them, although to my mind this compounds the problems of putting. For those interested, the average angle of lie of a putter is about sixty-five degrees. An upright lie is about fifty-five degrees and a flat lie would be around seventy-two degrees.

One factor that is seldom considered but that is of importance to certain golfers is the degree of loft built into the face of the putter. Obviously the putter has to have some amount of loft in order to get the ball rolling. If it were perfectly straight, unless you kept the club perfectly perpendicular at impact, you would drive the ball into the green, resulting in a quick pop-up and a bouncy ride to the hole. Lofts can range from as little as two degrees to as much as four. I have heard that Calamity Jane, Bobby Jones's famous single-blade Schenectady putter, had a loft of eight degrees, which is amazing when you consider that this is very close to the loft on a 2-iron. The only explanation I can think of for this is that in the old days the greens were considerably heavier, and Jones needed the extra loft to get the ball rolling on the taller grass.

The degree of loft on your putter would seem, in fact, to depend a great deal on the type of greens you play regularly. On fairly thick greens you might want something in the neighborhood of four degrees, while on the close-cut greens a putter with as little as two and a half degrees of loft might be more suitable.

The shaft of a putter can be made of steel, fiberglass, hickory, or even graphite. As far as I'm concerned, the shaft material is of little importance as long as the shaft is reasonably stiff. In order to avoid trying too many materials, if you settle for a lightweight steel shaft (the one on most putters) you should be well satisfied.

A final consideration is the weight of the putter. There is a great deal of variation among blades, mallets, and center-shafted styles, and your final criterion should be feel. It's worth noting, however, that on fast greens, such as those at the Apawamis Club in Rye, New York, site of the annual United States Seniors' Golf Association tournament, a heavy putter can be dangerous, since it inhibits the soft touch needed on slick greens. For fast greens a light putter is usually best. Conversely, on slow greens or large greens, where you need some power behind your putts, the heavy putter is helpful. One of the best putters alive, Ben Crenshaw, claims that his own putter "travels well," since it is neither light nor heavy and can adapt to the varying speeds of greens he encounters on the PGA Tour.

Ball Selection

The United States Golf Association has prescribed certain conditions in the construction of a golf ball. In order to be legal in America the ball must not be larger than 1.68 inches in diameter (the smaller British ball is 1.62 inches), it must be no heavier than 1.62 ounces, and it must have a maximum initial velocity of no more than 255 feet per second. Within these limits the constructions and materials vary widely.

There are now two major types of golf ball construction—the two-piece ball and the wound ball. In addition, a couple of companies market a completely solid ball, all in one piece. To make things more complicated, the wound ball can have either a balata cover or a Surlyn cover, and each of those types can, in turn, have either a solid center or a liquid center. Add to this the fact that most brands of golf ball come in two or three different compressions and you begin to appreciate the breadth of choice available to you as a golf ball purchaser.

Surlyn, a plasticlike material patented by the Du Pont company about ten years ago, is now the most popular golf ball cover in America. It is harder than its competitor, balata, and thus it is tougher to cut. Balata is a rubber product made from the sap of a tropical tree. Although it is easier to cut than Surlyn, it is also easier to maneuver and provides better click and feel; at least that's what the pros and top amateurs say. As to the centers of the balls, there is little real difference between the wound ball with a solid center and the wound ball with a liquid center. In actuality, they both have solid centers, cores of a rubber material known as polybutediene. These cores are about the size of a marble, but in the case of the liquid-center ball the core is hollowed out and filled with water or glycerine or some "secret solution" proprietary to the manufacturer. Incidentally, none of these liquids is harmful to you, lest you have believed all these years the old wives' tales about golf balls being filled with blinding acid and the like. The Titleist and ProStaff lead the wound ball market.

The two-piece ball always has a Surlyn cover. The rest of the ball is made of a different blend of that same polybutediene. As you can imagine, this ball is simpler and less expensive to manufacture than the wound ball. It is also claimed to provide a bit more distance when it is struck at lower impact velocities, perhaps 5 yards longer when you hit it with a 5-iron. The Topflite is the best known example of this construction.

So which ball do you buy? What it comes down to is "whatever feels good is best." However, there are certain basic guidelines. If you're not a pure striker of the ball, you might be happier with the Surlyn ball, either in a two-piece or wound construction. You won't cut it easily and you'll thereby save money. On the other hand, if you have, let's say, a 9 handicap or less you may find more response in the softer-covered balata ball. For a complete listing of the available brands of golf balls, categorized according to construction and material, consult "What Do You Know About Your Golf Ball?" an article in the May 1977 issue of *Golf* magazine.

What about compression? Well, what you may have heard is true. No matter how strong a golfer you are, you will hit a 100-compression ball farther than you will hit a 90-compression ball and a 90-compression ball farther than an 80-compression ball, if you strike them all at the same impact velocity.

Why, then, use anything but 100 compression? First of all, those 100-compression balls are often in a balata cover that is easy to cut, so they're not always the most economical choice. Second, there is that feel factor. Weaker golfers often feel that the high-compression ball feels like a rock at impact, and this is probably due to the fact that the 100-compression ball is more difficult to compress than the 90 or 80. It should feel harder, because it is intended for golfers who swing hard. Therefore, you may not be happy with the tighter 100 ball; it may not give you the feel you require with your own swing. Finally, if you play in a cold climate, the 90-compression ball is definitely preferable to the 100. When the temperature goes down, the characteristics of a ball change, and the 90 changes less markedly than the 100. Even the pros will go down to a 90 on cold days, so you should consider changing to an 80. In any case, the 90 seems to be the most practical ball for the majority of golfers, in terms of total playability and distance in all kinds of conditions. On the other hand, if you desire distance, then you should certainly give the 100-compression balls a try, even if that feel is not there. After all, most of us are willing to take a bit of sting in the hands if it means watching the ball carry an extra 10 yards. So experiment with wound balls and two-piecers, balatas and Surlyns, liquid and solid centers, 100s and 90s until you find that one golf ball that performs best for you.

When the Going Gets Wet

I can't leave this chapter without passing on a couple of homespun hints that have worked well for me in wet weather.

If you wear glasses you know what a problem steaming-up can be, both in wet weather and in the cold. I have found that by cleaning my glasses with a drop of gin on each side just before I go out leaves a film that prevents fog from forming and seems to shed the rain a good deal better than if they were left alone. Some friends have taken, at my suggestion, to carrying a little gin right in their golf bags. Not only does it enable them to repeat this treatment during the course of the round, it sometimes comes in handy for internal use when things go badly.

Gripping the club is always tough on a wet day, and I've found that the best thing to wear is cotton gloves, the kind worn by ushers at a funeral. No matter how wet they get, these cotton gloves always give you a reasonably good grip.

Health and Conditioning for Senior Golfers

No matter what your age, good health and physical conditioning are important if you want to enjoy the game of golf. I have talked with Bill Hyndman, Dale Morey, Curtis Pearson, George Pottle, Harry Welch, and others, all of whom place strong emphasis on staying in shape. Most of these senior golfers do special golf exercises every day. Fortunately, these exercises, many of which will be described in this chapter, are relatively simple and unstrenuous. They will, however, keep you both strong and limber—they'll build you up without knocking you out.

The single most constructive thing you can do to keep in shape is to walk at least half a mile a day, every day, at a good pace. This is especially advisable for those seniors who live in the colder climates where it's impossible to play golf year-round. When you do play golf you should make an effort to walk at least part of the eighteen holes rather than take a cart. In an average eighteen-hole round you'll walk over five miles, and that goes a long way toward keeping your legs toned up. Even if you have to take a cart for one reason or another, it's a good idea to walk one hole and then ride the next while your partner walks.

Swimming and running are also good for you, since they involve every part of the body. Excessive swimming, however, can do harm to your golf game. Lots of swimming can build up your chest, shoulders, and upper body to the point where it becomes difficult to make a full, free shoulder turn. That can strip away power.

For the same reason, you should probably refrain from doing any serious weight lifting, except the kind that concentrates on your legs and forearms. The spring grips available at most sporting-goods stores are particularly effective for building the forearms, wrists, and hands. An excellent way to test yourself with these grips is to squeeze a coin between the grips and see how long you can squeeze before letting the coin drop out. If you're diligent, you'll find that you'll be able to hold the coin longer and longer each day. If you don't want to invest in the grips, squeeze a squash ball or hand-ball—they're just as good.

Another way to strengthen your arms (and get some swinging practice at the same time) is to hit practice balls with your left hand only. This will build up your left arm and wrist while at the same time ingraining the "muscle memory" that will help you delay your wrist release into impact. Several pros use this practice method, and it can be helpful to any level of golfer, especially to senior golfers who are constantly battling against distance loss.

Another way to get the same result as the one-arm practice is to spend a lot of time hitting balls out of sand traps and heavy rough. You'll meet extra resistance at impact on these shots, and that will build up your arms and increase your ability to pull powerfully through the ball. It should also have a salutary effect on your scrambling game.

Of course, strengthening exercises should be balanced with exercises that improve the elasticity of your muscles. To that end, here are several stretching routines you can try before heading for the first tee.

You've probably seen it a million times, both on the practice tee and in print, but the old exercise of twisting back and forth with a club behind your back is still a great way to loosen up. And you'll be in good company if you use it; Jack Nicklaus never hits his first practice ball without doing this exercise.

Another good warm-up routine is to swing a weighted club.

There are a number of gadgets available (weighted head covers and lead rings to slip onto the shaft of your club) that will instantly triple and quadruple the weight of your clubhead. By swinging these heavy clubs you stretch your back and shoulder muscles and enable yourself to make a full free turn at the ball. You also can use this practice to develop timing and rhythm in your swing.

Herb Madigan, director of physical conditioning for the New York Athletic Club, has suggested another excellent warm-up exercise. Stand 3 feet from a wall, with your back to the wall and your feet about shoulder-width apart. Then, without moving your feet, twist to your right and place both hands flat against the wall behind you. Then twist back to your left and place both hands flat against the wall to the other side. Do ten such turns—five to the right and five to the left—and you'll be as limber as a fifteen-year-old when you step to the first tee.

Don't forget to loosen up the leg muscles, too. Stand against that wall again, about 3 feet away. Lean forward, keeping your legs stiff, and press your palms against the wall. Hold that position for seven seconds. Now bend your knees and squat down, putting all your weight on the soles of your feet. With your hands still on the wall, hold that position for seven seconds. Repeat this two-position exercise five times, and you'll be ready for the eighteen-hole walk I recommended earlier.

In addition to the strengthening and stretching exercises you might consider following a program in aerobics, a concept of exercise developed by Dr. Kenneth H. Cooper and used by the United States Air Force in its conditioning program. Dr. Cooper's definition of aerobics is this: "Aerobics refers to a variety of exercises that stimulate heart and lung activity for a time period sufficient to produce beneficial changes in the body." The whole program is spelled out in Dr. Cooper's book, *The New Aerobics*.

Of course, if exercises bore you, you can stay in pretty fair shape by cycling or playing tennis, paddle tennis, squash, badminton, or handball, or just by staying on your feet and active every day. Before attempting any sort of serious conditioning program you should take at least two precautions. The first of these is to check your present weight against one of the life insurance tables that indicate the proper weight for your sex, age, physical structure, and height. If you are more than 20 percent overweight you will probably have to do something about your weight before undertaking an aerobics conditioning program. In my own case, I was about 50 percent overweight when I first considered beginning a conditioning program. It was then that I went to Duke University and started a salt-free diet that helped me lose 38 pounds. For those who might be interested there are two weight-reduction plans available at Duke. Mine was given at the Dietary Rehabilitation Center and consisted of a 700-calories-a-day diet taken at a special dining area on the campus. I stayed at a nearby motel and participated in the program for about a month. During that time I was under daily medical and psychological supervision.

A three-day medical examination preceded my entry into the program. You must give up alcohol, or the weight just won't come off. Any cheating shows up at the daily weigh-in. You keep an elaborate diary to remind you of just what you are doing. When you leave the diary goes with you, along with a looseleaf notebook crammed with materials to enable you to continue the diet at home. I lost an additional 15 pounds in the first month following my return home. Now, after a year, I have put those pounds back on but have maintained my original 38-pound loss. I am still 30 percent overweight, and I plan to go back to Duke soon and stay until I get to my correct weight. I have changed my eating habits, and I feel that once I reach that ideal weight I will have little trouble maintaining it.

The reason I pass this along is that I am one of those many people who just can't diet at home. I have to go someplace where there is a complete program, along with other people fighting the same battle. I have lost literally thousands of pounds in a lifetime devoted to all kinds of diets, but until entering the Duke program I had never lost more than a few pounds at a time and was never able to keep them off for long. This is the only thing that has ever worked for me, and perhaps other chronic overweights may find it will work for them too. For further information, I suggest you write to Duke University, Durham, North Carolina 27706.

The other precaution that is absolutely necessary before entering a physical conditioning program is a thorough checkup by your physician. Only he can tell you how much you should try to do, and how soon. Without this prior knowledge, you risk giving yourself a heart attack. If possible, you should also have a test taken of your cardiovascular system at the hands of a "stress testing" machine. This is so important that the remainder of this chapter will be devoted to an article called "Give Your Heart to a Computer." The article first appeared in *The Winged Foot,* the monthly magazine of the New York Athletic Club, and it deals with the whole realm of stress testing and conditioning.

Give Your Heart to a Computer
by Fred G. Jarvis

Can a humble practitioner of afternoon badminton find success and acceptance in the electroded arms of a stress machine computer?

That was the question I put to myself and Dr. John Finkenstaedt, the Harvard conditioning whiz and cycling demon, who not so incidentally conducts a sophisticated testing program for Life Extension Institute.

I was curious just what level of fitness I was achieving by playing a racquet sport to the exclusion of other forms of exercise.

Stress testing is a cardiovascular barometer indicating the subject's ability to sustain hard exercise without producing heart irregularities and strain. It is rather like testing a racing car by running it flat out on a deserted track. You push your own engine in terms of pulse and blood pressure to reasonable upper limits and see how well—how smoothly—it responds.

At Life Extension Institute the response is quite precisely charted. The subject's body is wired into a computer panel which monitors pulse rate, takes blood pressure, provides a continuous oscilloscope printout, and does everything but show reruns of "Star Trek." While the light flashes on the control panel, you, the subject, walk on a simple treadmill whose speed and angle of incline are slowly elevated. You begin with a treadmill walking speed of 1.7

mph and a ten-degree incline. If you make it through the six stages (each is of three minutes' duration with no resting) you will be marching at a brisk 5.5 mph on a 20-degree incline. Frankly, it sounds easy and looks easy. Even if you are slightly daunted by the space-age equipment, electrodes, printouts, blinking lights, etc., it still boils down to walking. And everyone can walk.

That's what I thought. The first two stages are similar to leisurely strolling; the final stage is a "quick march." I was puffing in Stage Six, but I felt no different than I do during a prolonged rally in badminton. And the Stress Test had been preceded by a full three-hour physical and, consequently, no breakfast. So I was hardly at my sharpest and I felt I should have done much better.

Imagine my shock when Dr. Finkenstaedt casually dropped this fact: over 80 percent of the men who take the Stress Test do not get through Stage Two! At Stage Two they are strolling at 2.5 miles per hour on a 12-degree slope. And they can't do it. That's moving a bit faster than a Fifth Avenue window shopper.

This single fact is a stern indictment of the failure of the American male to take physical conditioning seriously with the concomitant loss of health, vitality, efficiency, and just about any other quality that depends on a smoothly functioning cardiovascular system.

Moreover, the Stress Test is a splendid diagnostic tool for red flagging subjects with potential or actual heart trouble. The traditional resting EKG discloses five to seven percent of the subjects as positive (bad news). Doctors experienced in the field know that this reading should be over 55 percent, so nearly 50 percent of the men with some form of heart problem go undetected with the electrocardiogram at rest. That's not much of an average. With the Stress Test, however, 50 percent of the test results show abnormalities; 15 percent potentially dangerous arrythmias and other previously undiagnosed electrocardiographic abnormalities; 20 percent latent hypertenses not detected before. So the Stress Test is a super detective with a diagnostic pattern almost exactly congruent with the suspected percentage of heart problems in the population as a whole.

The great problem with Stress Testing is that it is not generally available. The equipment at Life Extension Institute cost over $50,000 and was booked six weeks in advance when I visited Dr. Finkenstaedt. It requires a doctor and nurse to monitor the results. With heart disease America's number-one killer it is obvious that more Stress Testing is a necessity, a national obligation. The question is, to quote the song, where or when. For information on Stress Testing in your area, I suggest you write to *Runner's World* magazine, P.O. Box 366, Mountain View, California 94848, or consult your own physician.

The greatest problem in Stress Testing is translating the warning of the generally poor results into beneficial exercise programs. Of the 80 percent who fail to complete Stage Two, how many can be motivated or even frightened into changing their lifestyle or exercise patterns to the point where their physical condition shows a marked improvement?

Dr. Finkenstaedt believes that exercise, to be beneficial, should be conducted three to four consecutive days a week for at least thirty minutes. He is a firm believer in a five-minute warm-up and a five-minute warm-down. Heart rate should be driven to at least 85 percent maximum for the exerciser's age group and at least 20 minutes of higher activity sustained. Do you fit into this category? Obviously you could reach the desired upper level of exertion by jogging, walking, cycling, swimming, and playing one of the racquet sports or handball or hand tennis. Other sports require special skills or extra conditioning and strength—say judo, fencing, wrestling, gymnastics, figure skating, skiing. Basketball as played in the gym certainly requires tremendous stamina as does water polo. You could readily make up your own list of beneficial exercises. As long as you are extending your heart rate for a minimum sustained time, you're doing something of long-range value. It's obvious that some of America's favorite sports like bowling and golf don't fill this requirement; they are marvelous activities but their practitioners should supplement them with something more taxing in a cardiovascular sense.

I play badminton four times a week for an hour or so, and that apparently is enough exercise to produce an excellent Stress Test result. Dr. Finkenstaedt, who practices what he preaches, cycles about 100 miles a week—much of that total on the weekend. He's also a great advocate of walking—as much as possible, as far as time and weather permit.

The real point of this article is to find one lost sheep and bring him into the fitness fold. We need the maverick whose once trim physique is softened around the edges, who tells himself he has no time to exercise, and who is a prime candidate for a host of physical problems. We want to get him before a coronary does. Send him off for a Stress Test and physical, get him into the gym, in the pool, on the courts, and make a young man of him again. If we can convert one man, then this article will have been worth any amount of time and effort.

Dr. Finkenstaedt summarized the benefits of regular exercise for me and I'd like to pass them on as a parting shot. If you're a "waverer" perhaps they will reinforce your resolve.

Benefits of Regular Exercise

1. The heart and cardiovascular system can sustain a higher level of physical activity with less work output.
2. It enhances our ability to utilize glucose and fatty acids which are metabolized at a faster rate during exercise.
3. The adrenalin output is decreased during periods of emotional stress.
4. The clotting time of the blood is lowered, thereby decreasing the possibility of having a blood clot leading to a pulmonary embolus or heart attack.
5. Last, but not least, the individual feels better, eats better and has a better outlook on life in general.

Instruction for Seniors

Golf, more than any other game, is a process of "maintaining." One must maintain a straight left arm and an anchored head throughout the swing, the proper grip pressure and swing tempo, and patience and concentration from shot to shot. For seniors the business of maintaining becomes even more difficult. As age begins to take its toll we must work hard to maintain our distance off the tee and our accuracy from the fairway, as well as our resourcefulness and our nerves and confidence on the green. It is the goal of this chapter to help you do some of that maintenance work. The instructions that follow are written specifically for senior golfers who have played the game and know the basics but who are now in need of a few pointers to help them stave off the effects of age. Chances are, not every tip will be useful to you, but surely a couple of them will help you to maintain or even lower your handicap.

Maintaining Power

The first priority on the maintenance list of most senior golfers is power—distance off the tee. Fortunately, there are several things you can do to make sure you don't lose yardage as you get older. In the chapter on equipment, lightweight shafts were mentioned. If you have the type of smooth swing that works well with graphite, then you should certainly give it a try. By swinging the lighter shaft you'll be able to handle more weight in the clubhead end of the club, and that will help you to hit the ball farther. In the chapter on health and conditioning, several exercises were discussed. These exercises were geared to keeping your arm and leg muscles strong and your back and shoulder muscles supple. There is no question that if you can stay in good condition you will get the most out of your swing.

But there are other ways of preserving and extending power, and most of them are quite simple, requiring no mastery of a secret method or any great exertion on your part. One sound concept has to do with your feet, and the position they are in at address. Perhaps the most powerful position for the senior golfer involves a right foot that is absolutely perpendicular to your target line and a left foot that is fanned out about 45 degrees from heel to toe. By keeping your right foot perpendicular to the target line, you will set up a resistance, making the right leg act like a gate post around which your swing (the gate) can open on the backswing and close on the downswing. This perpendicular position also helps to guard against a lateral sway in the backswing, one of the major power leaks in poor swings. By having the left foot fanned out as much as 45 degrees you will enable yourself to move freely through the ball on the downswing, as you transfer your weight toward the target. Experiment with these foot positions, especially the left foot, until you find the stance that is comfortable and gives you the best combination of accuracy and power.

Another trick has to do with the way you tee the ball. For drives you might try teeing the ball a little higher. It is important to have the ball hang up in the air (assuming you're not fighting a wind) and the easiest way to do that is to increase the tee height of the ball. Of course, you can get the same effect by positioning the ball well "up" in your stance, playing it just to the left of your left instep. That way you'll be sure of catching the ball on the upswing.

Often a golfer, no matter what his age, will cheat himself of potential power just because he does not adhere to the fundamentals. The first of these is a "don't"—don't rush your swing. Seniors are particularly guilty here. We often become anxious to hit the ball, and lunge at it instead of letting the swing and clubhead do their jobs. It's important to make a smooth takeaway and a long wide extension of the club. You can practice this type of takeaway by using an extra tee. Place it in the ground about a foot or so in back of your ball. When you take your driver back, try to make the club pass over the tee. The wide takeaway will extend the arc of your swing, and the greater your arc, the more power you'll have. Also, it will help you guard against a quick pick-up of the club, a potentially destructive move in any golf swing.

As you continue the backswing, your main objective should be to get as full a shoulder turn as possible. Here, I like to use my chin as an indicator. If I can get my left shoulder to pass under my chin, I know I've made an adequate turn of my upper body. Again, as you move to the top of the backswing, and begin your forward swing, remember your timing. Don't rush things. If you make a slow, smooth swing, you will give yourself more time to coil up, to store power at the top of the backswing. By "just letting it happen" on the way down, you will overcome the impulse to hit with your hands, to "hit from the top," as the fault is commonly called.

Another power move you should work at is the extension past impact, where your hands and the club reach outward toward the target. Senior golfers often tend to pull the club back in to them quickly after they hit the ball, instead of following through. If you are making this aborted finish you are not accelerating through the ball as vigorously as you should be. So work on extending outward, right at the target, after you hit the ball. Then, be sure you finish the swing with your hands high, and your belt buckle facing straight ahead.

Perhaps the best way to extend your distance is through the use of an intentional draw. As most of us know, this right-to-left spinning shot rolls farther than either a straight shot or a shot that goes left to right. It's not an easy shot to hit if you don't hit one naturally, but if you're really looking hard for more distance, and you're willing to do some practicing, then you should definitely give the draw a try. Simply set up in a closed stance, with your left foot a few inches forward of your right, and your hip and shoulder lines facing several feet to the right of your target. Keep your clubface square to your target. This alignment will force you to strike the ball with a glancing blow that will make it spin from right to left.

If the closed stance doesn't help you hit a draw, try strengthening your grip a bit, by rolling both hands about a quarter-turn to the right on the club. If you have a standard grip, the Vs formed by your thumbs and forefingers should point approximately at your right ear when you take your regular grip. With the strong grip those Vs should point to the edge of your right shoulder. The strong grip will encourage extra hand action as you come through the ball. In effect you will be closing the club face through the impact area, and this will promote the right-to-left draw spin. If, after several practice sessions using the closed stance and the strong grip, you still can't get a right-to-left shot, you should give up trying and concentrate on the other ways of boosting your distance.

Maintaining Accuracy

Accuracy in golf is dependent upon two factors—direction and distance, and often the second of these is neglected. It is surprising how many seniors I play with do not know how far they hit the ball with each club. The only place to find this out is on the practice tee in the early morning. Hit a few practice shots to warm up. Then, starting with your pitching wedge, hit a dozen shots and take their average distance. Be sure to allow for bounce and roll and count only the carrying distance. You will have to do this for every club in the bag. It may take several mornings, but it will be worth the effort. A word of warning: Don't pace off the distances unless you know the exact size of your pace. Mine is 30 inches, so I know that when I have made 150 paces I have walked 125 yards. In order to figure this with real precision you might invest in one of those 200-foot measuring wheels, available at most hardware stores.

Once you know your capabilities, don't ever overestimate them. By the time you're a senior golfer you should have outgrown the impulse to stretch an 8-iron into a 9-iron. If you want real accuracy you have to play within yourself, even if it means gripping down on a longer club and shortening your backswing a hair. And don't just go by pure yardage; consider wind velocity and direction, the presence of hazards, as well as pin position and your own ability during a given day. Some days you're simply stronger than others and you should keep this fact in mind as you make your club selections.

Once you've chosen your club you still have the direction half of accuracy to worry about. This is mostly a matter of alignment. Many senior golfers, especially those whose eyesight isn't what it once was, have trouble lining up with a target that is 200 or more yards away. To solve that problem try this tip. As you sight your line from the target to the ball, look for a spot just a couple of feet in front of the ball—it can be a leaf, a discolored patch of grass, or whatever—and try to use that as you set your final alignment. Jack Nicklaus has used this alignment method for several years.

Imprecision of execution is often the result of imprecision in concentration. To help you focus, don't just look at the ball, pick a precise *area* of the ball and concentrate on it. One trick that works is to place the ball on the tee with the label at the point where you want to make contact. Hitting the ball while you concentrate on a specific spot can really help you keep your head down and swing smoothly. It works well on the putting green, too.

Incidentally, if you have trouble keeping your head down, here's another effective training method. Go out to the practice tee and hit several shots in a row without looking up to see where they went. That way you can groove yourself into a head-down frame of mind. If you can hit, say, fifteen shots in a row without ever looking up, then you have probably cured any tendency to peek at your shots.

At one time or another all golfers are bothered by a slice. And while a slice can be caused by a number of things, it is often the result of an outside-in swing across the ball. In the path of a proper swing the clubhead comes from the inside out to the ball and then moves back inside directly after impact. It is thus an in-to-out-to-in swing rather than an outside-in or inside-out swing. A good way to practice this type of swing calls for the aid of a 2×4 plank. Simply lay the plank parallel to your target line, about 2 inches outside your ball. Then just take your swings at the ball, making an effort not to hit the plank. If you take the correct swing at the ball your club will miss it every time.

You won't hit close to the target unless your shots are always target-oriented, even on the practice tee. This is especially important on the middle and short iron shots. If there are no flags on your practice area, create your own targets, either by lining up with distant objects or by constructing your own targets. A shag bag makes a good target, and it's also convenient to switch the distance of the shag bag as you pick up the balls. But do use a target. Hitting iron shots without a precise goal is useless exercise, no matter how straight those shots may seem to be.

The most important shot to practice is your wedge shot. Put your target out about 20 yards and hit about ten balls. Now move the target out another 20 yards and hit another ten balls, and so on until you reach your maximum distance for the wedge—about 75 yards for most seniors. These wedge pitches are among the most important shots in golf, and they're exactly the shots we never seem to practice.

One more tip on alignment. A mistake that 90 percent of amateur golfers make is to line up to the right of their intended target. Too often, we stand so that our "toe line," the line that connects the tips of our toes, is pointing directly at the pin. This is *wrong*. That line should point parallel to the line that extends from the ball to the target. If you point your toe line at the pin, then you will make one of two mistakes. Either you will hit the ball squarely and have it end up slightly to the right of where you want it to go, or you will make a subconscious adjustment in your swing and end up cutting across the ball, causing a slice. So be sure to set the toe line parallel to your target line. It will feel as if you're aligned to the left of your target, but in actuality you'll be right on line.

There are a couple of ways to improve your accuracy without ever picking up a club. One is to keep your eyes open as you play your round. On most courses you'll have an opportunity or two to see the greens of holes you will be playing later in the round. When that happens, take note of where the pin is—whether it's in the front or back, and where it is in relation to the major slopes on the green. It will help you later on.

One thing that few golfers take into account when hitting approach shots to the green is the slope of the green. Often they just compute the distance to the pin. But if you also consider that the green banks sharply upward, you'll realize that your iron shot won't take much roll and it will have to be hit almost stiff if it is to get close. On courses where the greens slope downward, away from the golfer, the opposite will be true. You'll probably be able to take a little less club and expect some roll downward toward the pin. If the actual yardage in such a case dictates a 5-iron, you'll be better advised to use a 6-iron.

Maintaining a Sharp Short Game

Probably the best bit of advice for senior players has to do with their philosophy in chipping: whether to use one club and vary the way you hit or use several clubs with the same type of hit. Jack Nicklaus, Bobby Locke, and others have had great success using just one club, and you can too. It's much easier to practice with just one club than with several, and if you rely on one club you can "get to know" it as you would a putter. Which club you choose is up to you, but I would recommend something in the 7-iron to pitching wedge range. Clubs with less loft than the 7-iron are not versatile enough, and the sand wedge with its thick, heavy flange is both difficult and dangerous to use for the finesse shots.

When you do practice with your club, try to make it hit all sorts of shots—high ones, low ones, stoppers, and runners. You can vary the trajectory of the shot by moving the ball up or back in your stance. Be sure to grip down on the club, for optimum feel.

One good way to practice on a putting green is to try to chip balls directly into one of the holes, on the fly. There's no better practice for your hand-eye coordination.

Most of us know the basics of the chip shot, so I won't go into them here, but each of us at one time or another just loses his touch completely around the green. If this should happen in the middle of a round, try this trick. Make believe your chipping club is a putter, and stroke your chips exactly as you would a putter. Many people do this all the time, and with great success.

After power, putting is probably the number-one "maintenance" area for senior golfers, and the hardest for us to conquer. Look at the pros—Hogan, Snead, Palmer—all of them could have had winning careers for years longer than they did were it not for putting. Their nerves went bad and so did their scores.

There are a few things you can do to protect yourself against the jerky stroke that increases as you get older. One common method is to straighten out the index finger on your right hand so that it points directly down the shaft of the putter. This provides an extra brace for the stroke and adds firmness to your move through the ball.

If you don't want any flippy-wristedness in your stroke, you can try weakening your grip—rotating your hands a quarter turn or more to the left. This makes it difficult to open and close the blade of the putter as you go back and through the ball. Another way to get the same results is through use of the old Leo Diegel method of putting, where you extend both elbows straight out from the shaft. This encourages a totally stiff-wristed stroke, where the putter is taken back by a movement of the arms and shoulders. The LPGA's Laura Baugh uses the Diegel method to great effect.

There's also the split putting grips, such as the one Sam Snead uses, where you brace the club up against your leg or stomach with the left hand, which acts as a hinge, and then grip down on the metal with your right hand, which then pulls the club back and pushes it through the ball. This is a little awkward at first, but it did help Sam (and many other seniors) save shots on the green.

If a different grip doesn't help you, then you might improve your putting if you take your eye off the ball. Henry Cotton has a tip in which he says line up the putt, take your grip and stance, and then, instead of watching the ball, watch your right thumb as it goes back and forth. He used this procedure himself for a while. At one point he even painted his thumbnail silver as a reminder.

If your thumb doesn't hold your interest, try watching the cup. Several golfers do this. After all, there are a number of precedents in other sports. Consider basketball, hockey, horseshoes, even pitching pennies. It's the target that is the object of concentration. So try looking at the hole as you stroke the putt. It may improve both your nerves and your accuracy.

Then there's the other side of the coin—where you never look at the putt at all. This is a practice method. Try hitting several 20-foot putts, and after you hit them, keep looking straight down as you try to guess where the putt ended up. This is a great way to develop touch. When you're always guessing "stiff" and you're right, the practice will have paid off.

A striped practice ball can be of help on the putting green. Line it up so that the stripe is facing your target, perpendicular to the blade of your putter. After you stroke the putt, take note of the way the stripe rolls. If you are cutting your putts the stripe will cant to the right; if you are pulling them it will cant to the left. If you're hitting them straight the stripe should roll on a straight line with no cant at all.

A final trick that is illegal during the round can nonetheless be helpful during practice. On the short putts place the putter behind the ball and shovel the ball into the hole. This is an excellent way to develop a feel for the accelerating stroke that is common to all good putters.

In putting, confidence is really the name of the game. And you can have great confidence on the putting green but nothing on the course. The idea is to try to instill some of the pressure of the course on the practice putting green. One good way to put this pressure on yourself is to try to make ten putts in a row. Start with one-footers, then move up to two-footers, finally three-footers, and maybe even four-footers. It's tough and it may take a while, but when you get up to those putts number 8, 9, and 10, believe me, you'll feel some pressure, and you will be getting some good practice for the real thing.

Once you do get confidence on the practice green you should try to extend it to the course. A good way to do this is to conclude all your practice sessions by hitting several short putts. Chances are you will make most of these, and the sight and sound of several putts going into the hole can give you a positive feeling for when you get to the first green. In any case, remember that the more time you spend on the practice green, the less time you'll spend on the real ones.

Maintaining Your Cool in Scrambling Situations

I've found that senior golfers have a reputation, whether de-served or not, of being canny masters of the scrambling shots.

Perhaps because we are not known as long hitters, this label as up-and-down experts is attached to all older golfers. Well, whether it's true or not, it *should* be true, because most of us have been in hundreds of trouble situations over the years, and we ought to have a good ability to save strokes. Unfortunately, all too few of us know how to scramble. I hope a few of these hints will help you with your next difficult lie.

On the subject of sand play, I have only one hint, which, if followed, will make you a better player than 90 percent of today's golfers. The hint is to practice. No one, but no one, practices sand shots, and if you spend just a few minutes in a practice trap you'll improve your game vastly.

Any golfer who wants to be a consistently good sand player has a very important decision to make when he begins to learn the trap shots. How is he going to regulate the distance of the shots? Will he hit the same distance behind the ball on every shot, and swing softly for short shots and firmly for long shots? Or will he swing with the same tempo on all shots and control the distance the ball flies by controlling the amount of sand he takes—a little bit for long shots, more for short shots? Either method will work—just be sure you choose one of them and don't vacillate between the two. You can probably make the ball do more tricks if you vary the amount of sand, but that's a little harder than the varying-swings method, which allows you to penetrate the sand in the same place every time. Log some practice time in the bunker and you'll undoubtedly choose your own best alternative.

Perhaps the most feared kind of mishit from a sand trap is the skull—a shot that normally zings across the green and ends up in a worse lie than the one from which it originated. My own pro, Mike Burke, has given me an excellent tip that virtually guarantees that you won't skull a shot from the sand. Simply address the ball as you normally would, with an open stance, the blade square to your target or slightly open. Then take your grip, placing your left hand on the very end of the club so that only your index finger and thumb are actually gripping the club. Then go ahead and hit the shot as you normally would. You'll find that, by keeping the last three fingers of your left hand off the shaft, you'll allow the right hand to dominate the shot, to pound down and under the ball, and explode it out every time.

Perhaps the worst sort of fairway lie is the hilly lie, where your stance is uneven and uncomfortable. You'll see all kinds of advice for these lies—some of it worthwhile, some of it just confusing. I've found that the one useful thing to remember when I have a hilly lie is to bend the uphill knee. That is, when I have an uphill lie I'll flex my left knee and when I have a downhill lie I'll flex my right knee. This simple adjustment sets me up for a well-balanced swing and a solid move through impact.

The lie where you're on a side hill and the ball is above your feet demands careful, abbreviated swings in order for you to keep from falling off the shot. There is also a tendency, when the ball is above your feet, to hit a draw, a right-to-left shot. I've found that a good way to counteract that is to open the clubface a hair at address. The opposite is true on the side hill lie with the ball below your feet. One tends to hit a fade from this lie, so it's wise to close the blade just slightly at address, to offset the left-to-right spin you'll be putting on the shot.

Rough can cause the senior golfer a lot of trouble, since most of us do not have the powerful pile-driving swing it takes to propel a ball from heavy grass. But there are a couple of adjustments you can make that will help you cope with rough.

When your ball nestles into a deep patch of grass the idea is to get your club down on it with as little interference by the grass as possible. What that means is that you must have a steeply vertical angle of attack on the ball, so that the club will not have to slash its way through the long blades. It's relatively simple to make this kind of swing; all it takes is a couple of minor adjustments in your address position.

First, play the ball about 2 inches farther back in your stance than normal—back in midstance. Secondly, stand about an inch closer to the ball than you would for a regular fairway shot. These two stance changes will encourage you to take the club up quickly on the backswing and make an upright pass at the ball. The club will come down sharply into the weeds and burst your ball out. This is far better than the "sweeping" technique. And don't be afraid to use this swing with a 4-wood or 5-wood. The fairway woods, with their heavy heads and round, smooth soles, are ideally suited to hitting shots from the rough.

Jack Nicklaus was once asked whether he thought it was a big advantage for him to be strong in the legs, so that he could slam the ball out of the rough. He replied that his leg strength had little to do with his ability in heavy grass, that it was the vertical swing that gave him the edge. So take encouragement—while few of us have the power of Nicklaus, we can all develop this vertical swing from the rough.

Heavy grass around the green presents a problem, especially if you are close to the hole. You either take a great whack at the ball and knock it across the green or you let up and the ball hardly moves. The trick is to hold the hands close to the body, break the wrists sharply and drop the edge of the club right on top of the ball. After a few tries you'll get the hang of it, and you'll find that the ball will pop right out, often with amazingly good results. It is a desperation shot, but if the ball is buried in that clinging grass there is nothing much you can do anyway. Just have confidence and give it your best try.

A tip on keeping cool during a big match comes from my friend Gene Hill, a fine competitive trap shooter and a good golfer when he gets time to play regularly. Gene says to take three deep breaths whenever tension mounts. Make them deep and exhale very slowly. It is easy to do this so no one will notice even just before that all-important putt. Athletes in other sports do it all the time but few golfers ever think of it. It is often an additional help to exhale just before a shot and to hold your breath until you hit the ball.

Especially for the Ladies

There are certain areas of golf that require a separate discussion for men and women. One of these is equipment. During a recent conversation with Peggy Kirk Bell, author of *A Woman's Guide to Better Golf* and perhaps the most knowledgeable female teaching professional in the game today, several interesting points were raised. Peggy feels that it's extremely important for female golfers to have the right equipment, more important than it is for males, since women, being less powerful, need all the help they can get from their clubs.

Peggy runs a women-only golf school at Pine Needles Lodge and Country Club in Southern Pines, North Carolina, and most of the women who attend her sessions have either hand-me-down clubs or cheap ones bought at discount. These clubs are usually wrong for them and, according to Peggy, there is no way they are going to improve their golf swing significantly until they get a set of clubs that fits their size and strength. Peggy feels that women whose husbands have fancy sets of clubs should give a little more consideration to the kind of clubs they themselves play.

Peggy is all for graphite shafts for women, feeling that anything that makes the club lighter will help women to hit the ball farther. Her other bugaboo with regard to equipment has to do with the sand wedge. At her school she tries to get women to abandon the 9-irons and pitching wedges they normally use from traps, and invest in a good sand wedge with a medium flange and a rounded sole. After going through her four-day school very few women leave without having mastered the sand shot.

Another knowledgeable woman golfer is Mrs. Albert Bower, the 1975 USGA Senior Women's Amateur Champion. She feels that most women play with clubs that are too short or too light for them. She is 5 feet 8 inches tall, and she uses men's clubs with a regular flex and a swingweight of C-9. She emphasizes that bigger women should not hesitate to buy men's clubs if that is what fits them best. Mrs. Bower also feels that the grips that come on women's standard clubs are too small for most women who, though they have small hands, have long fingers.

Mrs. Bower makes an interesting point about the difference between male and female golfers. She believes that the male torso is much more rigid than the female. Because of this, men have to work in order to get a big shoulder turn. Women, being more supple, should therefore guard against overswinging. It is true they need a big arc in order to get distance, but some women take the club back so far that they lose control of the club at the top of the backswing. The club should be taken back as far as possible but the grip must remain firm throughout the swing. Otherwise the club will have to be regripped during the downswing, and this can lead to all sorts of problems.

She also believes that women should have a strong grip, with the Vs of both hands pointing at the right shoulder. She uses the Vardon grip, in which the little finger of the right hand overlaps the knuckle on the forefinger of the left hand. With the stronger grip, women will be able to have increased hand action through the ball. This should increase their clubhead speed, and so their distance.

Both Mrs. Bower and Peggy Kirk Bell believe that women have an edge in golf in that rhythm is an important ingredient in the swing. Women often are good dancers, or at least have some experience in dance or ballet, and this serves them well in golf. The other ingredient in a golf swing is power, and Peggy Bell has some definite thoughts on that. "In most cases the women I teach have almost no strength in their left hand and arm. And a woman golfer, no matter how well she is taught, must have some strength in that left side. I encourage my students to do as many things as possible with the left hand, to think of themselves as being left-handed for a certain period of time during which they do everything left-handed. At the same time I explain to them that hitting a golf ball is not all strength—it takes coordination, flexibility, rhythm, and balance. Just trying to hit hard doesn't work. I tell my women not to hit the ball but to swing through it. I want them to swing the clubhead. I don't care whether they move off the ball. Too many of them are afraid to sway and as a consequence they stay on their left side and chop down. I want them to move those feet. Back on to the inside of the right foot and then back to the left. It is better to sway a little than to hit the ball only with your arms. It's especially important for the smaller person to develop leg action as well as a full shoulder turn."

Most of all, if you're a beginning golfer, don't be afraid to go to a qualified pro and get started on the right track. In golf it's important to take your first step in the right direction. After that, it's just a process of maintaining.

5

Courses for Seniors

by Robert Trent Jones

Senior golfers are no different from their younger counterparts in that their ranks include players of every ability. No better examples of this fact need be cited than Sam Snead, Julius Boros, Art Wall, and John Barnum, all of whom won tournaments after reaching the half-century milestone. Another case in point is A. L. "Jim" Miller of Chicago. Still one of the most active members of the Chicago District Golf Association, Jim is a member of the "men-only" Bob O' Link Golf Club in Highland Park, a great course perennially rated as one of America's one hundred best. Jim doesn't consider it remarkable that his game hasn't deteriorated as he has grown older. "Sure, I'm still shooting my age," he said recently. "Hell, it isn't very hard when you're 84!"

The point is, senior golfers do not require any special consideration save one—yardage—in the design of a golf course. As a senior golfer myself, I still savor the challenge of playing a well-designed, interesting course that demands thought and shotmaking from tee to green and intelligence in discerning the line, speed, and slope a putt must negotiate to achieve its ultimate target.

The best golf course for a senior is the best golf course for everyone. It is the strategic concept of design that offers the individual golfer the choice of how to play a particular hole. The strategic concept differs from the penal and the heroic designs in that it does not impose shotmaking demands of exceptional distance and confining accuracy, but instead leaves solely to the player the decision of how to play a hole commensurate with his capabilities.

A strategic golf course is a thinking man's golf course. Such a prerequisite fits right in with the philosophy of senior play. People who have been playing a game for most of their adult lives have, through the experience of competition and a thorough knowledge of their shotmaking capabilities, the ability to think their way around a golf course. It is within their ken to determine when to lay up short of a hazard, when to go for the flagstick or to shoot for the fat part of the green, and when to lag a "commercial" putt to avoid three-putting. These are decisions that confront all golfers, but most seniors have been over the road so many times that such a decision is almost automatic.

The penal concept of golf course design is the most restrictive. It flourished during the early part of the century when the British influence was strongest. A penal golf course was designed to exact heavy penalties for a wayward shot. It was rife with hazards—indiscriminate, cavernous bunkers, steep-sided "chocolate drop" mounds, deep rough growing all but through the green, and small, severely contoured putting surfaces which, lacking today's modern watering methods, were punishingly fast.

Playing a penal course, a golfer was forced to follow a line of play from tee to green dictated by the architect. To deviate was to suffer. Such courses demanded disciplined shotmaking, minimized strategy, and reduced the game to an almost mechanical performance. In effect, when playing a hole, a golfer was not required to think—only to obey.

Commenting on the penal concept, Bobby Jones once remarked, "I have found that most of the courses in America may be played correctly the same way round after round." The penal design, fortunately, is no longer in vogue, but vestiges of it remain on many older courses and on particular holes wherever golf is played.

The Pine Valley Golf Club in Clementon, New Jersey, is a superb example of the penal design. Since it came into being shortly after World War I, Pine Valley has achieved an enviable reputation as "the World's Toughest Golf Course." It is. Pine Valley can best be described as one vast bunker punctuated by islands of grass that serve as teeing areas, fairways, and greens. Although the fairways and greens, with one or two exceptions, are reasonably expansive and attainable targets for competent players, the course can impose horrendous penalties if one strays into its vast areas of unraked sand, its girdle of pine trees, and its water hazards that must be carried on four holes. The requirements of Pine Valley are so unrelenting that I would suggest that any golfer incapable of carrying a drive 170 to 180 yards with a modicum of accuracy forego the "pleasure" of testing it. I have—despite the fact that I've been a member for years.

Until recently Pine Valley was a course unique in the annals of golf. But in the early 1960s I was commissioned by the Northern California Golf Association to build a course at Pebble Beach. The ultimate result was Spyglass Hill, which incorporates over its first five holes many of the characteristics identified with Pine Valley. Spyglass stretches along the rugged shoreline of Monterey Bay. Before turning inland it offers golfers an exciting sequence of holes with ribbons of green fairways threading through sand dunes rampant with wild seaside vegetation and the notorious ice plant. The five greens that comprise the ultimate targets for aspiring golfers repose, as at Pine Valley, as island way-stations in the progression of holes.

The Oakmont Country Club is another throwback to the penal design, although it has been modified considerably since Henry C. Fownes first unfurled it over the rolling site near Pittsburgh in 1903. Fownes was a disciple of the penal philosophy, and to enforce his dictates he put an estimated 350 bunkers in this course, each of which was filled with heavy sand and furrowed to a depth of 4 inches. Fownes coupled his proliferation of hazards with greens—cut to $3/32$ of an inch—regarded as the fastest in the world. His fundamental belief was simple: "A poorly played shot should be a shot irrevocably lost."

In recent years nearly half of the bunkers at Oakmont have been eliminated and furrows no longer are dug into the sand. But its present 187 traps and the undiminished speed of its greens continue to remind us of the demands an architect can impose.

The heroic concept is prevalent more in the design of a particular hole than in an entire course. It sets up a shot as an unmistakable challenge, one that forces the golfer to extend himself fully if he is to succeed. Such a demand offers the reward of an easy par or a birdie if the shot is executed, and a severe penalty if not. The heroic requirements, however, differ from the penal in that some sort of bypass is afforded a golfer who will not, or cannot, accept the challenge. Taking the easier route normally obviates any possibility of playing the hole to the established par.

One of the most sustained stretches of heroic holes is to be found at the Merion Golf Club in Ardmore, Pennsylvania, which, like Oakmont, has been a favorite of the United States Golf Association as a site for that organization's major championships. Merion's last five holes are regarded as one of the strongest finishes in golf. To play them in par (4-4-4-3-4) requires length, accuracy, and touch, as the modest greens are full of subtle breaks and rolls.

The most famous hole in that stretch, a hole of absolute heroic demands, is the 445-yard sixteenth, or "Quarry Hole." The green is 25 feet higher than the fairway where a drive will come to rest. To reach the putting surface a golfer must hit over the quarry, a wild tangle of junglelike growth and sand that intrudes along the line of play for 75 yards and terminates at the front edge of the green. The quarry also is a factor on the 230-yard seventeenth hole and the 458-yard eighteenth, but from an elevated vantage point it is not quite as intimidating as it is from below. The holes in which the quarry intrudes and the fourteenth and fifteenth all may be played by less direct routes, but to do so is to concede that par is virtually unattainable.

I should also like to cite three holes of heroic requirements that have one thing in common—the Pacific Ocean. To my mind, the eighth hole at Pebble Beach offers the most awesome second shot in golf. A well-struck drive there comes to rest on an escarpment some 40 feet higher than the green. But in between lies an arm of the ocean where surging waves crash against the rocky shore 100 feet below. Carrying this monumental natural hazard requires a shot of at least 170 yards to a small green cut into the face of the slope below. And if the dramatic features with which nature endowed this hole aren't enough, there is a necklace of four bunkers surrounding all but the front entrance to the green. For the timid and the ill-equipped who would forego the thrill of trying to play this 428-yard hole conventionally, there is another way around. But to follow it is to minimize any chance for par.

Nearby is the Cypress Point Club and the most famous par-3 in the world—the celebrated sixteenth—renowned in photographs and paintings. This is an absolute carry over wild water that crashes into two jagged headwalls confining the cove. A 222-yard hole, it requires a shot that must remain airborne for 210 yards or suffer the consequences. And the narrow margin for error between the "drink" and the green is covered with a tangle of ice plant, virtually inescapable. With the wind blowing, as it usually is, and the sea lions barking in the ocean beyond, the sixteenth is a fearsome assignment. It can be played, however, without the loss of a ball if you aim left toward the vicinity of a crooked cypress tree and pitch onto the green. Follow this route and a par-3 is possible—if you one-putt.

Equally dramatic is the third hole at the Mauna Kea Beach Hotel and Country Club in Hawaii, a course I created. The Pacific Ocean also serves as the major hazard for this 215-yard one-shotter. Both the tee and the green are about 25 feet above the swirling waters of Turtle Spawning Cove, but to traverse the intervening yardage you must carry the shot 180 to 200 yards, depending on where the cup is positioned on the 25,000-square-foot green set on a diagonal. Like the sixteenth at Cypress Point, wind, water, and the beauty of the setting make this hole one of the game's most spectacular challenges. And also as at Cypress, the ocean may be circumvented making a par-3 possible if the pitch-shot second is deadly accurate.

The strategic golf course mentioned earlier is a concept that has been prevalent in my work since I embarked on a career in golf course architecture more than forty-five years ago. I have followed this precept for good reason—a strategic course is "everyman's golf course." It incorporates in its design certain features and characteristics that guarantee a challenge and a pleasure for golfers of every ilk. It is a golf course with great flexibility, not only in regard to yardage but also in the kind of challenges that may be imposed in shots to the green.

I believe the vitality of the game of golf is that it offers men and women a personal challenge for combat. They attack the course and par. The architect creates fair pitfalls to defend the course against easy conquest. In a true sense, the game is one of attack and counterattack. New and improved instruments have been developed which, together with practice and skill, may "bring a course to its knees." The architect calls on his ingenuity to create a hole—or a course—that will reward only achievement.

Obviously, if either side in a contest is unfairly weighted, interest will wane. A penal golf course is just too tough an assignment for the rank-and-file golfer; a pitch-and-putt course is no challenge for a player of some competence. The happy medium is unquestionably the strategic course.

This concept has been with us since the inception of the game, although to a lesser degree in the early days. St. Andrews in Scotland, the birthplace of golf, is the very essence of strategic architecture. Once oriented to the Old Course with its crafty trapping, the undulations of its massive greens, and the endless tumbling of its fairways, a golfer realizes that when he plays a fine shot he is rewarded, when he doesn't he is penalized. Such a golf course encourages initiative and benefits a well-played, daring stroke more than a cautious effort. Of overriding consideration, however, is the fact that there must be planning and forethought before each shot.

Credit for adapting the strategic concept to the American scene should go to Robert Tyre Jones, Jr. It was Jones's belief, inspired by his love for the Old Course, that a great golf course must be a source of pleasure to the greatest possible number of players and that it must require strategy as well as skill if it is to hold a player's interest. It should give the average golfer a fair chance and at the same time require the utmost from the expert who tries to break par.

This theory was articulated by Jones in the creation of the magnificent Augusta National Golf Club course, which he coauthored with Scottish architect Dr. Alister MacKenzie. Jones had been impressed by MacKenzie's work in the design of Cypress Point in which the architect's main thrust was that of strategic play, not penal, which was prevalent at the time. Significant to MacKenzie's concept was the elimination of an overabundance of traps. Although Augusta National has an unchallenged reputation as a premier championship test, it incorporates a total of only forty-two bunkers—but each is significant.

The Jones-MacKenzie creation is a superb test, despite its minimal sprawls of sand and double the fairway acreage of an average course—approximately seventy to a normal thirty-five acres. It requires superior know-how and tactical consideration. "Position" means everything at Augusta. Its greens, generally plateau in nature, follow no general size or configuration and offer surfaces that are crowned, swaled, hollowed, terraced, and sloped. Couple these elevation changes with a deployment of putting surfaces at angles to the line of play, with extensions hemmed by sand, and with "fingers" that require pinpoint approach accuracy, and it is not too difficult to appreciate why you must "think" your way around Augusta.

Augusta National, for those fortunate enough to play it, comes as a pleasant surprise to the average golfer. Generally he scores better on it than on his home course. For conventional play in the 6,300 to 6,500-yard range, its broad fairways and reasonably receptive flag placements allow the 17-handicapper a maximum of leeway in plotting his way around the course.

When it is set up at its maximum length of 7,020 yards as the stage for the Masters, Augusta is a different course entirely. It is then that its minimal bunkering comes sharply into play and its broad fairways become a ruse that can mislead a golfer not wary enough to realize there is a limited area from which he should play his next shot. That subsequent shot is critical in that in "seeking out the pin" it must come to rest in an area that minimizes the complexities of contouring of the green surfaces so dominant at Augusta.

The genius of Bobby Jones which served as the inspiration for

Augusta National also is evident at the Peachtree Golf Club in Atlanta, which was instrumental to my career, as I collaborated in its design and construction. Peachtree differs from Augusta National particularly in the basic principle of green design. Each green has five or six definite pin positions of which four are of such quality as to be ideal for tournament play. These pin positions represent the target area for the better golfer, whereas the whole green represents the target area for the average golfer. The greens are undulating in character, but not as severe or as continuous as the slopes at Augusta. Peachtree's greens are large, in keeping with this principle of design.

The greens at both Peachtree and Augusta National are of the plateau type for the most part, but Augusta's greens normally slope from front to back while the Peachtree greens do not slope in any general direction, but take the nearest and most obvious outlet to all sides of the green.

Another Peachtree feature is the tremendous flexibility of the course, brought about by the extreme length of the teeing areas, which in some instances are close to 200 feet. The course can vary from 6,300 to 7,400 yards, yet proper positioning of the flagstick can guarantee that nothing tricky will confront the golfer.

In the development of Peachtree, Bobby Jones and I attempted to create several different target areas that represented the point of anticipated skill. To this end, much of the fairway bunkering was positioned on the basis of several drives which he hit from the teeing areas to ascertain initial targets, and, in the case of the long par-5 holes, the secondary targets prior to an approach to the greens. To my knowledge this was the last time Bobby Jones actively engaged physically in anything related to playing golf. Even then, he was in the early stages of a crippling disease that eventually incapacitated him.

Through the fairway we determined where hazards should be imposed and in the green area the deployment of sand and other such playing considerations of a modest punitive nature. As a result, if a shot did not come off as anticipated and the golfer was wide of his target, he found sand or trees. And failing to achieve the penultimate target—a particular segment of the green—his chip shot or approach putt became more difficult because of undulations and rolls, and getting down in two was more demanding.

On the other hand, the average golfer benefits to the extent that the whole green, which is unusually expansive, is his target area, and while he is not expected to find the pin area as frequently as is the better golfer, he also will not find the bunkers so accessible, as they are not as intimidatingly close as they would be on a smaller green. Our belief was that while the average golfer may not have the playing skill of a low handicapper, he may be quite adept when it comes to putting. The green is the one area of a golf course where the low and high handicappers tend to level off.

The concept and characteristics of the Peachtree course are those that can be found to a lesser degree on almost every other golf course of my design. The extended teeing areas, great expanses of green surfaces, and strategically located bunkers can be interrelated to test the skills and capabilities of golfers running the gamut from the scratch player to the maximum 40-handicapper. Only the strategic golf course can offer such infinite variety; to build a golf course of any other concept would be foolhardy.

For seniors who enjoy the challenge of new conquests, may I suggest the following courses which, in the main, were designed by me. The list all but spans the globe. Each is a different and distinct challenge, but all were created with profound respect for the sanctity of par and an inherent regard for shotmaking values. These are golf courses to be played and enjoyed, but remember—to paraphrase a well-known axiom—thought must be the father of the shot.

Outstanding Golf Courses Recommended for Seniors

American Courses

*All Seasons CC, Lake of the Ozarks, Missouri

*Arcadian Shores CC, Myrtle Beach, South Carolina

*Arcadian Skyways CC, Myrtle Beach, South Carolina

*Arthur Raymond GC, Columbus, Ohio

*Indicates a Robert Trent Jones, Inc., golf course.

*Aventura CC, Miami, Florida (36 holes)

Bowling Green GC, Milton, New Jersey

*Bristol Harbor CC, Rochester, New York

*Broadmoor GC, Colorado Springs, Colorado (54 holes)

*Cacapon Springs GC, Berkeley Springs, West Virginia

Canyon CC, North Course, Palm Springs, California

*Carolina Trace G & CC, Sanford, North Carolina

*Dunes G & CC, Myrtle Beach, South Carolina

*Elkhorn CC, Sun Valley, Idaho

*Golden Horseshoe GC, Williamsburg, Virginia

*Goodyear G & CC, Litchfield Park, Arizona (36 holes)

*Grand Teton Lodge GC, Jackson Hole, Wyoming

*Horseshoe Bay CC, Marble Falls, Texas

*Incline Village GC, Lake Tahoe, California (36 holes)

*Inverrary G & CC, Lauderhill, Florida (36 holes)

*Lakeridge CC, Reno, Nevada

*Lake Shastina GC, Weed, California

*Lyman Meadow GC, Middle-field, Connecticut

*Madeline Island GC, LaPointe, Wisconsin

*Mauna Kea Beach Hotel GC, Kamuela, Hawaii

Mid Pines C, Pinehurst, North Carolina

*Nassau County Park GC, Salisbury, New York (36 holes)

*Otter Creek GC, Columbus, Indiana

Pebble Beach GC, Pebble Beach, California

Pinehurst CC, Pinehurst, North Carolina (90 holes)

Pine Needles CC, Pinehurst, North Carolina

*Rail GC, Springfield, Illinois

*Silverado CC, Napa, California (36 holes)

*Spyglass Hill GC, Pebble Beach, California

*Steamboat Springs CC, Steamboat Springs, Colorado

*Stone Mountain GC, Stone Mountain, Georgia

*Stumpy Lake GC, Norfolk, Virginia

*Tanglewood Park GC, Clemmons, North Carolina (36 holes)

*Upper Cascades GC, Homestead Hotel, Hot Springs, Virginia

Foreign Courses

Banff Springs Hotel GC, Lake Louise, Canada

*Brasilia GC, Brasilia, Brazil

*Campo de Golf de Sotogrande,

Guadiaro, Spain (36 holes)
*Campo de Golf, Tres Vidas, Acapulco, Mexico (36 holes)
*Cancun Golf Course, Quintana Roos, Mexico
*Cerromar Beach GC, Dorado Beach, Puerto Rico (36 holes)
*Cotton Bay Club, Eleuthra, Bahamas
*Dorado Beach G & TC, Dorado Beach, Puerto Rico (36 holes)
*El Basque GC, Chiva, Spain
*El Rincon CC, Bogotá, Colombia
*Fountain Valley GC, St. Croix, Virgin Islands
*Golf '72, Karuizawa, Japan (72 holes)
*Half Moon–Rose Hall GC, Montego Bay, Jamaica
*Jaspar Park Hotel GC, Lake Louise, Canada
Lucayan Beach Hotel GC, Grand Bahamas
*Mazatlan GC, Mazatlan Sinaloa, Mexico
*Moor-Allerton GC, Leeds, England
*Navatanee GC, Bangkok, Thailand
*Nueva Andalucia GC, Marbella, Spain (36 holes)
*Pacifica Harbour GC, Deuba, Fiji
*Pevero GC, Costa Smeralda, Sardinia, Italy
Princess Hotel GC, Freeport, Grand Bahamas (36 holes)
*Royal Dar es Salaam GC, Rabat, Morocco (45 holes)
St. Andres GC, Bogotá, Colombia
Shamrock GC, Grand Bahamas
Tryall Club, Montego Bay, Jamaica
*Zihuatanejo GC, Guerrero, Mexico

How to Select a Golf/Retirement Community

If you plan to retire to a golfing community, there are several considerations to keep in mind. The first of these is climate, both the meteorological climate and the social climate. Do you want something in the general area where you now live, so that you won't have to give up your old contacts and friends? Are you seeking a place where the weather is always warm? Or would you rather retire to an area that is relatively warm but still has four distinguishable seasons?

The financial side must also be faced. Before you select a retirement community you must make a realistic projection of the annual income for your retirement years. This income should be adequate to allow you to live in a style that is equal to the other residents of the community you select. If your income is $25,000 a year, you don't want to move into an area where the homes start at $100,000 and everyone else's income is $50,000 or more. To me there is nothing sadder than trying to keep up with the Joneses when you just don't have the money to do it.

The ideal way to select a golfing community is to visit a number of them for a week or more, renting either a condominium or a villa similar to the one you might buy. That way you can enjoy some resort golf while seeking the answers to your questions about retirement golf. Most of these communities have such rental deals, and there are usually dining and recreational facilities available.

In many cases the community also contains a large resort hotel with full facilities. This fact is important to one of the decisions you must make when selecting a golf/retirement site. Do you want one where there is a large hotel, meaning you will have to share the community's facilities with tourists and vacationers? This has obvious advantages and disadvantages. If you like to meet and entertain people, you will have plenty of opportunity. People on vacation are always happy and proud to meet residents of the area they are visiting. On the other hand, this kind of setup often means more crowded conditions for you as a golfer, and this must be taken into consideration.

A few years ago, when these resort/retirement communities were first springing up, the plan was for the developer to come in, build a golf course, clubhouse, roads, and all facilities, sell you a lot and landscape the area where homes were to be built. Then eventually, when all the lots were sold, the developer would turn the property over to the homeowners, who would run the place or hire management to do so. In actual practice, this has seldom happened, for apparently it takes forever to sell all the real estate, and while those sales are going on there is good profit to be made by managing the property, clubhouse, facilities, and golf courses. Because of this fact, a few more questions must be answered.

If cost is a factor, as it is for most of us, what would the annual dues of the community be, and what guarantee is there that those dues would not increase substantially each year? Also, will there be a limit on the number of members included in the community? Many golf/resort communities, even those that have three or more courses, have crowded conditions on the links. When there are thousands of homes, there will be thousands of golfers—and this kind of place is always crowded. Many of these courses are also in poor condition because the heavy play makes it difficult to maintain the fairways and greens properly.

Another consideration has to do with the way you get around a golf course. Many seniors, seeking exercise, enjoy pulling carts most of the time, but this is not permitted at some courses. Of course, if you end up at one of these courses you can still get plenty of exercise, by sharing a motorized cart and walking every other hole while your partner drives. But if you have to rent the motorized carts from the pro shop, they will probably cost you at least $10 for eighteen holes. If you're an avid golfer, that can add up to a lot of money very quickly. The frequent alternative—purchasing your own cart—raises another set of questions. Can you keep it in your own garage, charge it, and drive it to the course? Or can you pay the pro a fee for keeping it charged?

A crucial question relates to the demographic makeup of the community in which you plan to live. Strong pressure has been brought to bear on the developers of retirement communities for the fact that, by limiting the age of the people who may buy or visit the community, they are interfering with one of the basic freedoms. Obviously, a private club has no such problems and can determine its own membership at will. However, when you come to an open development, it behooves you to look carefully at the age makeup of the residents. If you select a community that is youth-oriented, keep in mind that the golf courses will likely be swarming with youngsters. And delightful as kids are, they seldom add pleasure to a senior golfer who has a serious interest in golf. Other seniors, of course, may prefer an area where there is a good mix of young and old, in the thought that having young people around will help keep them young. To a certain degree this is probably true, but the whole subject certainly deserves your careful consideration.

Another factor to consider is whether you want to travel during part of the year and rent your property while you are not using it. If this is the case, you should pick your home in a resort community rather than in a retirement area. Of course, if you buy a large enough place you can have it both ways, by maintaining a permanent residence in part of the house and renting out the other half. The rent from resort homes, if they are in a successful area, can provide a substantial addition to your retirement income.

I said that you should be aware of just how many homes the developers plan to rent, in order to guard against overcrowding. Well, you should also be careful not to lock into a community that has not yet proved itself. There are many sad stories of senior citizens who bought homes in retirement communities that were disbanded because of insufficient response. It's nice to get in on the ground floor with a good price on your home; just try to be sure that the future of the community is sound. Try to talk to some of the local bankers about the financial situation of the development you're interested in. After all, as local bankers they are interested, and in many cases may own the place you are going to buy, with the real estate company merely acting as an agent. When you decide to buy, hire a local attorney, making certain that he is simultaneously retained by the owners or developers.

And if all else fails, consider this: Buy a small house in a quiet village somewhere where the weather is right for you; then join the local country club.

Chapter 7 lists a couple hundred possible golf vacations in the United States. Pay particular attention to the ones marked with an asterisk; these resorts also make excellent retirement spots.

7

Golf Vacations for Seniors—United States

With the leisure time you have as a senior, you will undoubtedly be taking more golf vacations. Very often a vacation destination will become a quest for a retirement haven. Accordingly, besides discussing good senior golf vacations, I will indicate some golfing communities that combine vacation with retirement.

If you're planning a golf vacation, your eventual destination may depend a good deal on where you're living now. If your home is in the North, you will most likely plan your vacation around a trip to the southern climes, where warmer conditions prevail. If you are a Southerner it may well be that you would plan your golfing trips during the heat of the summer, and head north, where the weather is cooler. In order to come up with a selection of places suitable for golfers living everywhere, I will cover the country region by region, discussing briefly some of the famous resorts that have long given golfing pleasure to vacationers, as well as some of the very new ones that seem to offer good value and pleasure for your vacation dollar. Note: Those resorts preceded by an asterisk have golfing communities that make good retirement areas.

The Northeast

The northeastern part of the United States boasts many fine golf courses, but good golf courses with complete resort facilities are not nearly as common in this area as one might imagine. The golf in this region is also very seasonal. Mid-May to mid-October is the ideal time to play golf in the Northeast. To my mind the nicest time is the period after Labor Day to the end of October.

Maine

In Maine there are just two courses that I would recommend going out of your way to play. Neither one of them is exactly a re-

Lining up a putt in central Maine.

A Maine course near Augusta.

sort, but there are accommodations at the doorstep of each. The first is the Poland Spring Golf Club. This course dates back to 1894, making it one of the first resort hotel courses in the country. The old Poland Spring and Mansion House are now closed, but you can still buy the famous Poland Water there, and the course is as good as it ever was. Measuring 6,400 yards with a par of 71, the layout includes a number of interesting holes, most notably numbers 4, 6, and 14. Green fees are about $4. Facilities include a dining room, golf carts, caddies, and rental clubs. Accommodations at the Poland Spring Lodge are European plan, about $20 double.

The other course, Kebo Valley at Bar Harbor, is a par-70 measuring 6,300 yards. It has a locker room, dining room and bar, golf carts, hand carts, and caddies. Green fees run $5 to $6, and there are discounts when you stay at certain hotels in the area. Bar Harbor has a great many excellent hotels and a couple of good restaurants. Good places to stay are the Atlantic Oaks or the Bar Harbor Motor Inn.

New Hampshire

In New Hampshire can be found some of the old-line resort-type courses and hotels. Recommended selections would include the Balsam at Dixville Notch, an older resort hotel situated on 15,000 acres with surroundings of great beauty. There are twenty-seven holes of golf, and two of the nines are designed by famed golf course architect Donald Ross. Although they are not examples of his best art, they are well worth playing. The 232-room hotel, open late May to mid-October, offers rooms from $33 per person, double accommodations with meals.

The Bald Peak Colony Club at Melvin Village, New Hampshire, has a really exciting 6,100-yard, par-70 course. The whole setup is run as a private club; it is necessary to be introduced, but it is well worth getting an introduction—not only because of the quality of the course but for the very fine accommodations. It is located along the shores of Lake Winnipesaukee and offers, in addition to golf, a great many outdoor activities.

At Portsmouth, New Hampshire, is another fine resort hotel with 245 rooms in an inn and cottages. Called Wentworth-by-the-Sea, it has a very good resort course. Open May to October, its length is only 6,200 yards, very definitely the kind of course seniors would enjoy. They have a nice clubhouse with lockers, dining room and bar; golf carts, hand carts, and caddies are available.

Massachusetts

Although Massachusetts has a number of fine golf courses, many that would especially appeal to seniors are private, and require playing with a member. Others are public courses, very crowded

The Berkshires provide a scenic backdrop
for the Jug End Barn resort's lovely course.

and frequently located in areas where there are no convenient hotels. There are, however, half a dozen very good resort courses that seniors should enjoy.

Clauson's Inn Golf Course, North Falmouth, is a comfortable resort course with many fine views. Complete resort facilities are available at the inn and the clubhouse with locker room and bar. Pull carts, riding carts, and caddies are available. Rates are about $30 per day per person, modified American plan.

*Jug End, South Egremont, is one of Massachusetts' newest resort courses, 6,250 yards, par-72. The course is open April to December. Trout fishing in May is an extra bonus. Dining room, bar, golf carts, and pull carts are available. Rates for two range from $18 to $34 each, modified American plan.

The par-72, 6,530-yard Taconic Golf Club, Williamstown, is one of the finest courses you will ever play, with lovely views of the Berkshire Hills on all sides. Owned by Williams College, the course is listed as semiprivate, but members of other USGA clubs are always welcome. Facilities include locker room, driving carts, pull carts, and caddies on occasion. The place to stay while playing here is the famous Treadway Inn in Williamstown.

Island Country Club, Martha's Vineyard, is a par-70 at 6,018 yards. The clubhouse has a snack bar, dining room, and bar. Golf carts, hand carts, and caddies are available. A moderately trafficked course, this is well situated along the seashore. Accommodations are at the country club.

Coonamesset Golf Club, par-72, 6,480 yards, is a very good golf course in a resort area with full facilities. Accommodations next door are at the Coonamesset Inn. The golf course and inn are located in Falmouth.

Chatham Bars Inn, Chatham, is a very good older resort hotel with a fine golf course in a scenic location. Clubhouse has full facilities, including golf carts, pull carts, and caddies. Full American plan at the inn is approximately $64 per day for two.

Eastward Ho Country Club is also in Chatham. This is one of the best courses in the country, but it is a private club and an introduction is necessary in order to play.

Vermont

Vermont has a number of golf courses in resort situations, but I must caution you that many of them are cardiac alleys in miniature. Beautiful scenery—but it is up and down some of the worst hills you will ever walk. There's no problem if you can rent a golf cart or if you're in top physical condition. The courses recommended here all rent motorized golf carts. Not only are Vermont golf courses scenically attractive, but most of them are uncrowded, and the golf couldn't be friendlier.

Lake Morey Country Club, Fairlee, par-69, 5,400 yards, is a pleasant resort course along Lake Morey. It is fun to play because it is set in a valley, and hills are no problem. The clubhouse has locker, snack bar, dining room, bar, golf carts, hand carts, and caddies. Accommodations at the Lake Morey Inn are immediately adjacent. The golf season ranges from May to October. Double accommodations at the inn are from $22 each, modified American plan.

Equinox Country Club, Manchester, par-71, 6,750 yards, is certainly one of the most scenic courses in the country and one of the most famous. Not all the holes are great, but there are a few you will never forget. The clubhouse has lockers, motor carts, hand carts, and caddies on occasion. As far as accommodations are concerned, unfortunately the famous old Equinox House is closed at the moment, but there are many good places to stay in Manchester. The Wilburton Inn, open mid-May to mid-October, would be a good choice.

Stowe Country Club, a private club, welcomes guests staying at hotels in this resort area. This 6,700-yard, par-72 layout is a beautiful course with plenty of ups and downs. The clubhouse offers lockers, lunch service bar, golf carts, hand carts, and caddies on occasion. All kinds of accommodations are within reasonable distance of the golf course. A good choice would be Edson Hill Manor, and the Lodge at Smugglers Notch.

Basin Harbor Club, Vergennes, is par-72, 6,000 yards. One of the few courses where hills are no real problem, this is a good resort course situated on a bluff above the shores of Lake Champlain. The clubhouse has a dining room, bar, golf carts, pull carts, and caddies on occasion. Accommodations are at the Basin Harbor Club immediately adjacent. Open mid-June to mid-October, this is one of Vermont's finest resorts, with full recreation facilities. Double rates are $30 each, full American plan.

Woodstock Country Club, Woodstock, now has eighteen holes, par-69, 5,960 yards. The course was redesigned by Robert Trent Jones who also added nine holes to give them the present layout. Part of the course is in the valley and the other nine swings through the hills. It's well worth playing. Accommodations are next door at the charming Woodstock Inn. European plan accommodations start at $28 double per day, with lower rates off season.

One of Vermont's better courses, Sugarbush Golf Club in Warren, par-72, 6,740 yards, is a hilly layout designed by Robert Trent Jones, who also added nine holes to give them the present layfull resort facilities, including fishing. Modified American plan ranges from $56 per day, double.

Connecticut

Connecticut has many fine golf courses, but unfortunately they are almost all private. There is really no resort in the state to my knowledge that has an eighteen-hole course of championship caliber. Banner Lodge in East Haddam has an eighteen-hole course, but I have never played it.

New York State

New York State possibly has more fine golf courses than any other state. I am sure that both California and Florida will dispute this statement, but I am talking primarily of well-kept courses of quality rather than numbers. Most of the fine courses are private; when it comes to resort golf courses, New York, in reality, has but a very few.

Lake Placid in the Adirondacks is as close to a golf center as anything in the state. There are two resort complexes, as well as a fine public golf course, all worth playing. The Lake Placid Club, a fine old resort, has for many years been run as a private club; it still has a membership that receives priority over reservations from the general public. This need not stop you, however, for the club is very

A tough lie in Vermont's Green Mountains.

large and there is usually room for all. This big hotel offers everything in the way of resort facilities on its 833-acre reserve. There are actually forty-five holes of golf at the Lake Placid Club. The lower eighteen is somewhat longer, but the upper eighteen is quite hilly, with a number of blind holes, and perhaps is the tougher course of the two. In addition, there is a very challenging par-3, nine-hole course that is a great deal of fun to play when you don't have the time for a full round. Lake Placid Club has a variety of accommodations, in cottages along the lake and in the main house. Rates, double, with full American plan are about $35 per day per person. You do not have to stay at the Lake Placid Club in order to play golf there. Guests are given preference, however, as to starting times on those rare occasions when the courses are crowded. Motor-driven golf carts, which were long banned on the upper courses, are now permitted, and this is a great blessing, as the Lake Placid hills are really something.

Craig Wood Country Club, par-72, 6,540 yards, is a municipal course open to the public at all times. It has a number of very good golf holes and is completely surrounded by many of the high peaks in the Adirondacks. Complete clubhouse facilities include bar, dining room, riding car rentals, pull cart rentals, and occasionally caddies. It is open May to November.

Whiteface Inn Golf Course is one of the best layouts in the state. This par-72, 6,490-yard course winds through the woods that conceal all holes but the one you are playing. It is in beautiful surroundings and a real challenge. Whiteface has a very lovely hotel and cottages with complete resort facilities for all types of outdoor activities. Rates, full American plan, are about $34 per person, double. Economical golf package plans are available.

By far the largest center of golf in New York State is the region where the Catskill Mountains are located. This is a built-up resort area with a number of hotels having golf courses open in the summer. The best of these is the Concord Hotel and Country Club, Kiamesha Lake, New York. This fantastic hotel boasts a golf course that is equally incredible. Called "The Monster," it is aptly named. From its 100-yard-long back tees, it measures just under 7,700 yards, but probably plays even longer. Jimmy Demaret, who was pro there a while ago, made suggestions to Leo Finger in the design of the course. Demaret has always called it the toughest golf course in the world, and many who have attempted it from the back tees would be quick to agree. The average senior can enjoy the course, however, if he confines his play to the front of the tee. The Concord has three other golf courses, but this is unquestionably the one you will want to play. The Concord has something like 1,200 rooms and every possible resort facility, but if you want to concentrate on the golf, I would suggest staying at the lodge immediately adjoining "The Monster." It has its own dining room and bar and is completely devoted to golf. Rates are $42 each, double, full American plan. Other courses worth playing in the area are: Grossinger's, near Monticello; Kutsher's, also at Monticello; and Nevele Country Club, at Ellenville, New York.

An eighteen-hole, par-72, 6,372-yard course is attached to the Otsego Hotel. This resort in Cooperstown with full vacation facilities is open from May to October. The golf course is a very good one with a great variety of terrain. The resort is situated on the shore of Lake Otsego and the lake comes into play on the golf course for a number of holes. There are complete country club facilities including locker room, bar, dining room, pull carts, and electric carts. American plan rates are about $30 per day, per person, double occupancy.

Saratoga Spa Golf Course, a fine eighteen-hole, par-72 course, stretches to over 7,000 yards on the back tees. It is set on mildly rolling hills and offers a number of challenges that make it well worth playing. In addition, there is a very exciting par-3 golf course. Accommodations are at the adjacent Gideon Putnam Hotel, one of the state's better-known and older resort hotels.

New Jersey

This is another state that boasts any number of well-known and fine private golf courses.

The only resort course in New Jersey is Great Gorge Resort Hotel at McAfee. This is a George Fazio course, an eighteen-holer at 6,820 yards and another nine at 3,300 yards, set in lovely rolling hill country. Three interesting quarry holes and a number of spring-fed lakes come into play.

The Atlantis Country Club at Tuckerton has a golf course that runs through the woods. Because of its narrow fairways and small

Seniors at the Deal Golf Club in New Jersey.

greens, it's quite a challenge. There is a motel with dining room and bar adjacent.

Pennsylvania

Pennsylvania offers ideal golf terrain, and some of the country's best resort courses are located in this state.

A selection of the better resort courses would certainly include Hershey Country Club. There are two very good golf courses here. The older West Course runs through hilly country and presents a number of challenges; the new East Course, when stretched to its full 7,300 yards, is one of the toughest in the country. Accommodations are at the Hotel Hershey in Hershey, home of the chocolate bar. The hotel also has a fun nine-hole course immediately adjacent. The courses are open April through October, with rates about $30 each in double room with full American plan.

*Shawnee Inn and Country Club is located in Shawnee on the Delaware. Twenty-seven holes of exciting golf are set along the Delaware River and surrounded by mountains. There are points on the golf course where, like Washington, you can try to cross the Delaware—it takes about a 230-yard carry at the narrowest point. Shawnee, which was famous as a golf resort long before Fred Waring and a group bought it in 1944, is even more famous today, a place everyone should play at one time in their lives. The inn is a pleasant place to stay and has full resort facilities.

Tamiment Resort and Country Club. Tamiment, played from the back tees, is over 7,100 yards. This is one of the most challenging courses you'll ever see and one of the best that Robert Trent Jones ever designed. Seniors will enjoy this layout only if they play from the front tees. Tamiment has full recreation and golfing facilities.

Buck Hill Inn and Golf Club, Buck Hill Falls, boasts twenty-seven holes of exciting resort golf set in the rolling hills of eastern Pennsylvania. Long known as the Inn at Buck Hill Falls, this is one of the great resorts of America. The inn has every possible resort facility, including two or three miles of beautiful trout stream available to the guests.

Other well-known resort courses in Pennsylvania are: Bedford Spring Hotel, 6,730 yards Bedford; Downingtown Inn and Golf Club, 7,050 yards, Downingtown; Pocono Manor Inn and Golf Club, 6,400 yards, Pocono Manor; and the Sky Top Club 6,365 yards, Sky Top.

Mid-Atlantic States and the South

Maryland

Almost all the courses worth playing in Maryland and in nearby Washington, D.C., are private. One exception is the 6,580-yard course at the Hunt Valley Inn and Golf Club in Cockeysville. They have eighteen holes worth playing, and a new nine should be ready by the time you read this. Just nine miles north of Baltimore, this is one of the few hotels near a large city open to the public, offering full resort facilities. Martingham Golf Club, near St. Michael's on Maryland's Eastern Shore, is also well worth playing.

Virginia

Virginia, surprisingly, has very few resort courses worth playing. I may have missed one or two of the newer ones, but the only two I can strongly recommend are the Homestead, in Hot Springs, and the Golden Horseshoe, in Williamsburg.

Everyone should make an effort to spend at least one vacation at the fabulous Homestead, quite possibly the greatest resort hotel in the world—and that is covering a lot of territory. They have three extremely exciting golf courses. The Upper Cascades, set in a mountain valley just a short distance from the hotel, is one of the most beautiful courses you will every play. A mountain stream, stocked with trout, flows through it and comes into play on a number of holes. Truly, it is an unforgettable course. The two other courses next to the hotel are not quite as challenging, but will long be remembered. The hotel, of course, has every conceivable recreation facility.

The Golden Horseshoe at Williamsburg can be a shocker from the back tees. The 6,753-yard course, surrounded by huge trees, is superb and it is enormous. It has been said that this was one of the

Golf on the Golden Horseshoe at Williamsburg, Virginia.

few times that Robert Trent Jones was given all the acreage he wanted in order to build the course he wanted. There is also a par-3 golf course at Williamsburg, and a new eighteen is being built to give the guests an easier course as a relief from the tough Golden Horseshoe. Accommodations are at the adjacent Williamsburg Inn, which offers true colonial hospitality. A new eighteen-hole layout has just been completed, scheduled for opening in 1978.

You can golf year-round in Virginia. The Williamsburg Lodge golf course is an all-time favorite.

Safely out.

West Virginia

As might be expected, this small state is not strong on resort golf courses but it does have one, the Greenbrier. Having said the Homestead is possibly the greatest resort in the world, I would have to say that the Greenbrier might very well claim the same title. Located at White Sulphur Springs, this big resort hotel offers three outstanding eighteen-hole tests. The most famous is called Old White, but they are all a real pleasure to play. The Greenbrier has long been famous as the home of Sam Snead, who for many years was chief pro at the resort. For some reason he is no longer associated with the hotel. The Greenbrier's accommodations and facilities are truly outstanding and well worth a visit.

North Carolina

With North Carolina we begin to get into a different situation for the vacationing golfer. In recent years because there is so much more vacationland in this fine state, a number of resort and golfing communities have opened. For the vacationer, North Carolina is certainly prime golf country. It also should be remembered that whereas the private clubs near metropolitan centers such as Washington, D.C., New York, Boston, Los Angeles, and Chicago, usually do not welcome visiting golfers to private clubs, there are few private clubs in North Carolina that do not extend their hospitality to a visiting private club member from out of state. North Carolina also offers a variety of golfing terrain stretching from the seashore through the Piedmont Plateau and on up into the mountains in the western part of the state.

It is impossible to cover all the fine courses in this beautiful state, but a few of the outstanding ones need to be described. The golfing hub of North Carolina is unquestionably *Pinehurst, set in the sand hills amid the rolling terrain of the Piedmont Plateau. Once the property of the Tufts family, Pinehurst is a golfing community spread over many square miles and featuring some 15 courses that wind in and out among tall cathedral pines. By far the most famous of these courses at Pinehurst is Number 2, designed by the late Donald Ross, a Scotsman who left his native heath in Dornach as a young man to seek his fame and fortune. He eventually settled in Pinehurst, where he designed a number of courses and became one of the great golf course architects with monuments to his skill as a designer everywhere in America. Pinehurst Number 2 is perhaps his most famous creation, a sterling test of golf ability. It is long if you play it from the back tees, but above all it is demanding. You must keep the ball on the fairway or wind up on a pile of pine needles behind a towering tree or in one of the many strategically placed traps. When you finally arrive on the putting green, you are faced with large rolling greens demanding your utmost skill.

The other four courses at Pinehurst are well worth playing, but remember that the cost of maintaining this huge complex means that at certain times of the year a number of the courses will not be in top shape. This is especially true during the winter when the bent grass dies and the winter rye takes over. The five courses, as well as the old Carolina Hotel and Holly Inn, are now owned by the Diamondhead Corporation, which has modernized the hotel, built other recreation facilities, but still maintains many of the old traditions. Today it is open all year and there are a number of golf packages offered at the hotel as well as at condominiums leased for from several days up to a month. It is a very busy place, and the old guard may find some disappointments.

Nearby in Southern Pines there are three outstanding examples of Donald Ross's genius. *Pine Needles Lodge and Country Club, 6,900 yards, offers not only a Donald Ross course but an extremely attractive country club and lodge. Golf packages, which can save you substantially, are offered throughout the year.

Just a good 2-iron from the entrance to Pine Needles is the Mid Pines Club with another famous Donald Ross layout. Using an electric cart, play Pine Needles in the morning and Mid Pines in the afternoon, and see which you think is the better of these two really fine golf tests. Mid Pines has good accommodations and serves meals. There is a bar where you can get a mixed drink, something rare in this area, but Pine Needles also offers this service. Julie Boros, one of the world's great senior golfers, was for many years pro at Mid Pines.

The Elks Club in Southern Pines also boasts a Donald Ross course, and, believe me, it is well worth playing. They have a dining room and bar, but there are no accommodations. Incidentally, you don't have to be an Elk to play here.

*The Country Club of North Carolina in Pinehurst has three of the best nines you will ever play. Its layouts have been ranked in the top twenty by all the major golf publications. The club is completely private; you must have an introduction by a member in order to play here. By all means make the effort; you will never regret it.

There are few courses in our country that offer a true linksland for golf, similar to the wonderful courses along the Irish, English, and Scottish coasts. *Seascape at Kitty Hawk, North Carolina, is one of them. Set among the high dunes overlooking the Atlantic Ocean, it provides an unforgettable golfing experience. The clubhouse and facilities are not elaborate, but you can get lunch, mix your own drink, and rent motor carts as well as pull carts. There are no formal hotel-type accommodations at the course, but the pro shop will be happy to help you lease one of the beautiful homes that overlook various holes on the golf course. Prices for these are modest, and they are all furnished. Incidentally, there are plenty of other regular motel and resort accommodations in and around Kitty Hawk.

A mile or two away and just as close to the ocean is another fascinating course, *Duck Woods Golf Club. This is a private golfing community, but privileges are readily extended to visitors who belong to other clubs. They also have furnished homes of various sizes located in and around the course, which you can rent for a modest fee. The course itself is entirely different from the Seascape, in that although you are next to the ocean, you would never believe it. The course winds in and out of huge cypress forests interlaced with creeks that come into play on many holes. The sense of remoteness as you get out on the course is almost overpowering.

Moving into the Great Smoky Mountains of western North Carolina, the golfer invades an entirely different area. Here we are in majestic mountain country and the golf is entirely different, but just as breathtakingly beautiful. Seniors should remember, because of the altitude and the many hills on some of these courses, it is well to ride a cart, particularly when playing in the summer months. In western North Carolina, as in all other sections of the state, we are faced with an embarrassment of riches when making a selection; thus, we can mention just a few of the wonderful golf experiences available in this region.

Seniors may not be thrilled to learn that the Beaver Lake Golf Course, Asheville, has the world's longest par-5—690 yards—or that nearby Black Mountain has what may be the game's longest hole, a 745-yard par-6.

The following courses all have golf accommodations immediately adjacent: Beaver Lake Golf Course, 6,550 yards, Asheville; Beech Mountain Golf Club, 6,470 yards, Banner Elk; Blowing Rock Country Club, 6,090 yards, Blowing Rock; Seven Devils Golf Club, 6,300 yards, Boone; High Hampton Inn, 5,900 yards, Cashiers; Maggie Valley Country Club, 6,430 yards, Maggie Valley; Grandfather Golf and Country Club, 6,885 yards, Linville. The last is a private course but they usually offer reciprocal privileges; it's well worth playing. It is a shame not to be able to list more of the superb North Carolina courses, but once you play a few I'm sure you'll agree that this is golfing country preeminent. It will be an exciting search to ferret out the rest of them on your own.

South Carolina

South Carolina, although it can boast of having the oldest golf club in America, must concede that many of its fine resort courses are of a much more recent vintage than those in neighboring North Carolina. This is not to say that they aren't as good, merely that they are not, in many cases, as traditional. Because golf is so recent in South Carolina, some of the best courses are found in clusters, primarily along the coast.

With twenty-six resort courses all located within a thirty-mile driving distance of downtown Myrtle Beach, the so-called "grandstand area" could surely claim that it has more resort courses within a small area than anyplace else in the world. I know of no other place that is so completely golf oriented as this strip along the South Carolina coast. When you learn that there are some 36,500 motels, hotels, and other accommodations, plus 9,000 campsites in this

Golf on the Grand Strand, Myrtle Beach, South Carolina.

area, you may begin to wonder if this is the place you would like to spend that perfect vacation. The ideal time to play these courses is in the late spring or the early fall. However, despite the great number of courses, they are often crowded at this time. During the summer, since many people are on a family-type vacation, the courses are not necessarily that crowded but the weather is extremely warm. For my own part, I have a secret ambition: I would like to get three senior golf pals and come here in late February or early March when everything is uncrowded, stay at one of the beautiful hotels on the water when the rates are half-price, and play all twenty-six of the courses. Meanwhile, to sample some of the feast, here are a half-dozen courses in this region that are worth anyone's time and money.

The best-known course in the area is the Dunes Golf and Beach Club. This 7,000-yard layout was designed by Robert Trent Jones, and it is a fine example of his best work. The most famous hole, watery number 13, is 555 yards long, but you may cut off as much of the water as you feel you can carry. The aim is to cut off enough so you can reach it in two, but my understanding is that Mike Souchak is the only golfer ever to have reached it in two. Regardless, it is a fun hole to play as are many of the holes on this course so near the ocean.

The Dunes is a private club that offers playing privileges to guests at certain motels and hotels in the area. A good nearby motel that offers such privileges is the Caravelle.

Another fine course is Pine Lakes International Country Club, the oldest course in this area; a little farther inland, it is reminiscent of the courses in Pinehurst, with big pine trees and several small lakes that come into play to make the course a scenic challenge. A few rooms used to be available here, but that is no longer true. However, there are plenty of good motels in the nearby area.

Arcadian Shores Golf Club, designed by Rees Jobes, opened in 1974. It is an interesting 7,070-yard layout with many lakes that come into play.

Immediately adjacent is the Hilton Inn. Robbers Roost Golf Club, Quail Creek Golf Club, and the Surf Golf and Beach Club are other clubs in the area.

Teeing up at the Sea Pines Plantation, Hilton Head Island, South Carolina.

A world-famous hole on Hilton Head Island, South Carolina.

Moving south along the coast we now head for Hilton Head Island and its many courses, but on the way there are three tempting courses set along the Carolina shore. These are Seagull Golf Club at Corlies Island, Fripp Island Golf and Country Club on the coast east of Charleston, the superb *Seabrook Island course. Moss Creek Plantation,* on your left just before you cross the bridge to Hilton Head Island, has a great new course by George Fazio.

The Hyatt on Hilton Head Island at Palmetto Dunes.

At Hilton Head Island,* the most famous golf test is the Harbour Town Golf Links, at Sea Pines Plantation. Designed by Pete Dye and Jack Nicklaus, it has been selected by many experts as one of the 20 best courses in the world. Sea Pines also has two resort courses which, though not up to Harbour Town, are certainly a pleasure to play. Sea Pines Plantation offers a number of packages throughout the year that can save you money off the established

Walk out your cottage door onto the golf course.

rates. You either stay in villas at various points on their compound or at the Hilton Head Inn.

*Port Royal Inn and Golf Club, one of the first resorts on the Island, has two very good courses and full American plan accommodations in villas overlooking the sea.

*Hilton Head Plantation has the Dolphin Head golf course, a fine layout designed by Ron Kirby and Gary Player.

Another fine resort on Hilton Head Island is the well-known *Palmetto Dunes, which offers a fine course by Robert Trent Jones as well as a newer but highly praised George Fazio layout. Palmetto Dunes is also nationally known for its golf school under the direction of Bob Toski. Accommodations are taken care of by the new multimillion-dollar Hyatt Hotel as well as by its rental villas.

The rest of the courses on the Island are primarily community golf projects but they are open to the public and all are worth playing. Perhaps since it is an island, playing golf at Hilton Head gives a sense of remoteness that is not found on the courses in the Myrtle Beach area.

Georgia

Georgia, despite being the home of Bobby Jones, who did so much for golf in the United States and throughout the world, has only a few fine resort courses for vacationing golfers. Georgia is prime golfing country, in the area along the coast as well as in the western part of the state; it is expected that in the near future we will see a number of new resort courses as well as golfing communities in this beautiful state.

Resort golf in Georgia begins just outside of Savannah at the *Savannah Inn and Country Club, a testing golf course laid out in a lovely setting.

Moving south about seventy miles down the coast from Savannah, we come to Brunswick, where a short causeway connects the mainland to world-famous Sea Island. Sea Island Golf Club now has thirty-six holes of golf and every hole is worth playing. The course, set in a spectacular area overlooking the water, is not only challenging but presents the kind of variety that makes golfing a pleasure day after day. There is a fine clubhouse with full dining

A beautiful golf hole on one of Georgia's "Golden Isles."

facilities, a locker room with complete equipment, including driving carts, hand carts after 3:00 p.m., and caddies. Accommodations are at the Cloister Inn, one of the great resort hotels in the area, with full recreation facilities. Another golf layout on Saint Simons Island should not be missed; it is primarily a golfing community, but has accommodations in an inn and rented condominiums. Saint Simons Island Golf Club, with a well laid out Joe Lee course, has many water holes winding through live oaks and cathedral pines.

A few miles farther south along the coast is Jekyll Island, once the home of the famous Jekyll Island Club. Years back a few millionaires who came each winter would enjoy its seclusion and tropical beauty; now Jekyll Island is completely public. The golf courses are under the control of the Jekyll Island authorities who maintain them in excellent condition. They are reminiscent of the courses in the Pinehurst area with tall pines sheltering each hole and a great deal of sand and water. The Buccaneer Motor Lodge makes a good place to stay while playing these courses; they have a special golf package plan available most of the year.

To reach Georgia's last outstanding golf resort complex, we must move inland to a point just north of Columbus, on Route 27. Here in Pine Mountain is situated the famous Callaway Gardens. Built by the late industrialist, Casen J. Callaway, this resort offers thousands of acres of not only gardens, but quail hunting, fishing, hiking, and three eighteen-hole golf courses, as well as a nine-hole layout. The courses are well maintained and are challenging, with a good many of them sporting water holes with no sand traps and plenty of trees. It was the late Mr. Callaway's theory that sand traps slowed up play; as a result at Callaway Gardens you know where your ball is at all times—it's either playable or in the water. Callaway Gardens is really a sensational place with a lot of appeal for senior golfers. There are accommodations at a Holiday Inn with full facilities immediately adjacent to the course. Cottages are also available for rent, weekly and monthly, at quite reasonable rates.

Alabama

Before moving on to Florida and its myriad golfing opportunities let's take a quick look at the few resorts situated in Alabama and Mississippi, which are of interest to the senior golfer. (Keep in mind that I am describing resorts that have accommodations either on the grounds or immediately adjacent.)

Alabama is justly proud of its great resort, the Grand Hotel, at Point Clear overlooking Mobile Bay. This twenty-seven-hole golf layout is called the Lakewood Golf Club and its three nines are well maintained, the holes winding through tall trees. The Grand Hotel has a couple of senior invitational golf tournaments each year, and you might want to write them if you are interested in playing.

South of Point Clear and reachable by a toll-free four-mile bridge and causeway is Dauphin Island. Here is located Isle Dauphine Country Club, one of the most unusual golf layouts in the state. This 6,590-yard course is reminiscent of linksland links golf in the British Isles. One nine of the course winds along the shore of the Gulf; the other nine is set among giant sand dunes, water, sand, and loblolly pines. The course is not long, but on a windy day it's about as tough as any you would play. Accommodations are at a nearby Holiday Inn that has golf privileges at the semiprivate course.

Mississippi

Mississippi resort golf is primarily near Biloxi where the Edgewater Golf and Country Club and the Broadwater Beach Golf Club are two courses that offer a modest and enjoyable challenge to seniors. The famous old Edgewater Hotel is gone, but the Buena Vista Hotel guests have playing privileges at the Edgewater golf course. The Broadwater Beach Inn Hotel is immediately adjacent to the course. This resort hotel has excellent accommodations in the hotel as well as in cottages and every possible recreation facility. Set among thirty-three acres of lovely landscaped woods, it is definitely worth a visit.

Kentucky and Tennessee

Kentucky has many private golf courses and a few public ones, but so far as I know there is no resort golf setup at this time, where one would go out of his way to plan a vacation round. Much the same situation prevails in neighboring Tennessee, but one resort

You don't have to be a "Colonel" to play golf in Kentucky

course I can recommend highly is the *Inn at Cobbly Knob, near Gatlinburg. This layout designed by Gary Player wanders through a valley surrounded by mountains, and offers several fine golf holes. Most of the rooms in the inn have a stunning view of the golf course from their porches. There is a dining room and bar, as well as full recreational facilities.

Florida

A bewildering number of golf courses crowd this state, and they come in all sizes and shapes. In order to make a recommendation, I'm going to eliminate most of the courses located in golfing communities or developments, and concentrate on the true resort courses that have either hotel or motel accommodations at the course. Even here there are so many that it is possible only to make a selection of a few that I know will give satisfaction.

By and large one thinks of Florida golf as being played on flat territory. In many cases this is true, but throughout Florida, and particularly in the central part, there are many low rolling hills that provide ideal golf territory and give one the chance to play up, down, and sidehill lies.

Perhaps the simplest way to check on some of Florida's better golf resorts is to start at the top of the Atlantic Coast ending with a couple of courses on the panhandle and in the Tallahassee area.

Outside of Jacksonville is Amelia Island Plantation. Amelia Island is a recently opened resort with three golf courses designed by Pete Dye, who was the architect of the famous Harbour Town Course at Hilton Head Island. I have not played the courses as yet, but friends on whom I can rely say they are a real pleasure to play. There are full resort facilities here, including dining room, bar, and accommodations in the lodge and in rental condominiums.

Down the road a piece is the famous Ponte Vedra Club, Ponte Vedra Beach. Here there are twenty seven golf holes along the ocean with many lagoons, and indeed one hole completely surrounded by lagoon, the famous Island Hole. This course is open only to guests staying at the Ponte Vedra Inn.

St. Augustine has one of the better courses in Florida in the Ponce de Leon Country Club. For some reason, this layout is often ignored by golfers in their search for the better Florida courses. Perhaps the reason is that for many years it was a private course known as the St. Augustine Lakes and Country Club, owned and run by the Florida East Coast Hotel Company through the famous Ponce de Leon Hotel in St. Augustine. The Ponce de Leon, incidentally, is closed as a hotel and is being run as a girls' school. This course is similar to some of the European links running through salt marshes with many lagoons coming into play. In the old days many championships were played on these links. If you get a chance, stop by and try it; you will not regret it. There is today a 200-room lodge complete with dining room, bar, and a number of other resort facilities. It is run by one of the heirs of Henry Flagler, who founded the old Florida East Coast Hotel Company.

Before swinging much farther south down the coast, there are a few good resort courses in central Florida that must be mentioned. The Plantation Hotel Country Club at Crystal River has a Mark Mahannah course that features a number of water holes as well as the type of design where there are no parallel holes. There is a resort hotel immediately adjacent.

Walt Disney World Golf Resort, at Lake Buena Vista near Orlando, has three golf courses, each designed to give you a different type of golfing experience. Walt Disney World may not seem to be the ideal spot to spend a golf vacation, but if you come in the spring or fall, when things are not crowded, they are really very pleasant courses to play. There's plenty of first-class accommodations immediately adjacent, and much of interest to be seen in the area.

In the same region is Mission Inn and Country Club at Howie-in-the-Hills, Florida. This pleasant resort has one of the hilly golf courses we mentioned earlier. You will not believe you are in Florida when you play this one. Plenty of good bass fishing abounds on both the course's lakes and those nearby.

*Deltona Golf and Country Club in Deltona is one of central Florida's best courses. The hills and pines here are reminiscent of the Carolinas. Good bass fishing is available also. Accommodations are at the nearby Deltona Inn.

Placid Lakes Inn and Country Club, Lake Placid, is so named

because at one time it was the winter headquarters of the famous Lake Placid Club at the Adirondacks in New York. The course is 6,700 yards and well worth playing. Good fishing is basic to this area too. Accommodations at the inn are $29.50 per day per person on the golf packages that include breakfast and dinner as well as un-limited golf.

Two more courses that should be mentioned before leaving central Florida are the Cypress Creek Golf Club in Orlando and Arnold Palmer's Bay Hill Club and Lodge, also in Orlando. I don't know how often you're going to run into Arnie here, but the courses are superb, having been designed by the late Dick Wilson, and you will enjoy the whole golfing ambience. Introduction by a member is required.

Actually there are so many courses in central Florida that probably the thing to do is to pick one of these mentioned as a head-quarters, and then go out and play the others. The Chamber of Commerce in Orlando will provide you with a complete list and al-most all of them welcome play from visiting golfers.

Moving over to the coast, the *Sandpiper Bay Club at Port St. Lucie has two well-known courses—the Saints (6,100 yards) and the Sinners (6,428 yards). This, up until a short time ago, was the St. Lucie Hilton Country Club. Apparently there has been a change in ownership, but it is still run as a full resort with complete facilities. Neither of the courses is great, but they are both typical Florida courses. A good deal of water and sand, occasional trees, and proper trapping make them a modest challenge and very pleas-ant for seniors to play. There is also an eighteen-hole, par-3 course.

One of the prime senior golf situations is the Breakers at Palm Beach, now open all year. There are special golf package plans dur-ing the summer, which can save you a lot of money. The rates in the winter are by no means cheap but are reasonable for one of the great resort hotels of the world. A number of senior organizations run competitions at the Breakers during the winter months, and I'm sure that if you write to the hotel in Palm Beach, they will be happy to tell you where to contact these organizations for an invitation. (Also, see Ray McCarthy under listings of annual senior tournaments and competitions in the Appendix.) The Breakers has two courses for play, open only to guests of the hotel.

Three other courses worth investigating in the Palm Beach area are the Palm Beach National Golf and Country Club and the Fountains of Palm Beach. The Atlantis Country Club in Lantana has accommodations in the adjacent Atlantis Inn.

Below Palm Beach is another of the world's great golf resort hotels—the Boca Raton Hotel and Club in Boca Raton. There are four eighteen-hole golf courses available to guests staying at this famous hotel styled in the Spanish tradition. Its history goes back to the early days of Florida when one of the first owners rode around the course in a Rolls-Royce while playing. Tommy Armour taught at least a generation of golfers to play here many years ago, and he was followed by that great senior Sam Snead who for many years was head pro. As a senior, if I had to select just one place to play in Florida, it might well be the Boca Raton Hotel and Club.

The World of Palm Aire at Pompano Beach is a resort community with five eighteen-hole golf courses, four of them regulation, and one executive. This well-known resort offers every possible facility in addition to a now very famous spa.

When you get into the Miami area there are plenty of courses to play, but very few of them offer the getaway feeling that one looks for on a golf vacation. One exception is the Doral Hotel and Country Club. Despite the size of this outstanding resort and its four golf courses, it still provides a golfing experience that is almost unique. The most famous track is, of course, the Blue Monster, which we all enjoy seeing on television when the PGA Tour gets to Florida. The Doral courses are all noted for their lakes; in fact, there are very few holes where water is not a factor. Full resort facilities are at the hotel; golf packages offered at various times throughout the year can save you a considerable amount of money.

Just below Miami, you may take the McArthur's Causeway to Key Biscayne and the Key Biscayne Golf Club. This municipal course ranks as one of the best in the country and it is definitely among Florida's six or seven best golf courses. Though it is a municipal course, green fees run about $8 and this keeps some of the play down. The course winds in and out of the Mangrove Flats that edge the bay. There are some really sensational holes you will never forget. The clubhouse has full facilities including rental carts, pull carts, and occasionally caddies. There is a beautiful bar that overlooks the bay and Miami; there's also a fine dining area. Starting times are offered most of the year, so if you're going to play be sure to get one. There are ample accommodations on Key Biscayne Island.

Very little resort golf of any consequence is found in the Florida Keys. One exception would be the Ocean Reef Club at North Key Largo where there are two exceptional courses. However, the club is private and you need an introduction. They have been known to extend reciprocal privileges to members of certain private clubs, so it might be worth a call if you don't know a member. Accommodations here are excellent.

Coming up the West Coast, our next stop might be at Marco Island, which has two good courses and complete resort facilities at a large hotel on the water.

If you get into southwest Florida, you're bound to hear about a golf club called the Hole in the Wall. Though I've not had the pleasure of playing it, I understand it is one of the better courses in the area, and a refuge for seniors. The average age of the members is about seventy. They are suspected of having large bank accounts.

A good new course is the *Wilderness Country Club, also in Naples, which sports twelve lakes to make things interesting. They have condominiums to rent and a clubhouse with full golfing facilities. If you're looking for golf far from the madding crowd, a most interesting course is to be found on the small island of Boca Grande, which you can now drive to from the mainland. The course is owned by the Gasparilla Inn, and it is an extremely pleasurable layout to play. It winds through tall oaks with cypresses, and then along the water with dogwoods here and there, lakes everywhere—a very pleasant course. Boca Grande is also a good place to be in May and June if you're interested in some of the finest tarpon fishing in the world.

Another course on the West Coast that has the same getaway-from-it-all feeling is at the South Seas Plantation on Captiva Island. When I played it, there were only nine holes, but they are certainly interesting enough to play twice.

Palm-Aire Country Club at Sarasota has a good Dick Wilson course available to guests at the lodge. One of the really superb courses in the Sarasota area is Long Boat Key; guests at the Lido Biltmore have privileges here.

In the St. Petersburg area, there is the fabulous Belleview Biltmore Hotel and Country Club. This noted resort hotel, so reminiscent of a more graceful bygone era, has three eminently playable golf courses, which are now owned and managed by U.S. Steel, and they do a first-class job. The Belleview Biltmore is annually the scene of three of the more important senior invitational events. The hotel is open only from January to the end of April.

Just a little north of here at Tarpon Springs is Innisbrook Resort and Golf Club. Though only about seven years old, if I remember right, this beautiful place has achieved fame almost overnight and is one of the most pleasant places in Florida for a senior to spend a golf vacation. The terrain is heavily wooded and hilly and, of course, there's plenty of water. The three courses offer tests of varying degree from relatively easy to very, very tough. Accommodations are in twenty-four lodges, all of them situated on one or the other of the golf courses.

If we continue on up north we finally come to that area known as the Panhandle, west of Tallahassee, the state capital, to the border of Alabama near Pensacola, Florida. Golf in the Panhandle area is not nearly as crowded as in the southern part of the state and is, therefore, less expensive. The Bay Point Yacht and Country Club in Panama City Beach has an excellent Willard Bird Championship layout; although it's private, reciprocal privileges are often offered, particularly if you stay at the hotel immediately adjacent.

In Pensacola is the *Perdido Bay Inn Country Club, a very good layout in a lovely setting. This is a resort community, but there is a fine and comfortable inn with full facilities and you can rent condominiums.

This concludes our Florida suggestions. As mentioned earlier we are merely scratching the surface and only listing places that I'm personally acquainted with. Certainly there is room for adventure in course collecting among the many places I have not mentioned. A good way to do this is to send to Tallahassee, Florida, for a complete list of all the golf courses in the state. Pick a central point and then explore the many golf courses from there.

By and large, Florida golf is not expensive; at the very top you will probably pay $40 per day per person, two in a room, including in many cases your greens fees, two meals, and a number of other privileges. At some of the smaller resorts, the same package will cost $25 to $30 per day, again including greens fees, two meals, and possibly even sharing a cart.

The Midwest

In the midwestern part of the United States there are few resort courses as we know them elsewhere. It is a situation, particularly in the northern areas along the lakes and in the woods, where a great many extremely pleasant outdoor vacation spots have resort hotels that feature fishing, boating, canoeing, and, in many cases, have either a nine- or eighteen-hole golf layout. In most cases these will not approach championship caliber, but can be very pleasant to play when you're on a vacation and golf is not the prime object. There are, however, a few very good golf resorts and we will take a look at these.

Ohio

When you think of Ohio, you automatically think of Jack Nicklaus, although Ohio really has no resort courses where a dedicated senior might plan to spend a vacation. There is one golf situation worth looking into, and, as one might suspect, Mr. Nicklaus is involved with this particular setup. Kings Island, near Mason, Ohio, is a 1,600-acre amusement park with a number of different turn-of-the-century themes. It also has the Jack Nicklaus Golf Center, a championship course that hosts the Kings Island Open played in the fall, as well as an interesting executive course. The course is open to the public and there are plenty of accommodations immediately adjacent, including the Kings Island Inn, which is part of the complex, and the new Kings Point Hotel, completed in the spring of 1976.

There is considerable talk about Jack Nicklaus's new private club, Muirfield, which celebrates Jack's winning of the British Open in Muirfield, an important milestone in his career. The course in Dublin, Ohio, is in no way patterned after the famous course in Scotland, but obviously it embodies a great deal of Jack's strongest feeling about golf course architecture. The course is completely private and you must be accompanied by a member when playing.

Arkansas

Arkansas is hardly in the midwest, but we'll include it in this section anyway. One interesting resort setup for seniors in Arkansas is the famous Arlington Hotel and the Hot Springs Golf and Country Club immediately adjacent. There are two extremely good golf courses; the Arlington is the more famous of the two courses, but they are both well worth playing. There is also an executive-length course. Accommodations are at the Arlington Hotel, which is noted for its health spa and horseback riding.

Two other courses in the area at which golf privileges are available are the Belvedere Country Club and the Hot Springs Village Country Club. All in all, Hot Springs is an ideal place for a senior to spend a golfing vacation.

Illinois

Illinois has a number of private courses, but little or no resort-type golf. A senior who finds himself or herself on a business trip in the Chicago area need only look in the Yellow Pages under golf courses to find at least twenty places that extend privileges, many of which are outstanding layouts.

Indiana

Indiana has one famous golfing resort, French Lick. Here there are two courses, the Hill and the Valley, and as might be suspected, the Hill course offers a more difficult challenge than the easy but enjoyable Valley course. The huge Sheraton Hotel has complete resort facilities including 485 rooms.

Wisconsin

Wisconsin offers the Scotsland's Golf Club at Oconomowoc, a complete resort hotel with a summer theater and health spa. This

6,700-yard golf course features water and a stream on a number of holes. At Lake Geneva, Wisconsin, is the Playboy Club—Hotel Golf Club, which has two outstanding golf courses designed by Pete Dye. Full resort facilities are available at Bunnyville West.

Michigan

Michigan's golf is confined primarily to summer and early fall play, and it offers a number of good resorts that are well worth any senior's time. Most interesting, perhaps, is the Boyne Highlands Golf Course in Harbor Springs, where there are two courses. The one by Robert Trent Jones features gigantic tees that offer five different course lengths ranging from a maximum of 7,200 yards down to 5,600 yards. This is one place where you can't complain about the length of the course; you can choose whatever length appeals to you. The resort is one of the best in the country, offering every possible facility, with a luxurious Old English atmosphere. In Boyne Falls, there is another well-known golf resort, Boyne Mountain, with eighteen-hole Alpine Golf Club layout as well as a par-3 course. The resort features full recreational facilities in an attractive alpine style.

In Bay City is the Bay Valley Inn with a course designed by Desmond Muirhead, which runs up and down hill with many water holes. Complete resort facilities are at the Bay Valley Inn. In Bellaire, an outstanding resort, Shanty Creek, offers eighteen holes of golf in addition to all of the standard resort facilities.

Last we come to one of the most famous resort hotels in the world, the Grand Hotel at Mackinac Island. This unique resort in a stunning location does not really have an outstanding golf situation, but the hotel itself is so interesting that every golfer with imagination owes it to himself to visit it at least once. There are nine holes of pretty fair golf adjacent to the hotel, and another nine holes at a municipal course.

The West

The term "West" can cover a lot of territory and that is exactly what we are going to do. We will start with Texas and work our way up and down and across to the West Coast, finally finishing up in Hawaii. A lot of the courses in this area are quite different in character from those in the eastern part of the United States. Courses in the Southwest are usually barren of trees. A lot of them are in open sun country where wind is a factor, and water is used over and over to provide difficulties for the erring golfer.

Texas

Texas has a number of private golf courses, but unfortunately courses of the resort type are few and far between. The state has so much wonderful land for golf that more resort courses will surely sprout up in the near future. For the moment, however, your first choice may as well be Pecan Valley Golf Club in San Antonio. This club is private, but privileges are extended to members of other private clubs. It is a challenging course and was the site of the 1968 PGA tournament. San Antonio is one of the most historical and interesting cities on the continent, and as such is well worth a visit on its own merits, to say nothing about the fine golf to be found at Pecan Valley. There are no accomodations offered directly on the course, but downtown San Antonio is loaded with good places to stay and is only about fifteen minutes away. Hotel Menger is a good choice; not only is it historically interesting but it offers golf packages in conjunction with Pecan Valley.

Brownsville, on the Mexican border, boasts that it has the "southernmost golf course" on the mainland of the United States, and this may be so. In any event, Brownsville is a resort town that enjoys excellent weather all year. There are two resort courses of more than casual interest: the Valley International Country Club, two meandering eighteen-hole courses with water on a great many holes, as well as a nine-hole par-3 layout. Fairway cottages and an inn offer good accommodations. The Rancho Viejo Country Club has a golf course cut through a grove of citrus trees. An adjacent inn and fairway cottages offer accommodations.

*Waterwood National in Huntsville has a Peter Dye course built in the Scottish tradition. It was carved out of the East Texas pine forest. Lake Livingston, one of the South's best bass lakes, comes into play on a number of holes, and, at one point, you have a 200-yard carry to a peninsula-type green. There are complete resort facilities with accommodations in cabanas or lodge rooms.

Galveston Island is an area that has three or four golf courses in a subtropical atmosphere and a number of hotels that offer packages. The famous Flag Ship Hotel on Galveston Island has golf packages.

Laquinta Royal Motor Inn in Corpus Christi has an eighteen-hole golf course adjacent, as well as an executive nine-holer. They offer inexpensive golf packages during the off-season.

Wyoming

Wyoming has two resorts worth investigating. The Jackson Hole Golf and Country Club, in Jackson, was rebuilt by Robert Trent Jones. Surrounded by the high mountains of the Grand Teton Range, Jackson is a resort center with many excellent motels and hotels. The Old Baldy Club in Saratoga has an eighteen-hole course located in this exclusive resort. I have no firsthand knowledge of the quality or character of this course.

Colorado

Colorado is ideal golfing territory during the summer months; the golf in the mountain area is cool, pleasant, and bug-free. Though some of the courses are set in valleys, seniors golfing in the mountain states would be well advised to get a motorized cart. It is not only a question of walking up and down hills but also of altitude. Many of the courses in Colorado are set from 7,000 to 9,000 feet above sea level, and this can be a strain on the heart, while it also adds 30 yards to your drives.

One of the most famous golf resorts in the world is the Broadmoor at Colorado Springs. The Broadmoor actually has two golf layouts. The East Course is perhaps the best known, having been used as a "venue" for many championships. It was designed by Donald Ross and then redone by Robert Trent Jones a few years back. The West Course is considered the sportier course and is probably more fun to play, with many memorable par threes. A new course that may be open when you read this was designed by Arnold Palmer. The Broadmoor has a number of senior events during the summer; if you are interested, write them for information. The Broadmoor sits on 5,000 acres and has every possible recreational facility.

Colorado has some of the world's most beautiful courses.

Vail is nationally known as a ski resort, but it also has a good golf course that seniors will enjoy playing, the *Eagle-Vail Country Club. A new course, designed by Vonhagge and Devlin, it features a number of natural water hazards as well as elevated tees. There are ample accommodations in the nearby resort village of Vail.

*Tamarron is a relatively new development in Durango, set up by the same company that built Innisbrook in Florida. There are condominiums for rent and a lodge on the golf course.

Steamboat Village Inn, at Steamboat Springs, is another place where skiing is more popular than golf, but the inn has a Robert Trent Jones course set among lovely scenery and where natural water comes into play on many holes. The resort has all types of recreational facilities.

New Mexico

New Mexico is just getting started as a golf state and I'm sure that many new courses will be built there in the future. Today its most famous golf course is at Cloudcroft, adjacent to the Cloud Country Lodge. This is reputed to be the highest golf course in the United States with a number of points on the course over 9,000 feet high. Seniors interested in collecting unusual courses should try this one. The lodge is a pleasant place to stay.

Utah

Utah is also just getting started with the golf boom. The Wasatch Mountain State Park Golf Course, near Heber City, has three nine-hole layouts set in a valley between the mountains with a number of lakes and trout streams running through the courses. A lodge on the course offers full accommodations.

Idaho

Idaho's chief interest for senior golfers would be at Sun Valley, long famous as a ski resort established by the Harrimans many years back. It does have an eighteen-hole course open during the summer and, in addition, a nine-hole short course. The world-famous resort has every facility and is well worth a visit.

Arizona

Arizona has long been popular with senior golfers. The excellent dry weather provides an ideal climate for our aches and pains as we get older. There are only a relatively few days in the southern part of Arizona when one cannot play golf. Many golfers think of desert courses as being flat and uninteresting. This is certainly true of some layouts, but southern Arizona has some courses with rolling fairways and trees, which are a reminder of Eastern clubs.

Golf in Arizona is clustered in three areas: Phoenix, Tucson, and, above all, Scottsdale. There are some seventy courses in the state that welcome public play, but I will confine my selections to those places that have accommodations immediately adjacent. The

Arizona Biltmore in Phoenix is probably the best-known resort hotel in the state and indeed one of the most famous in the West. Designed by Frank Lloyd Wright and open since 1929, this resort offers all kinds of recreational facilities, including tennis, horseback riding, skeet and trap shooting, and an absolutely magnificent swimming pool. The course is a typical desert one surrounded by high mountains. Since it is very flat and often hard, it is certainly not a difficult course to play.

A number of other resorts are situated beneath famous Camelback Mountain or actually on its slopes. If you stay at the Paradise Inn or the Jokake Inn, you have a commanding view of this mountain. Guests at the Jokake play at the Paradise Inn Valley Country Club. If you stay at the Paradise Inn, you have the privilege of starting your day's round from your hotel room at whichever hole is closest to your room. The golf house will deliver your cart to the appropriate tee. The Paradise Inn course is a combination of a number of pedestrian holes combined with three or four sensational ones, including a 135-yard hole that is exactly 135 feet high. Another hole has a drop of 200 feet before the ball hits the fairway. There are full resort facilities at the Paradise Inn.

Nearby is Camelback Inn, a desert course in luxurious surroundings. This, incidentally, is located in Scottsdale, a suburb of Phoenix. The Marriott's Camelback Inn is a well-run resort with first-class facilities, including heated pools, hot therapy pools, and luxurious rooms. There is an eighteen-hole par-3 course.

Almost next door to Camelback Inn is Dell Webb's Mountain Shadow, another resort hotel with all of the amenities. The golf course is only 2,900 yards long, but it is a shrewd test of golf and a lot of fun to play. The surrounding scenery is nothing short of sensational. The sand in its traps is said to have been imported from Capistrano. If that isn't carrying "coals to Newcastle," I don't know what is.

Scottsdale Inn and Country Club is another resort with a course that seniors enjoy. In the Phoenix-Scottsdale area are two other courses that are worth a visit. The first of these, the San Marcos Hotel Course, is located in Chandler. The Wigwam in Litchfield

Park is another first-class resort offering the ultimate in luxury in beautiful surroundings and every facility for outdoor sports. Golf is on three desert golf courses of varying degrees of difficulty. This is a true desert oasis with plenty of water holes being featured. The course is owned by the Goodyear Golf and Country Club, which, though private, extends privileges to guests at the Wigwam. There is a private-club atmosphere about this whole setup. If you come, just make sure you're not driving a car with Goodrich tires.

To me, the first choice for an Arizona golfing vacation would be the Skyline Country Club in Tucson, situated on the side of the lovely Santa Catalina Mountains. Guests at the Skyline Inn are presented with stunning views of the city of Tucson in the valley below them. The course itself is as interesting as any you would play in the Southwest, with a great deal of privacy on some of the holes, water judiciously used, and, of course, the beautiful mountains always in the background.

The Tucson National Golf Club is more of a private club than it is a resort; however, properly introduced members of other private clubs are welcome to play and to stay at its comfortable motel-lodge accommodations on the course. The clubhouse itself has an old-fashioned locker room and bar facilities. Food is good, too. Tucson National is famous as the annual site for the Dean Martin Tucson Open. The course, to my mind, is less interesting than the one at Skyline, but it certainly will have its appeal for many seniors.

Nevada

I wonder if anybody ever went to Nevada just to play golf. Most folks go to gamble or to get a divorce, but Nevada does have some fair golfing opportunities. Lake Tahoe, at an elevation over 6,000 feet, offers two of the most beautiful courses to be found anywhere. Most of us think of Lake Tahoe as being in California, which, of course, the western part of it is; however, the two courses mentioned here are on the Nevada side. *Incline Village Golf Club is in Incline Village, a resort development along the shores of Lake Tahoe. The 6,800-yard Robert Trent Jones course offers a variety of complications and beautiful conditions, and is a joy to play. Nearby at Stateline, Nevada, is the Edgewood Tahoe Golf Club, a course designed by George Fazio.

A few years back, if you wanted to go golfing in Las Vegas, you could do it very inexpensively. The theory many hotels operated on was to charge relatively little in order to encourage you to come to the area and spend your money at the gaming tables. This, however, is now a thing of the past; hotel prices are in line with well-known resorts throughout the country. The courses in the area are not tremendously exciting, but if you're here for other purposes, play them. If you play in the early morning, you usually have them to yourself.

Resorts with golf courses on the premises are Dunes Hotel and Country Club, Desert Inn Country Club, and Sahara Nevada Hotel and Country Club. The Paradise Valley Country Club may well be the best tournament course in the area. It was the site in 1973 of the United States National Senior Open Golf Championship. The Tropicana Hotel and Country Club also has a course that is as good as any in the area. Bandleader Louis Prima has a resort course here called Fairway-to-the-Stars, which is reputed to have some interesting novelties. Much the best course in Reno is Robert Trent Jones's Lake Ridge Golf Club; there are no accommodations directly on the course.

Washington

In the western part of Washington virtually year-round golf is enjoyed on courses as green as anywhere in the world. This is due to a climate sheltered by the mountains to the west and warmed by coastal currents. Although the rainfall is frequent in the fall, it is mainly a misting type of rain that is gentle to both the golfer and the courses. At the moment there is not a great deal of resort-type golf in Washington. There are a number of public courses and some private clubs that extend reciprocal privileges, but it is not a state where one would normally plan a golf vacation. Alderbrook Inn, in Union, is a pleasant resort hotel that offers a golf course worth playing. Ocean Shores, at the southern end of the Olympic Peninsula, has been converted into a seashore resort community in the last few years. It has an eighteen-hole golf course in a scenic location and a number of very good motel-type accomodations.

Oregon

Oregon is much the same as Washington, with golf virtually year-round near the coast, but just a few resort-type courses. The most famous is Salishan Golf Links at Lincoln City. This handsome resort overlooking the Pacific is one of the best resort links you will play. Hewn out of the rocks along the coast, it features rolling, winding fairways hemmed in by shrubs and evergreens. A number of holes border the ocean and offer stunning views. Salishan, open year-round, is something quite unique and well worth a visit.

The Village Green at Cottage Grove is another outstanding resort offering a variety of recreation facilities. Nearby is the Hidden Valley Golf Course, a public course not to be missed. Just a short drive from downtown Portland in the Mount Hood National Forest is Bowman's Mount Hood Golf Club. This resort offers a variety of accommodations, most of them overlooking the golf course. The course itself is laid out in an alpine meadow and offers a number of holes in which lakes and streams come into play. Many of the holes are surrounded by gigantic fir trees. The view of Mount Hood is always in the background.

California

California may not have quite as many resort courses as Florida, but no state can offer the variety to be found here. There is seaside golf, there is mountain golf, and there is desert golf, with lots of variations in between. California is the kind of golfing territory that offers touring seniors an opportunity to sample an infinite variety of courses.

The Silverado in Napa is an outstanding Fred Harvey resort set on 1,200 attractive acres. There are full resort facilities and thirty-six holes of golf. Both courses are extremely good golfing tests. The Kaiser International Open is played here in October.

San Francisco boasts one of the best municipal courses in the country. Harding Park is a slightly rolly, tight course partly encircled by Lake Merced, with incredible flora. Vacationers should play it during the middle of the day when it is uncrowded.

One of the most famous courses in the country is the Olympic Club at Lakeside, which has two exceptional courses. The Lake Course is famous for unusual happenings in the U.S. Opens that were held there. It was here that Billy Casper came from seven shots back to tie Arnold Palmer and later beat him in a playoff. A few years before, the great Ben Hogan was beaten in a playoff by an unknown golfer named Jack Fleck. The Lake Course wanders around Lake Merced and is surrounded by enormous firs. The Ocean Course, as befits its name, is much more open and the wind comes into play here. The club, although private, usually welcomes visitors from private clubs outside the state.

The San Francisco Golf Club has a very good course but you won't find it easy to play here. Guests must be accompanied by a member.

Some ninety miles down the coast we come to the famous Monterey Peninsula. Golf on the Monterey Peninsula is quite different from golf in the adjacent Carmel Valley. It is hard to imagine a more rugged coastline than that of the Monterey Peninsula with the courses seemingly just clinging to the rocks as they are exposed to the caprice of the winds. Here the land comes down to the sea and most of the courses feature views of the ocean with impenetrable roughs of ice plant. When you stray off the fairway, there is just no recovery; take the penalty. On the inland side you will find completely shut-off holes surrounded by firs and cypress.

There are about ten notable courses to play in this region, but with the exception of Carmel Valley and Pebble Beach, there are no resort hotels directly on the golf courses. This is no problem in the Monterey area, as the whole region is filled with lovely places to stay and all the golf courses are open to the public with the exception of the very private Cypress Point.

Carmel Valley at Carmel offers an almost desert type of play. The two courses have quite a number of exotic trees and are incredibly lush. Anyone visiting the area will want to play all the courses, so here is a brief listing: Laguna Seca Golf Club, Monterey; Pebble

Beach Golf Club, Pebble Beach, which is of course, on the Monterey Peninsula where one can stay at the Del Monte Lodge directly on the golf course. Del Monte Golf Club, Monterey, is the oldest of the Peninsula seacoast courses; the Del Monte Hyatt House is near the course. Spyglass Hill Golf Club, Pebble Beach, is considered the toughest of all the Peninsula courses. Designed by Robert Trent Jones, it is one of the courses used in the Bing Crosby Pro-ams as are Pebble Beach and Cypress Point. Monterey Peninsula Golf Course is another fine course adjoining Spyglass Hill.

There are two principal courses in the Carmel Valley. One, the Carmel Valley Golf and Country Club, is a rather flat course, quite pretty and beautifully maintained; there are water holes, and despite what seems to be an easy-looking course, it just isn't so, as the seniors found when the USGA played their championship here in 1975. Quail Lodge, rated one of the best small resort hotels in the country, is immediately adjacent and guests have the privilege of playing the Carmel Valley course. Rancho Canada Golf Club, also in Carmel Valley, has two eighteen-hole layouts. They are on the flat side and not extremely difficult, but still a pleasure to play.

Driving south along California's magnificent Pacific Coast highway, we come to San Luis Obispo and here at Avila Beach is the San Luis Bay Inn and Golf Course. This lovely hotel overlooking the bay has excellent accommodations and resort facilities. The golf course is extremely interesting, part of it running along the valley with occasional long carrys over water, and another part running into the hills surrounded by apple trees and beautiful flora.

On your way to Avila Beach, you pass, if you're driving along the coast, the Hearst Estate at San Simeon, certainly one of the most interesting attractions in the entire state. If you are visiting here, the Morro Bay Golf Course, which overlooks the sea and has a view of San Simeon, is a real test. The Golden Tee, opposite the golf course, is a good place to stay.

North of Santa Barbara is a corner of Denmark, the town of Solvang, which is a picturesque-style community including four windmills. Since many Danes have settled here, this is an excellent place to sample Danish pastries. Also located here is one of the most famous guest ranches in the country, the Alisal, which in recent years has added an eighteen-hole golf course. Incidentally, the 10,000-acre working ranch has up to 3,000 head of cattle.

A short jog inland from Santa Barbara is the Ojay Country Club, which has a beautiful golf course set in the valley surrounded by high mountains. The luxury accommodations offer air-conditioned rooms in the inn as well as in seventy-two cottages.

The senior golfer in search of the ideal golfing vacation would do well to bypass Los Angeles. There are plenty of municipal courses in this area, but as might be expected they are usually crowded. In addition, there are a number of private clubs, but these all require an introduction and in some cases the accompaniment of a member. In other words, no resort situation exists in the immediate Los Angeles vicinity.

On the other hand, San Diego County is a great golfing center. This is the kind of place where it is best to spend at least a week at one of the resorts or the many fine motels in the area and play a selection of the seven courses offered. One of the best-known courses in the region is Rancho Santa Fe, long the scene of Bing Crosby's annual clambake before it was moved up to Pebble Beach; among the best courses in the state, it is a "must" for anyone visiting the area. There are accommodations at the Rancho Santa Fe Inn and in a number of cottages scenically located on the course.

Now that the Crosby has moved north, the most famous course in the area is Torrey Pines, where the annual San Diego Open is played. A portion of this layout wanders along the Pacific Ocean, and wind is a big factor. There are actually two courses, and both are extremely good, but it is the South Course that is seen on television during the Andy Williams San Diego Open. There is an inn with full resort facilities located adjacent to the golf course. The Torrey Pines Inn and Golf Club is at La Jolla.

In the region north of San Diego is the Apple Valley Inn, a pleasant resort for families, with eighteen holes of scenic golf that most seniors will enjoy.

El Cajon is a golf center by virtue of two country clubs: Cottonwood Country Club, which has two eighteen-hole courses, and the Singing Hills Country Club, which has two eighteen-hole courses and an eighteen-hole par-61 short course. Courses are primarily for guests staying at the Singing Hills Lodge. Also in the area is the famous La Costa Spa. This much publicized resort at Rancho La Costa has a Dick Wilson course, as well as complete resort facilities in the ultimate of luxurious surroundings.

At Escondito is the Rancho Bernardo Inn and Country Club, which has two courses in a lovely valley and a resort with full facilities. Here also is the Circle R Golf Resort.

In San Diego itself is the Stardust Country Club, which is private, but guests at the Master Hosts Inn have privileges. This by no means exhausts the possibilities in the San Diego area, but it will be enough to get you started. The San Diego Chamber of Commerce has a free list, with descriptions of all the courses in the area.

A good approach to Palm Springs, our last golf center, is by way of Route 79 with a stopover at Warner Springs. Warner is one of the state's older resorts, having been established in 1900. It has 3,000 acres with excellent accommodations, and a good eighteen-hole golf course at your doorstep. Farther along on 79 going north there is Gilman Hot Springs and the Massacre Canyon Inn. This resort sits on 500 acres at the foot of the San Jacinto mountains. There are twenty-seven holes of resort-type golf. The course, though flat, is surrounded by trees, and a couple of lakes add interest to the play.

Palm Springs has something in the neighborhood of twenty golf courses, none of them more than a few miles apart. Two or three of the better ones are completely private; however, the rest of them, even if they are private, usually extend golf privileges to visiting members of private clubs. Three of the better courses have accommodations adjoining, and guests are permitted to play. Just a reminder: Palm Springs enjoys an ideal climate except during the summertime, when the heat is so intense that all golf must be confined to the very early morning or late evening.

The Canyon Hotel offers accommodations overlooking the North Course, one of the best golf courses in the area. They have another fine layout in the South Course, but this one is primarily for owners of condominiums in the development.

Mission Hills Golf and Country Club in Rancho Mirage has what seems to me the best golf course in the entire Palm Springs area. It was laid out by Desmond Muirhead, and resembles a links course in the British Isles more than anything else. It is long, windswept, and water comes into play on a great many of the holes. From the back tees, it is an absolute monster, but few ever play it from there. Each year it is seen on television as the site of the Colgate Dinah Shore Tournament. There are condominiums for rent and guests at the Palm Springs Spa Hotel may play here.

Bermuda Dunes Country Club has twenty-seven holes of typical desert golf with the course meandering over rolling sand dunes. The club, though private, is open to members of other out-of-state private clubs. Accommodations are on the premises.

Indian Wells Country Club is seen on television each year during the Bob Hope Classic. It also accepts play from out-of-town club members and has accommodations at the course.

La Quinta Country Club is one of the really outstanding desert courses in the area. Laid out at the foot of lovely mountains bordered with lush tropical plants, it is extremely attractive and yet provides a severe test. The clubhouse is one of the most beautiful in the whole desert area. La Quinta is private, but if you stay at the La Quinta Hotel with its cottages you will be given guest privileges.

The final suggestion before we leave California is the Catalina Island Golf Club located on the island of Santa Catalina reached by daily boat service from San Pedro, a port of Los Angeles. This course, built in 1895, is one of the earliest in southern California; though only nine holes, it is double-teed to give a variety of play the second time around. The course is not long, but is scenically exciting with extremely narrow fairways and many doglegs. Las Casitas has thirty-four cottages opposite the golf course with full resort facilities.

Hawaii

Hawaii offers some of the finest golf in the world and play is year-round. You must make your reservations well in advance if you want to stay at one of the famous resorts that offer an outstanding golf course. There are as many as seventy courses on the five islands; many of them are public and can be crowded. There are a number of private clubs that require an introduction by a member or in some cases extend reciprocal privileges to members of private clubs in other states. However, if you are going to enjoy the golf, it is essential to stay at one of the resorts. This means making your plans well in advance, especially during the winter months.

Hawaii, the largest island in the chain, is called the Big Island. At Kona is the Mauna Kea Beach Hotel, one of the great resorts of the world. Situated along the oceanfront with a lovely beach this stunningly beautiful golf course was designed by Robert Trent Jones. The hotel also offers hunting, riding, and fishing. Rates are by no means modest, but if you want the best in the islands, this is it. When you play the course, be sure to use the front tees.

Other clubs on Hawaii worth investigating are Keauhou Kona Golf Club, which features black lava rough; and Seamountain Ninole Golf Club in Pahala, a new course with lovely ocean views in a resort development. Another resort development is Waikoloa Village Golf Course near Kamuela, a new Robert Trent Jones course in a resort community.

Oahu is the capital island and to my mind it has been over-promoted to the extent that most of the courses are mobbed. However, the Makaha at Makaha has two good golf courses with Bermuda greens on one and bent greens on the other. A little secluded, these courses are scenically situated on the lower slopes of the Waianae Mountains.

Kuilima Golf Club at Kahuku has a George Fazio course in a new Dell Webb resort complex.

On the island of Maui is the Royal Kaanapali Golf Club at Kaanapali Beach, another outstanding course designed by Robert Trent Jones, with lots of water, trees, rolling terrain and, of course, wind—truly a spectacular situation. They also have a 4,300-yard executive-length course. A new resort course on Maui is the Golf Club at Kapalua; this course was designed by Arnold Palmer and takes advantage of its situation along the ocean to offer fine views and a tough test any time the winds blow.

Golf Vacations for Seniors: Foreign

Canada

For some reason we in the United States seldom think of planning a golf vacation in Canada, except for such outstandingly publicized golf courses as Banff in the Canadian Rockies. This is a mistake because for many of us who live in areas of the United States where it gets hot during the summer, a Canadian golf vacation can provide pleasantly cool golf in beautiful surroundings and on courses that are in peak condition. There aren't many such resorts in Canada, but the few I am about to mention will prove rewarding for any senior who would like to travel north of the border.

The Maritime Provinces have several interesting courses with the Keltic Golf Course on Cape Breton Island in Nova Scotia the most sensational. This stunning golf course is set along the highlands of this island overlooking the sea. The official name is Cape Breton Islands Golf Links. The course, laid out by architect Stanley Thompson, is owned by the Canadian government and is situated in the Cape Breton Island National Park. Each of the holes has great definition and most are separated by tall trees. The views of the bay are magnificent, but the course is really one of the most difficult on the continent and seniors playing it should be in no rush to finish a round. Enjoy the magnificent scenery, and if things get too rough,

One of Nova Scotia's fine courses.

head back to the clubhouse. There are excellent accommodations immediately adjacent in the Keltic Lodge.

At the other end of the island of Nova Scotia is a famous course called the Digby Pines; not nearly as long as Keltic or as difficult, it does wind through the familiar tall pines and it is a course where many seniors will want to take a motor cart in order to enjoy their outing. Next door is the Digby Motor Hotel, which offers good accommodations.

Pine trees guard a New Brunswick green.

Green Gables on Prince Edward Island, one of Canada's best courses.

Another famous resort, St. Andrews-by-the-Sea, in New Brunswick, is reached easily by way of Calais in Maine over Route 1. St. Andrews has two courses owned by the Canadian Pacific's Algonquin Hotel. The Hotel Course is the more difficult of the two, and it offers lovely views of adjacent Passamaquoddy Bay, famous for tremendous tides that swing as much as 30 feet on each tidal change. The courses are a must for Donald Ross collectors. Seniors will find the course tough enough at 5,800 yards rather than at its full 6,400 yards. Adjacent is the Algonquin Hotel, a full resort hotel with all facilities.

The tiny province of Prince Edward Island offers a course called the Green Gables Golf Course, named after the classic novel, *Ann of Green Gables,* because the author lived here. The cottage, which overlooks the fourth hole, is a great tourist attraction. The course itself has nine holes inland surrounded by the inevitable towering pines, and then moves on to a great variety of seaside links terrain, a fearsome test when the wind blows, which is often. There are good accommodations immediately adjacent.

The time to enjoy Quebec's resort courses is from June to November, with September and early October a wonderful time to experience the Canadian autumn. There are a number of resorts in the Laurentian Mountains, just a short drive from Montreal, but easily the most famous is the Gray Rocks Golf Club at St. Jovite. This beautiful course combines rolling hills surrounded by high pine trees with the peaks of the Laurentians all around. Gray Rocks Inn, on the course, offers excellent accommodations.

Eighty miles west of Montreal and situated on the north shore of the Ottawa River is one of Canada's most interesting golf opportunities. This is the course of the Chateau Montebello, formerly one of the most exclusive resorts in the world, the Seignory Club. The course, set in a valley amid pine trees, is an extremely pleasant one with a number of water holes. The hotel, owned by Canadian Pacific, is one of the most complete resort hotels in Canada and the accommodations are outstanding. The Chateau is open year round and on its many thousands of acres offers hunting with guides in the fall and fishing for brook trout in many remote lakes in the spring, summer, and fall.

Eighty-five miles west of Quebec City, high on the shores of the St. Lawrence River, is Canada's most famous resort, The Manoir Richelieu overlooking Murray Bay at Pointe-Au-Pic. This majestic resort hotel with two thousand acres of beautifully landscaped grounds has a fine golf course which, though not long, is extremely rugged and hilly; so steep are some of the hills that at two points escalators are needed to get you to the next tee. During the

Golf in Ontario, Canada, with Lake Muskoka in the background.

summer the Manoir also offers trout fishing in remote lakes that have
separate accommodations for anglers. This certainly is a resort you
must visit at least once in a lifetime.

Ontario has many superb golf courses but none in a resort

Enjoying a round at Falcon Beach Golf Course in Manitoba, Canada.

situation where a senior would spend a serious golf vacation. The famous London Hunt and Country Club in London, Ontario, is one of Canada's outstanding courses, and on occasion extends privileges to visiting golfers. It is well worth the effort to play.

The Canadian Rockies overlook this course in Jasper National Park.

Manitoba, like Ontario, has several courses that the tourists can play, but none around which a golfing vacation would be planned.

Alberta has one the most spectacular of Canada's golf courses—the Banff Springs Hotel Golf Club at Banff. This course is another of Stanley Thompson's creations, and is unforgettable. Part

A hot day in British Columbia.

of the course is literally torn out of the mountains. Although there are many memorable holes, perhaps the most outstanding is the famous par-3, 178-yard sixteenth, called the Devil's Cauldron. Here, from an elevated tee, the golfer must carry across a lake, with winds an important factor. The Banff Springs Hotel has every facility a resort hotel should have. Nearby is the famous Chateau Lake Louis, also a Canadian Pacific Hotel and a beautiful place to stay.

Just a few hours north of Banff is another superb resort hotel at Jasper, the Jasper Park Lodge. The course here is as good as any in Canada and is famous for its friendly bears that have been known on occasion to retrieve balls from shallow lakes and trees. If you are in this area, you will not want to miss staying at least a couple of days at this unusual golf course.

Golf in British Columbia is virtually a year-round affair especially in the Vancouver area. Most of these courses are private or semiprivate, and all extend playing privileges to visitors. If you are visiting that lovely city, Victoria, there are two golf courses well worth playing. The Victoria Golf Club, established in 1893, is British Columbia's oldest course and is characterized by a good deal of water and many bunkers. The Uplands Golf Club has been the site of the Canadian Senior Championship and, though not long, is an interesting challenge. While in Victoria I would recommend your staying at the Canadian Pacific's ivy-clad Empress Hotel, which you should enjoy very much. British Columbia's only real resort/golf situation is at Fairmont Hot Springs where the Fairmont Hot Springs Resort offers excellent facilities, as well as a very good eighteen-hole golf course. Finally, although the course is only nine holes, the Harrison Hotel, in Harrison Hot Springs, another of Canada's great resorts, offers a memorable challenge. This is another spot in Canada where you can combine wonderful fishing with your golf.

Bermuda

When you decide to plan a golf vacation away from our native shores, the major factor has to be weather. For the most part, tropical golf may not offer the type of challenge that our courses provide on the mainland. However, where the breezes are balmy and the sun is hot, for seniors fleeing the cold winter the golf has to be good.

Approximately 640 miles off the North Carolina coast lies Bermuda with some very pleasant courses to play at almost any time of the year. Bermuda's weather during the winter months may at times be a little less than ideal, but to my mind this is more than compensated for by the uncrowded situation in the island during these months. Bermuda's most famous golf course is definitely the Mid-Ocean Club at Tucker's Town, and it is probably the best course on the island. Unfortunately, the club is private, and although you need not be accompanied by a member, you must have an introduction by a member. There is a large nonresident membership and if you search among your acquaintances you may find that some of them are members of Mid-Ocean. The club has excellent accommodations, dining room, and bar on the course.

The Belmont Hotel Golf and Beach Club at Warwick has a very pleasant, not overly long course that is a joy to play.

The Castle Harbor Golf Club at Tucker's Town is an up-and-down-hill affair with sensational views of the ocean, but you'd better keep it down the middle; anything off the fairway can roll for what seems miles in the wrong direction. The hotel on the course here has full resort facilities with golf packages offered virtually year round.

The Princess Hotel in Southampton has a par-3 course that you will not forget in a hurry. It is up and down, and all around, and in and out with beautiful seascape views and little ponds coming into play—really a tester. The hotel is a modern, up-to-date resort. If you are staying here or in the area nearby, Riddel's Bay Golf and Country Club, although private, offers visiting golfers of other clubs privileges on a daily, weekly, or even monthly basis. The course has many holes along the water and elevated tees that require carrys across lagoons. The whole setup is extremely pleasant, and you may want to make it your personal club in the area. The course is a little difficult to walk, but now they have plenty of motorized carts and I would advise most seniors to rent one.

The only other course on the island worth a serious golfer's attention is the Port Royal Golf Course. This public course gets a considerable amount of play from both native islanders and visitors and it can get quite crowded. However, it is a course well worth playing and certainly would rank as one of the two best on the island. Designed by Robert Trent Jones, it runs along the ocean where a number of holes feature greens almost built into the cliffs, which can provide some frightening shots.

Puerto Rico

For many years the Dorado Beach Hotel in Puerto Rico, part of the Rockresorts chain, was really the only place to enjoy a golf vacation on that island. It still is a number-one choice because the location is outstanding and the hotel is still beautifully run. The courses by Robert Trent Jones are among the best he has ever done in tropical areas. There are actually two full eighteens that feature holes surrounded by enormous palm trees and other lovely foliage, with much water coming into play in the form of lagoons as well as the sea. The lagoons, incidentally, are a fine place to catch baby tarpon. If you'd like to be right next to the courses ask to have your reservations at the villas nearest the Su Casa.

Nearby Cerromar Beach Hotel is a newer setup also owned by Rockresorts and Eastern Airlines. It has two eighteen-hole courses, designed by Robert Trent Jones, with much water coming into play again. The new hotel there has full resort facilities.

Also nearby is a course that used to be run by the Hilton Hotel chain known as the Dorado Hilton. The course stretches to over 7,000 yards, and although not tremendously exciting, certainly is a difficult one to score well on. It is now called the Dorado Del Mar Hotel and Golf Club.

At Las Crobas on the eastern end of the island is situated the fantastic El Conquistador Hotel and Club, which sits high over the sea with a cable car to take you down to the beach and up-and-down course the likes of which you'll seldom see. It can be fun to play but be sure and take a cart; this is no place to try to walk (in fact, I think they make you take a cart).

At Luquillo is Puerto Rico's newest course, the Hyatt Rio Mar. You see a lot of advertising for this course with Chi Chi Rodriquez as the pro. He was pro for many years at the Dorado Beach. I have not seen the course myself but understand it winds through a rain forest and plays along the beach at times; probably well worth investigating.

Another relatively new course is Marriott's Palmas Del Mar at Humacao, a course designed by Ron Kirby and Gary Player that runs through coconut groves and cane fields.

Virgin Islands

In the Virgin Islands the only real course worth playing is Fountain Valley Golf Club in St. Croix. There is no hotel on the course itself, but nearby hotels have playing privileges. The well-known Buccaneer Beach Hotel at Christiansted has built an under-6,000-yard course that I have not had a chance to see.

Dominican Republic

In the Dominican Republic is the famous Cajuiles Golf Club at La Romana. This much publicized course designed by Pete Dye has five or six unforgettable holes running along the sea. The hotel has full resort facilities, and many of the accommodations overlook the golf course. There have been reports that the service at the hotel is not what it should be. Let's hope by the time you read this it has improved again.

Barbados

Barbados, which seems to be everyone's favorite island, has a worthwhile golf opportunity in the Sandy Lane Golf Club at Sandy Lane Estate. This course, which was nine holes for many years, has been expanded to provide eighteen holes of golf featuring lovely views at every turn.

Jamaica

Jamaica's best golf is offered by the Tryall Golf and Beach Club. Tryall is actually one of the best courses to be found in the entire tropics. It has the atmosphere of a fine old planters' estate and the course winds through some of the most beautiful trees and flora you will see. The first nine holes are hilly and inland, and the other nine runs along and comes down to the sea in a series of fine golfing moments. This course, when it is in good condition, and it is not always so, is a joy to play. The adjoining accommodations and beach club provide every facility for a getaway type of vacation.

Montego Bay is the center for the rest of Jamaica's resort golf with four resort hotels offering golf, which while good is certainly not top-drawer. One of these courses is the Rose Hall Golf Club with its adjacent hotel Rose Hall, which opened in 1974. Ironshore Golf and Country Club definitely has some worthwhile holes but it also has some that are best forgotten. It has no accommodations on the property, but there are many resort hotels in the area. The Half Moon–Rose Hall Country Club has a very long and narrow Robert Trent Jones course running basically east and west, and exposed. Here when the wind blows you have one big headache. Runaway Bay Golf Club at nearby Runaway Bay has another rugged layout that is not always in top condition but boasts a number of good holes. Hotel accommodations are immediately adjacent.

There are a number of good courses in the Kingston area but none of them is of a resort nature. If you're in the area, the Caymanas Golf and Country Club near Kingston is certainly one of the best layouts on the island. It is a private club, but introductions can be arranged.

The Bahamas

A few years back, the Bahamas were not often thought of as a destination for a winter golf vacation, but in recent years this has changed as a number of courses have been built not only on Nassau and Grand Bahama but in remote areas on the out islands. Nassau, the capital of the Bahamian Islands, is located on the island of New Providence. Its best golf course is probably the Lyford Cay Club, a private club founded a few years ago by a Canadian millionaire. It is absolutely necessary to have an introduction to play. There are good accommodations adjacent. The course itself is one that seniors will enjoy very much and it is worthwhile searching among your friends to see if you can't get an introduction to play here.

The Coral Harbour Golf Club is a very good and testing layout with many trees, water, and wind coming into play. Unfortunately, the hotel is closed as of this writing but the course itself may still be played. The Sonesta Beach Hotel has folded and is now known as the Ambassador Beach Hotel and Golf Club. It has a fairly playable course. Paradise Island, which is the gambling center, has an attractive golf course with the usual design to be found in the tropics. There are a number of hotels nearby. After Lyford Cay, the best course on the island is probably at the South Ocean Golf Club. This beauty was designed by Joe Lee. Nine holes of it are set along the ocean and they provide as fine a golf as you will find anywhere in the islands.

Grand Bahama, though a relatively small island, has a heavy concentration of golf courses in its small area. A lot of folks go to Grand Bahama Island for the gambling, and I guess it is only to be expected that on this island we would find one of the few par-3s that are lighted for night play. This one is located at the Bahama Reef Country Club, which also has a regulation golf course. It is the custom on the island to stay at any one of a number of hotels and then play the courses on the island that are open to the public. These are: The Grand Bahama Hotel and Country Club at West End and the Lucayan Country Club at Freeport with a Joe Lee course that is pine bordered and has rolling fairways rather reminiscent of golf in the Pinehurst area.

The Cotton Bay Club in Rock Sound, Eleuthera, once private, is now open to guests staying at the Cotton Bay Club, which also was formerly a private club. The course wanders along the water and is a fine test of anyone's golf; it's one of the top courses in all of the Bahamian area—well worth a visit. Some may remember it from one of the early CBS match plays on television, in which Arnold Palmer played. There are full accommodations and other resort facilities at Cotton Bay and also excellent bone fishing in the nearby waters.

Great Harbor Club at Great Harbour Cay in the Berry Islands is definitely a place to play. If you want to get away from things, this is certainly a remote spot.

The Treasure Cay Golf Club on Great Abaco was designed by the late Dick Wilson and it is another course that is far above the norm for most tropical layouts. As on many Dick Wilson courses, greens are elevated and guarded by lakes and sand traps. The course is beautifully located and the adjacent hotel, the Treasure Cay Beach Hotel, has full resort facilities.

The British Isles

To my mind, if you're going to plan a golf vacation abroad, the British Isles would be the first place to go. Scotland, of course, is the home of golf and this is reason enough to visit that area. But the chief reason is to play the many marvelous links courses that are set along natural, rolling, treeless land left by the receding seas hundreds of thousands of years ago. These courses can never be thought of as having been designed by man. Nature designed them ages ago, man came along and put a few greens in strategic places, dug up some of the turf and left the sand to catch your ball, and finally planted a little grass here and there to provide target areas. They are like no other golf courses in the world. The first two or three times you try them, the sense of desolation and loneliness, particularly on a rainy day, can be so overpowering that you may form an instant dislike to this kind of golf. I urge you to bear with it; like scotch whiskey, it is something that grows on you.

While most British courses are private, there are few that will not extend playing privileges to visiting Americans if they present proper credentials to the captain of the course. This gentleman is usually a retired military person and he winds up being all things to all men at these clubs. He is elected by the membership but receives a salary in many cases. You should bring with you, if you plan to play, a letter of introduction from the chairman of the golf committee at your club. It is never difficult to get a game of golf if you are by yourself in the British Isles; golfers are most friendly and the captain will usually be able to fix you up with a game.

There are just so many fine courses to play that, if you are planning a serious golfing tour of any length, it would be wise to write the British Travel Service, which has offices in many major United States cities, for their golf booklet that describes various courses and indicates conditions of play. Most of the courses we are about to list in England have been venues for the British Open on at least one occasion and some of them many times.

You normally don't go to England on a golf vacation to play the courses around London. However, there are many good ones, and any hotel can direct you to the ones you can reach during the day and come back to your hotel at night. The Selsdon Park Hotel Golf Club, which is easy to reach from London, has a fine restaurant and lounge bar and full accommodation right at the course. Perhaps the best-known golf club in the London area is Walton Heath. This one is set high up above sea level on land covered with heather and though it has the feel of a links course, it is not actually one. The best way of reaching Walton Heath is by rented car from London. There are two fine courses here, and the bar and restaurant are open from nine in the morning to nine at night. Incidentally, before you play any of these courses, be sure to call ahead and make your arrangements. It is disappointing to arrive at a course and find that there is a tournament scheduled for that day, which prevents any sort of outside play.

The Royal North Devon Golf Club on the shores of Bideford Bay is in an area that provides uncrowded golf, and the course itself is one of the great ones. There are no accommodations at the club, but I would suggest calling the club to recommend a hotel for golfers in Bideford.

Prince's Golf Club, at Sandwich, is one of England's most famous seaside links and offers twenty-seven holes of superb golf in beautiful surroundings. There are accommodations in the club's Dormy house, one mile from the course. Also near Sandwich is the famous Royal St. George's Golf Club, a must on any collector's list. Walter Hagen won an Open here and Bobby Jones led an American team to victory in the first Walker Cup match played in England. No accommodations directly on the course but nearby Deal and Sandwich have plenty of good places.

In the Liverpool/Blackpool area are Formby Golf Club, Royal Birkdale Golf Club, and the Royal Lytham and St. Anne's Golf Club. These are courses of great historic interest on which many championships have been played through the years. Both Royal Lytham and Formby have accommodations in a Dormy house attached to the club. These accommodations are in most cases for men only. Actually, a good deal of the golf in the British Isles is for men only. Not that there aren't courses where wives will be extremely happy, but in many, traditionally and physically, the clubs are not set up to serve women.

Your itinerary may not include Wales but a good excuse to visit that country would be the Royal Porthcawl Golf Club in Porthcawl. This is a true links course with views of the ocean from every tee and a tremendous challenge when the wind blows. This is a long course from the back tees but seniors can enjoy it by playing it well forward. And by all means use the small ball; not only does it handle better in the winds that roar and blow about your ears but on the dry sod it rolls farther than the slightly larger American ball. Please remember when you're considering whether to play the long or short tees that there are no motorized golf carts at the average British course. Lords and ladies of high degree pull their own carts on occasion or if caddies are available you may take them, but no motor carts will be found.

The golf courses in Scotland listed below all have accommodations on the course. This list is only touching the surface of Scottish golf, but the courses mentioned make wonderful headquarters and once there you can easily check with the hotel as to other courses worth playing in the area. If you fly into Prestwick, near Glasgow, you are immediately plunged into the heart of some of Scotland's finest golf. If you are wise, you will have long since made your reservation to stay at the famous Turnberry Hotel, whose two outstanding courses, Ailsa and Arran, are known throughout the golfing world. Turnberry is reached by a relatively short drive in a rental car along the Firth of Clyde south of Glasgow. Once there, if you can tear yourself away from the two magnificent courses at your doorstep you may make forays to play Prestwick, the scene of so many championships in the early history of golf, and Troon, which was next to your airport when you landed. Alternatively, if you can get reservations, and these must be made long in advance, you could stay at the very good Marine Hotel, overlooking the eighteenth hole at Troon, and using this as a base, take daytime rides to the courses previously mentioned. However, being next to the airport may not be a good idea—there are many planes landing and taking off and to my mind they spoil Troon as the ideal place to stay. Your next head-quarters might well be Gleneagles, an unforgettable place. The two courses, the Kings and Queens, are as delightful as any you will play. They are inland courses a little reminiscent of golf in the northeastern United States but still quite different in subtle ways. Gleneagles has a third short eighteen and a fun par-3 course that winds in and out of the grounds. There is also trout fishing on the premises, and fishing for salmon and trout in the nearby waters can be arranged at the hotel. Golf can be relatively inexpensive in Scotland, but this is not true of Gleneagles. The luxury offered here is superb and you pay for it. Gleneagles is the ideal place for playing St. Andrews, as it can be reached by a short drive. A little longer trip can be made to Carnoustie.

If you prefer to plunge yourself into the midst of golf's antiquities, then you should stay at St. Andrews and the hotel to stay at there is the old Course Hotel. There are four courses at St. Andrews, with stunning views of the Old Course from the bar at the top of the hotel. Many of the bedrooms also overlook the courses. All of the courses at St. Andrews are open to the public and you must reserve starting times well in advance, particularly if you wish to play the historic old course. The Royal and Ancient Club right at the first tee of the Old Course is a completely private club and well worth a visit. But you must have an introduction if you would like to see the interior of this famous club. You are required to wear tie and jacket if you plan to go to any of the inner rooms. Also, no women are allowed in the inner rooms. However, there is a private club for women in St. Andrews, so they have their own place to enjoy an after-golf tea.

If you have the time to drive north into the highlands there is one other course in Scotland I would recommend without reservation; this is Dornoch, certainly one of Scotland's greatest golf courses. This magnificent course, which runs along the sea, is so typical of what a truly great links should be. It is well worth making a pilgrimage to Dornoch; not only will you never forget the course, but you will be treading the ground where one of our greatest golf architects, Donald Ross, was born and grew up as a boy. Golf is very inexpensive at Royal Dornoch, as are the accommodations at the Royal Golf Castle Hotel immediately adjacent to the course.

From Dornoch you may make daily trips to a number of fine golf courses, perhaps the most famous of which would be Nairn. Another course that should not be missed is the home of the Honorable Company of Edinburgh Golfers, Muirfield. Muirfield, though near the sea, is faintly reminiscent of northern United States courses. It is among the best maintained layouts in the British Isles. Greywalls, a fine small hotel adjacent to the course, is the place to stay. You should have a letter of introduction from your club and you must contact them well in advance in order to make arrangements to play.

There are a number of other courses worth playing while in this area, including three at Gullane and a beautiful seaside links at North Berwick. Incidentally, if you can't get into Greywalls it is easy to play all of the courses making Edinburgh your headquarters, and Edinburgh, unlike Glasgow, is a beautiful city, well worth visiting.

I have saved Ireland for the last because, being of Irish ancestry, it is my favorite. I have made a number of trips to Ireland, both to fish and golf. Dublin makes an ideal base to play one of the finest courses in the British Isles. It is called Portmarnock, and it has been the scene of much championship play. Set along the seaside, it is a true links. It is only a short taxi drive from your hotel, or you can drive out in a rented car. The pro here is one of Ireland's most famous golfers, Harry Bradshaw, and when you call to make your reservation you might see if he is available for a playing lesson. Even famous pros at many places in Ireland give playing lessons for a modest sum. It is a wonderful introduction to a course like Portmarnock. Within the confines of the city of Dublin is the Royal Dublin Golf Club, and this, too, should not be missed by the visiting golfer.

Another choice for a golfing vacation in Ireland would certainly be Killarney. This absolutely beautiful and world-famous resort has three lovely golf courses that wander along the lake giving one the feeling at times of seaside golf. If you're staying at Killarney the Hotel Europe right on the lake would be a good choice if you enjoy a completely modern hotel with all of the amenities. Something in the grand hotel manner is the Great Southern Hotel. Both are handy to the golf.

While in Killarney, plan a day's outing to nearby Bally Bunion, thought by many, including golf writer Herbert Warren Wind, to be Ireland's finest course. It is an unforgettable experience if you catch it on the right day. Nothing could be more beautiful than these holes along the sea, winding up and down through valleys and finally ending up in a magnificent return to the club, where people sitting in the bar by a huge window can bear witness to all of your final mistakes.

Another course that must be put on any collector's list is Lahinch. This course along the sea has many holes that are unforgettable, particularly a par-3 in which you shoot over a large hill and have no idea what happened to your shot until you finally crest the hill and start on down. A good hotel that makes excellent headquarters for seeing some of the famous sites in the Lahinch area is the Aberdeen Arms Hotel right along the sea in Lahinch.

Before you leave Ireland you must play a wonderful course very close to the airport at Shannon, and so you might make this your first stop or your last stop on the way home. It is the Waterville Golf Links, a stunningly beautiful layout along the coast that is a stern test when played at its full 7,200-yard length. This was once a nine-hole course that was redesigned and an entirely new nine holes added to make eighteen holes. A lovely hotel overlooking the lake there called the Waterville Lakes Hotel is very near the course. It also offers salmon fishing in the famous Butler pool and in a number of lakes and streams nearby. There are also some magnificent sea trout to be caught in the lakes.

Appendix

Recommended Reading List

The following books by no means constitute a definitive list and are intended to be merely a starting point for seniors who may or may not become collectors. A few of the books may be out of print, but a visit to one or two good secondhand book stores should turn them up. Anyone who is interested in serious collecting should try and obtain *The Library of Golf* by Joseph Murdoch, published in 1968 by the Gale Research Company, Detroit ($12.50). In the list that follows, books are listed alphabetically by author.

Armour, Tommy
Tommy Armour's ABC'S of Golf (New York: Simon & Schuster, 1967). Armour is the author of several other golf books but this simple and readable book is probably the best. The ABC format makes it easy to check on any phase of your game.

Bell, Peggy Kirk (with Jerry Clausen)
A Woman's Way to Better Golf (New York: Dutton, 1966). A very good book for the woman golfer. Senior women can really help themselves with this book—almost as good as going to Pine Needles Lodge to take a lesson with Peggy amid the North Carolina pines.

Boomer, Percy
On Learning Golf (London: John Lane, 1942). One of the classic teachers
of the game. The "Bible" for a great many well-known people. The
late Duke of Windsor, who was an admirer, wrote the foreword.

Boros, Julius
How to Play Golf with an Effortless Swing (Englewood Cliffs, N.J.:
Prentice-Hall, 1964). A good book for seniors who would do well to
emulate one of the sweetest swingers of all times.

Burke, Jack
The Natural Way to Better Golf (Garden City, N.Y.: Hanover House,
1954). To me one of the best books on instruction for beginners and a
great brushup for golfers of all ages and abilities.

Cotton, Henry
Henry Cotton says . . . (London: Country Life, 1962). A book of thoughts
to take with you out on the course for the day's round—one at a time,
of course. A delightful book with funny illustrations by Roy Ullyett.
Cotton has written a number of books and they are all worth owning.

Dante, James, and Leo Diegel
The Nine Bad Shots of Golf and What to Do About Them (New York:
Whittlesey House, 1948). If you are a seasoned golfer and run into
those problems that plague all of us at one time or another, this book
may be able to put you back on the track.

Darwin, Bernard Richard Meirion
Tee Shots and Others A collection of columns written for the *London Evening Standard* by Mr. Darwin. A good sampling by this best of all
golf writers. When he died at age eighty-five he had written some two
dozen books and every one of them is worth reading. Most, unfortunately, are now collector's items.

Grimsley, Will
Golf: Its History, People and Events (Englewood Cliffs, N.J.: Prentice-
Hall, 1966). My favorite quick reference book, it has a special section
by Robert Trent Jones. Covers an astonishing range from the beginnings of golf to 1966.

Hagen, Walter J. (as told to Margaret Seaton Heck)
The Walter Hagen Story (New York: Simon & Schuster, 1956). Hagen was

golf's most colorful character. This book is a joy to read especially today when so many pro golfers are sedate businessmen.

Hogan, Benjamin William (with Herbert Warren Wind)
The Modern Fundamentals of Golf (New York: A. S. Barnes, 1957). There are a lot of "bests" but to my mind this is the best book on instruction ever written. The illustrations by Anthony Ravielli are just as important as the text.

Jones, Ernest (as told to David Eisenberg)
Swing the Clubhead (New York: Dodd, Mead, 1952). One of the classic books on golf instruction. Jones was an English pro who lost his right leg in World War I. Not to be put down by this tragedy, he soon proved that he could still play golf in the low seventies while standing on one leg. To do that he had to "swing the clubhead," which is the whole point of the book.

Jones, Robert Tyre Jr.
Golf Is My Game (Garden City, N.Y.: Doubleday, 1960). Written by Jones himself and a joy to read. Of special interest is the portion of the book telling how the Masters was developed with a description of how to play each hole.

Keeler, O. B.
Down the Fairway: The Golf Life and Play of Robert T. Jones, Jr. (New York: Minton Balch & Co., 1927). This is the standard work on the life and times of Bobby Jones by his "Boswell" O. B. Keeler.

Locke, Arthur D'Arcy
Bobby Locke on Golf (London: Country Life, 1953). This interesting autobiography was written by Locke when he was thirty-four. He is very frank in his opinions and this makes for a most readable book. Contains excellent illustrated instructions.

Mathieson, Donald Mackay
The Golfer's Handbook, published continuously since 1898 in Glasgow, Scotland. Founding editor, Donald Mackay Mathieson. Current editor, John B. Duncan. Easily the most successful annual ever published; 1,100 pages of useful and interesting information. Aimed at the British golfer, but all golfers will enjoy it.

Price, Charles
The World of Golf: A Panorama of Six Centuries of the Game's History

(New York: Random House, 1962). Price manages to bring a new approach to the history of the game. The illustrations are especially interesting and unusual.

Runyan, Paul

Paul Runyan's Book for Senior Golfers (New York: Dodd, Mead, 1963). An excellent instruction book for senior golfers.

Sarazen, Gene

Better Golf After Fifty (New York: Harper & Row, 1967). A book well worth reading by any senior golfer. *Thirty Years of Championship Golf: The Life and Times of Gene Sarazen,* by Herbert Warren Wind (Englewood Cliffs, N.J.: Prentice-Hall, 1950). A readable biography about a most colorful golfer.

Scharff, Robert, Editor

Golf Magazine's Great Golf Courses You Can Play (New York: Scribner's, 1973). The most complete guide to golf courses around the world ever published.

Snead, Sam (with Al Stump)

Education of a Golfer (New York: Simon & Schuster, 1962). Snead's golf experiences make for some of the richest humor in all of golf. If there is one book I miss from my library it is this one. I wish whoever borrowed it would return it.

Steel, Donald et al., Editors

The Encyclopedia of Golf (New York: Viking Press, 1975). The first comprehensive encyclopedia of golf. The illustrations are very effective.

Ward-Thomas, P.A. et al., Editors

The World Atlas of Golf (New York: Random House, 1976). A stunning book that examines 170 of the world's best and most famous courses in forty-three countries. Contains three-dimensional maps in color and colored photos of golf courses.

Wind, Herbert Warren

The Story of American Golf: Its Champions and Its Championships (New York: Knopf, 1975). This is the third edition of this masterpiece by America's greatest golf writer. The first edition was published by Farrar, Straus in 1948 and the second by Simon & Schuster in 1956. If you don't already own this beautifully written book, treat yourself to a copy.

As an admirer of Herb Wind, I own most of his books. Two I would especially recommend are *Herbert Warren Wind's Golf Book* (New York, Simon & Schuster, 1971), a superb collection of his writings that have appeared in the *New Yorker* and various other magazines and books, and *The Greatest Game of All, My Life in Golf,* by Jack Nicklaus with Herb Warren Wind (New York, Simon & Schuster, 1969).

Golf Magazines

Golf Digest (monthly). Annual subscription: $9.50. Address: *Golf Digest,* 363 Greenwich Avenue, Greenwich, Connecticut 06830.

Golf Journal (10 months). Official publication of the U.S. Golf Association. Annual subscription: $3.50. Address: *Golf Journal,* P.O. Box 2015, Radnor, Pennsylvania 19089.

Golf Magazine (monthly). Annual subscription: $7.94. Address: *Golf Magazine,* P.O. Box 2786, Boulder, Colorado 80302.

Golf World (weekly newsmagazine). Annual subscription: $13.95. Address: Golf World Company, Inc., 600 Access Road, Southern Pines, North Carolina 28387.

Senior Golfer (quarterly). Annual subscription: $2.50. Address: P.O. Box 4716, Clearwater, Florida 33518.

Golf Oddities

Long-Lived and Low-Scoring Seniors In Chapter One some of the remarkable accomplishments of the late Ellis Knowles were mentioned. Henry Hopple, a Los Angeles freelance writer, in the April 1975 issue of *Golf* tells of another astounding oldster. "George Miller first attempted to hit a golf ball in 1932 and after missing the ball repeatedly he became determined to master the game. Within a few months he broke 100. That sounds like every player's first months of the game. The only difference here was that George Miller was age 55 in 1932. Several years after he broke 100, Miller snapped 90 but it was not till after he retired at age 65 that Miller really got on his stick. . . . The Anaheim, California, resident scored a 71 on a 6100 yard course at age 76. When a 'mature' 92, he shot three consecutive 79's on a regulation course. Now 97, Miller plays four times a week and shoots in the low 90's. After an illness at age 92, his doctor advised he ride a golf car. Miller's comment: 'I hated to give up walking but as long as I didn't have to give up golf it was okay. I know I owe my good physical condition to golf.' "

Perhaps the oldest golfer who ever lived was Nathaniel Vickers who celebrated his one hundred and third birthday on October 9, 1949, and died the following day. He was the oldest member of the United States Senior Golf Association and until 1942 competed regularly in their events and won many trophies in the various age divisions. When one hundred years old he apologized for being able to play only nine holes a day. He attributed his longevity and good health to heredity, contentment, and to taking it easy. He was never known to hurry.

A notable veteran who died in February 1963 was Willie Auchterlonie, aged ninety-one. He was the last "home" Scot to win the open championship—in 1893 at Prestwick. He founded the firm of D. and W. Auchterlonie, the famous St. Andrews club makers, and was an expert in fashioning handmade clubs. He had been professional to the Royal and Ancient Club since 1935.

The tenth Earl of Wemyss played a round on his ninety-second birthday, in 1910, at Craigielaw. When eighty-seven the

Earl was partnered by Harry Vardon in a match at Kilspindie, the golf course on his East Lothian estate at Gosford. The venerable Earl, after playing his ball, mounted a pony and rode to the next shot. He died in June 1914 in his ninety-sixth year. The eleventh Earl, also a keen golfer, succeeded to the title when over seventy. He used to say, "Everybody's father dies but my father."

A remarkable New Zealand veteran golfer, Jimmy Drake, at age eighty-nine, and while a regular playing member of the Miramar Club, Wellington, returned scores lower than his age on 192 occasions. The first time he accomplished the feat was when he was seventy. He lost count of the number of times he equaled his age, saying, "That's no challenge; the challenge is to better it."

South African golfer William Edmonds celebrated his eightieth birthday in 1968 shortly after completing a medal round of 74. This was the 150th occasion he had shot his age or better. His achievements include having won the South African Seniors Championship five times, the last time at the age of seventy-one. At the age of seventy-two he won the Maritzburg Club Championship with rounds of 74 and 75 on a par 72 course.

An unusual golf tournament was held at the LaGorce Country Club, Miami, Florida. It was called the Super Seniors. To compete in the event you had to be at least eighty. The six spry gentlemen who made up the field totaled 523 years. They were Lou Maul, ninety-two; Bill Kennedy, ninety; Russell Osborne, eighty-eight; Bob Shreve, eighty-six; Dr. Paul Putzki, eighty-four, and the baby of the competition Fred Weidersum at eighty-three. The winner was not announced but the important thing, of course, was for them to make the tournament.

Senior Associations

American Seniors Golf Association
P.O. Box 241
Palm Beach, Florida 33480

Eastern Seniors Golf Association
7 Red Oak Road
Bronxville, New York 10708

Francis H.I. Brown International
Team
World Senior Golf Federation, Inc.
The Broadmoor
P.O. Box 1356
Colorado Springs, Colorado
80901

Great Lakes Senior Golf Association
Heights-Rockefeller Building
2483 Lee Boulevard
Cleveland Heights, Ohio
44118

Heart of America Senior Golf Association
7139 Lafayette
Kansas City, Kansas 66109

International Seniors Amateur Golf Society
1841 Broadway
New York, New York 10023

New England Senior Golf Association
Woodland Golf Club
Auburndale, Massachusetts
02166

North and South Senior Invitationals

Pinehurst Country Club
Pinehurst, North Carolina
28374

Southern Golf Association
P.O. Box 9151
Birmingham, Alabama 35213

Southern Seniors Golf Association
P.O. Box 1629
Winter Park, Florida 32789

Trans-Mississippi Seniors
5353 Nall Avenue
Shawnee Mission, Kansas
66202

Tri-State Seniors Golf, Inc.
P.O. Box 3834 C.R.S.
Johnson City, Tennessee
37601

United States Golf Association
Liberty Hill Corner
Far Hills, New Jersey 07931

U.S. National Senior Golf Association
2140 Westwood Boulevard
Los Angeles, California 90025

U.S. Seniors Golf Association
60 East 42nd Street
New York, New York 10017

U.S. Senior Women's Golf Association
c/o Sam Watkins
80 Clapboard Ridge Road
Greenwich, Connecticut
06830

Western Golf Association
    1 Briar Road
    Golf, Illinois 60029

Western Senior Amateur Association
    John Olson
    25115 Kirby, Sp, 335
    Hemet, California 92343

Western Seniors Golf Association
    5446 Michigan Road
    Indianapolis, Indiana 46208

Senior Men's Amateur Championship*

CHAMPIONSHIP TROPHY

Presented in September, 1955

by Frederick L. Dold

Member of the United States Golf Association Executive Committee 1950–54

HISTORY

1955—The Senior Amateur Championship was established in 1955 as a result of a remarkable growth in senior golf. Many senior associations had come into being on district, state and sectional levels. The oldest was the United States Seniors' Golf Association, which limited itself to 850 members and had a substantial waiting list. Most of the others also were similar to private clubs, conducting fine, enjoyable tournaments for their members. However, as there was no one event open to members of all USGA Regular Member Clubs, the USGA was requested to inaugurate such a competition. Thus was born the USGA Senior Amateur Championship, for members of USGA Regular Member Clubs who were 55 years old and had handicaps not exceeding 10. One hundred twenty-eight players qualified through 18-hole sectional rounds, and they competed in another 18-hole qualifying round at the Championship site for 32 places in match play. The match play comprised one 18-hole round each day. The first Championship was held at the Belle Meade Country Club, Nashville, Tenn., in September-October, and J. Wood Platt, 56, of the Saucon Valley Country Club, Bethlehem, Pa., defeated George Studinger, of San Francisco, Calif., 5 and 4. Platt was two under par in the final. The medalist was Martin M. Issler, of West Orange, N.J., who equalled par with a 72. Sixteen tied for the last ten places at 79. The Championship attracted 370 entrants. Thirty states and the District of Columbia were represented among the qualifiers.

1956—Frederick J. Wright, Jr., 58, of the Oakley Country Club, Watertown, Mass., a member of the 1923 Walker Cup Team, became the second Senior Amateur Champion by defeating J. Clark Espie, 57, of Indianapolis, Ind., 4 and 3, at the Somerset Country Club, St. Paul, Minn., in August, 1956. The medalist was Weller Noble, 65, of Berkeley, Calif., who scored a two-over-par 72, the same score which won the medal the first year. The defending champion, J. Wood Platt, lost in the third round to Wright. The entry dropped to 282. The qualifying limit fell on 10 players tied at 79, also the same score as in the first year, for the last four places. In the first round Thomas C. Robbins, of Mamaroneck, N.Y., went 21 holes to defeat Perry T. Taylor, Huntington, W. Va., the longest match thus far.

1957—In a reversal of the 1956 final, J. Clark Espie, 58, of the Hillcrest Country Club, Indianapolis, Ind., defeated Frederick J. Wright, Jr., 59, of Watertown, Mass., 2 and 1, at the Ridgewood Country Club, Ridgewood, N.J., in September-October. Wright had beaten Espie, 4 and 3, the previous year

*Reprinted from USGA Record Book.

and seemed on the way to another victory when he was 3 up with seven to play. J. Wood Platt, of Philadelphia, Pa., the first Senior Amateur Champion, lost in the first round to Christopher A. Carr, of Hamburg, N.Y. In the same round, Judd L. Brumley, of Greenville, Tenn., went 22 holes to defeat William L. Goodloe, Sr., of Valdosta, Ga., the longest match of the Championship thus far. The medalist was Thomas M. Green, Jr., 55, of Seattle, Wash., who scored a two-over-par 73. Eleven tied for the last nine places at 81. Runcie B. Martin, of Duluth, Minn., the oldest qualifier at 72, was playing 52 years after his first appearance in a USGA Championship, the Amateur of 1905.

1958—Thomas C. Robbins, 65, of West End, N.C., became the oldest winner by defeating John W. Dawson, 55, of Palm Desert, Calif., 2 and 1, at the Monterey Peninsula Country Club, Pebble Beach, Calif., in September-October. J. Clark Espie, 59, of Indianapolis, Ind., the defending Champion, established a new qualifying record with a one-under-par 71 but lost to Robbins in the semi-final, 2 up. Eight players tied at 79 for the last of the 32 qualifying places. Chick Evans, 68, of Chicago, who had played in all four Championships, was among the qualifiers for match play for the first time and won two matches.

1959—J. Clark Espie, 60, of Indianapolis, Ind., the 1957 Senior Amateur Champion, won again at the Memphis Country Club, Memphis, Tenn. His margin in the final was 3 and 1 over J. Wolcott Brown, of Manasquan, N.J. In the five Senior Amateur Championships played thus far, Espie had won twice, been runnerup once, and a semi-finalist and qualifying medalist on a fourth occasion. In 1959 he participated in the longest match to date in the history of the competition—a 24-hole match which he won in the second

round against Larry E. Stage, of Lafayette, Ind. Other semi-finalists were George Dawson, of Chicago, and William E. Norvell, Jr., of Chattanooga. Defending Champion Thomas C. Robbins, of Pinehurst, N.C., missed qualifying for the Championship flight of 32. There was a record entry of 391. There was no handicap limit for the first time.

1960—Michael Cestone, age 55, of Montclair, N.J., competed in his first Senior Amateur Championship at Oyster Harbors Club, Osterville, Mass., and finished as winner. He defeated David Rose, 56, Cleveland, Ohio, on the 20th hole of the final match. Other semi-finalists were W.B. McConnell, Kennett Square, Pa., and Edward E. Lowery, San Francisco, Calif. Weather played a large part in the Championship. Steady rain fell on the day scheduled for the qualifying and continued into the following day, causing cancellation of play both days. The USGA Senior Championship Committee was forced to extend the Championship one day, to Sunday, and to schedule two match rounds on Saturday. The Committee also decided to permit the use of automotive transportation on the double-round day. Defending Champion J. Clark Espie, Jr., Indianapolis, Ind., withdrew because carts were to be permitted in the emergency, contrary to the published conditions of the tournament. The medalist was S.S. Rockey, of Los Angeles, who shot 74. A record 517 entered the Championship—126 more than the previous high.

1961—Dexter H. Daniels, 56, of Winter Haven, Fla., won the Senior Amateur Championship on his first attempt. He defeated Col. William K. Lanman, Jr., also 56, of Golf, Ill., in the final match by 2 and 1. The finalists, after five days of play under serene weather conditions at the Southern Hills Country Club, Tulsa, Okla., were confronted with a wind that blew in gusts up to 35 miles per hour. Daniels, 2 up after nine holes, maintained that lead through the 16th and ended the match by halving the 17th. Joseph Morrill, Jr., of Great Barrington, Mass., was the medalist at the site of the Championship with a 74, three over par. Seven men who advanced to the quarter-final round of the 1960 Senior Amateur Championship competed at Southern Hills but only Richard H. Guelich, of Hamburg, N.Y., was able to equal his 1960 performance. Guelich lost to Lanman in a semi-final match.

1962—Merrill L. Carlsmith, a 56-year-old Hawaii attorney, won the eighth Senior Amateur Champsionship by defeating Willis H. Blakely, Portland, Ore., by 4 and 2 in the final match at the Evanston Golf Club, Skokie, Ill. Carlsmith saved his strongest efforts for the last two matches. He was even par against Michael Cestone, the 1960 Champion, through 17 holes of their semi-final, and only one over par in the final against Blakely, the Oregon State and Pacific Northwest Senior Champion. The field, considered by many to be the strongest ever for this event, produced 31 scores of 77 or better in the qualifying round. Previously, the lowest score required to qualify had been 78. Dexter H. Daniels, the defending Champion, was automatically awarded the 32nd qualifying berth. He lost in the first round to David E. Rose, the 1960 runnerup. Three men playing in the event for the first time shared the qualifying medal with scores of 72—James M. Johnson, LaDue, Mo.; William S. Terrell, Charlotte, N.C., and Henry L. Robison, Albuquerque, N.M. There was a record entry of 525.

1963-Merrill L. Carlsmith became the first to defend the title successfully. He defeated William D. Higgins, of San Francisco, by 3 and 2 at the Sea Island Golf Club, Sea Island, Ga. This

was the third appearance in the Championship for the 57-year-old Hilo attorney and the first for Higgins, also 57. Egon F. Quittner, Jenkintown, Pa., defeated Maurice R. Smith, Charlotte, N.C., in the first round after nine extra holes, an overtime record for the Championship by three holes. A new all-match play format was introduced. The original entry of 494 was reduced by Sectional Qualifying to 128 for the Championship proper. On Friday, the only day of two rounds, players were permitted to use automotive transportation for the first time under original conditions of a USGA Championship.

1964—William D. Higgins, 58, of San Francisco, won the Senior Amateur Championship at the Waverley Country Club, Portland, Ore., after losing in the final round in 1963. Higgins birdied the last two holes to win his semi-final match 1 up from David Goldman, Sr., of Dallas. He then beat Edward Murphy, a member of the host club, by 2 and 1 in the final. Merrill L. Carlsmith, of Hilo, Hawaii, was seeking an unprecedented third straight Championship but he was defeated in the second round by Ralph Swan, of Vancouver, Wash., by 3 and 2. The format was altered to 36 holes of qualifying at the site to determine 32 players for match play. A.L. (Jim) Miller, 71, turned in a remarkable qualifying performance to earn the medal with scores of 74-76—150.

1965—Robert B. Kiersky, 57, of Oakmont, Pa., won the 1965 Championship at the Fox Chapel Golf Club, Pittsburgh, by defeating George C. Beechler, Prineville, Ore., on the 19th hole. On the extra hole, 405 yards long, Kiersky two-putted for his par, but Beechler was short with his second and took three to get down. Two former Champions lost in the semi-final round. Merrill L. Carlsmith, winner in 1962 and 1963, lost to Beechler; William D.

Higgins, San Francisco, the 1964 Champion, was ousted by Kiersky. Curtis Person, Sr., of Memphis, Tenn., won the medal with 149.

1966—For the second consecutive year George C. Beechler reached the final of the Senior Amateur Championship and met defeat. Dexter H. Daniels, of Winter Haven, Fla., defeated Beechler, 1 up, at the Tucson National Golf Club, Tucson, Ariz. Robert B. Kiersky, of Delray Beach, Fla., the defending Champion, lost in the third round to Merrill L. Carlsmith, the Champion in 1962 and 1963. Two former Amateur Champions played: Jack Westland, the 1952 Amateur Champion from Pebble Beach, Calif., lost in the first round to Curtis Person, Sr., of Memphis, and Richard D. Chapman, of Oyster Harbors, Mass., was defeated in the third round by David Goldman, Sr., of Dallas. Person was medalist with 143.

1967—Raymond Palmer, 55, of Lincoln Park, Mich., won the Championship the first year he was eligible. In the final he defeated Walter D. Bronson, of Chicago, 3 and 2, at the Shinnecock Hills Golf Club, Southampton, N.Y. Dexter H. Daniels, of Winter Haven, Fla., the defending Champion, failed to qualify for match play. Palmer, J. Wolcott Brown, of Sea Girt, N.J., and David Goldman, Sr., of Dallas, tied for medalist with 153.

1968—Curtis Person, Sr., 58, of Memphis, Tenn., barely qualified for match play, then won the Championship. He defeated Ben Goodes, 55, of Reidsville, Tenn., 2 and 1, in the final. Person escaped a sudden-death playoff for a qualifying position by one stroke. He won his first match in 20 holes, and his next three by 1 up. Person was 1 down after 13 holes against Goodes, won the 14th with a birdie, and the 15th

and 17th with pars. John Tullio, Aurora, Ohio, was qualifying medalist with 146. The entry of 674 set a record.

1969—Curtis Person Sr., of Memphis, Tenn., became the second to defend the title successfully when he defeated David Goldman, Sr., of Dallas 1 up, at the Wichita Country Club, Wichita, Kans. Merrill Carlsmith, of Hilo, Hawaii, won successive Championships in 1962-63. For the second consecutive year Person barely qualified for the match play phase. His 36-hole score of 154 was one stroke under the cutoff. The par-3 holes made the difference in the final match; Person won three of the four with pars. The match was even after 10 holes, then Person won the par-3 11th when Goldman missed the green. They halved the remainder of the holes with Person holing a 10-foot putt for par on the 18th. Goldman was medalist with 146.

1970—Gene Andrews, 57, of Hacienda Heights, Calif., won the Senior Amateur Championship by defeating James Ferrie, of Indian Wells, Calif., 1 up at the California Golf Club of San Francisco, South San Francisco, Calif. Andrews was a former Amateur Public Links Champion and a former member of the Walker Cup Team. He was the first Senior Amateur Champion ever to have won another USGA Championship and was the second former Walker Cup player to have done so. Frederick J. Wright, Jr., the 1956 Champion, was a member of the 1923 Walker Cup Team; Andrews was on the 1961 Team. Andrews won two extra hole matches, defeating Robert E. Cochran, of Normandy, Mo., in 21 holes, and Ernest Pieper, Jr., of San Jose, Calif., in 19. He won the final by sinking a 27-foot putt on the final hole. Curtis Person, Sr., Champion in 1968 and 1969, lost in the first round to Truman Connell, of Delray Beach, Fla.

Bruce N. McCormick, of San Gabriel, Calif., was medalist with 147. The entry of 683 was a record.

1971—The 1971 Championship drew what was probably the best field in the history of the Senior Amateur, and it was won by Tom Draper, 57, of Troy, Mich. Draper defeated Ernest Pieper, Jr., of San Juan Bautista, Calif., 3 and 1, in the final at the Sunnybrook Golf Club, Plymouth Meeting, Pa. Among players competing for the first time were William Hyndman, III., who had been runnerup in the British Amateur in 1969 and 1970. Hyndman was eliminated by Pieper in the semi-final round. Draper was 1 up on Pieper after playing fine bunker shots on holes 9, 13, and 14, birdied the 152-yard 15th with a tee shot 30 inches from the hole to go 2 up, and hit a 2-iron 15 feet from the hole on the 183-yard 17th to win another hole and end the match. Draper also defeated Bob Cochran, of St. Louis, a former Walker Cup player and medalist at 148, and Truman Connell, of Boynton Beach, Fla., the current American Seniors Champion. Draper also defeated William F. Colm, of Pebble Beach, Calif., in the semi-final round. Gene Andrews, the 1970 Champion, did not defend.

1972—Lewis W. Oehmig, 56, of Lookout Mountain, Tenn., won the Senior Amateur Championship at the Sharon Golf Club, Sharon Center, Ohio, in his initial attempt, beating Ernest Pieper, San Juan Bautista, Calif., on the 20th hole. This was the second successive year that Pieper was runnerup. Tom Draper, the defending Champion, failed to qualify for match play. Three of Oehmig's victories were extra hole matches, and in each he had to make a sizable putt or hit his approach shot close to the hole. He defeated Edward L. Meister, Jr., Willoughby, Ohio, on the 19th hole with a birdie; Bill Hyndman, Huntingdon Val-

ley, Pa., on the 20th hole by making an 18-foot putt, and Pieper on the 20th hole of the final round after squaring the match on the 18th with a 30-foot birdie putt. He also beat Timothy Holland, Wenham, Mass., 3 and 2, in the first round, and Curtis Person, Memphis, Tenn., 2 and 1, in the quarter-finals. Person, Senior Champion in 1968 and 1969, played an impeccable round against Oehmig, but caught him on a day when Oehmig shot 15 pars and two birdies. Hyndman was the medalist with 71-74-145. In his match with Oehmig, Hyndman was two up after nine holes and was still one up after 17. On the 18th Oehmig hit his approach shot four feet from the flagstick and made his birdie, sending the match into extra holes. On the 20th he holed an 18-footer to beat Hyndman. Senior amateur golf continues to burgeon, with the result that there were 617 entries and another strong field.

1973—Bill Hyndman, of Huntingdon Valley, Pa., won the Senior Amateur Championship at the Onwentsia Club, Lake Forest, Ill., defeating Harry Welch, of Salisbury, N.C., 3 and 2. This victory marked the end of a long quest for Hyndman; for more than 20 years he had been one of the country's leading amateurs, but he had never won a national Championship. He had been runnerup in the 1955 U.S. Amateur and had been the finalist in the British Amateur Championships of 1959-69-70. He had also reached the semi-finals in the Senior Championship, 1971 and 1972. The defending Champion, Lew Oehmig, of Lookout Mountain, Tenn., was eliminated in the first round. Hyndman defeated Bill Trepsas, Stoughton, Mass., 2 and 1 in his first match, then Ernest Pieper, San Juan Bautista, Calif., by the same margin in his second round. Two down to Pieper after nine holes, Hyndman scored four birdies in eight holes for a four-under par 31 to win the match. Pieper was even with par on the second nine. In the final round against Welch, Hyndman won the first hole and was never behind. Hyndman either won or saved holes on three occasions by brilliant recoveries from greenside bunkers. It should be noted that Hyndman and Welch played their match—16 holes—in only two hours and 20 minutes. Co-medalists were the new Champion and Sam Friedman, of Fort Walton Beach, Fla., with scores of 73-74—147. A total of 633 entered the Championship.

1974—In his first year of eligibility, Dale Morey, of High Point, N.C., defeated Lew Oehmig, the 1972 Champion from Lookout Mountain, Tenn., 4 and 2 in the final round at the Harbour Town Golf Links, Hilton Head Island, S.C. The defending Champion, Bill Hyndman, of Huntingdon Valley, Pa., was eliminated by Oehmig in the third round, 1 up. Morey was runnerup to Gene Littler in the 1953 Amateur Championship, but he had never won a national Championship. On his way to the final he defeated Ed Meister, of Willoughby, Ohio, 6 and 5; Wally Sezna, West Chester, Pa., 4 and 3; John Humm, Rockville Centre, N.Y., 2 up, and Ed Tutwiler, of Indianapolis, Ind., 3 and 2 in the semi-final round. Tutwiler was medalist in the 36-hole qualifying with 144, two over par. Morey won four of the first five holes in the final round but Oehmig rallied and was only 1 down playing the 14th when he became involved in one of a series of Rules decisions that were made during the Championship; in all, Oehmig was party to three such incidents. His tee shot on the 152-yard, par-3 hole came to rest within the confines of a water hazard. Electing to play the ball, because he was not entitled to relief without penalty, Oehmig took his stance and then soled his club. The Referee had no choice but to invoke Rule 33-1b and advise Oehmig that he had lost the hole.

Morey won the 15th hole to go 3 up with three to play. Oehmig conceded the 16th hole and the match to Morey. The entry of 743 set a new record.

1975—William F. Colm became the third Californian to win a USGA Championship during 1975. He defeated fellow-Californian Stephen Stimac, 4 and 3, in the final of the Senior Amateur Championship at the Carmel Valley Golf and Country Club, Carmel, Calif. Players from 37 states competed for the Championship, but when it was over the winner was the man who traveled the shortest distance—nine miles. Colm lives in Pebble Beach. He had been a semi-finalist in 1971, and in 1970, 1972, and 1973 he lost in the first round. Stimac, of Walnut Creek, had never advanced beyond the second round before, but he kept winning his matches, including his quarter-final round against former Walker Cupper Bob Cochran, which went 19 holes. . . . There were 737 entries.

1976—Lewis W. Oehmig, 60, of Lookout Mountain, Tenn., who had won the Tennessee Amateur in 1937 at 21, won the Senior Amateur Championship for the second time. He defeated John Richardson, 55, of Laguna Niguel, Calif., 4 and 3 in the final round at the Cherry Hills Country Club, Englewood, Colo. Oehmig won previously in 1972. The Championship was played in erratic weather—showers and temperatures in the 50s. Oehmig and Dale Morey, High Point, N.C., were the only former Champions among the 32 who qualified for match play. Richardson eliminated Morey in the second round. Richardson also defeated Ed Tutwiler, of Indianapolis, the medalist. In the final round, Richardson, 1961 California Amateur Champion, consistently outdrove Oehmig, but he could not match Oehmig's short game. In the 15 holes played, Oehmig had five conceded birdies and lost only two holes. He three-putted both. On the first hole Richardson drove into a lateral water hazard and conceded a birdie to Oehmig. Holes two and three were halved, then Richardson drew even when Oehmig three-putted the fourth green. Oehmig then won four holes in succession to go 4-up. Oehmig three-putted again, on the 10th, to stand 3-up. Both players birdied the 11th and halved the 12th, 13th and 14th with pars. The match ended on the 15th, where Oehmig put his tee shot within 12 feet of the hole and Richardson hit into a bunker. The entry reached 833, breaking the record 743 set in 1974.

Senior Women's Amateur Championship*

CHAMPIONSHIP TROPHY

Presented in October 1962 by the

United States Golf Association and Friends of Senior Golf

HISTORY

1962—The inauguration of a Senior Women's Amateur Championship followed the same pattern that preceded the start of the Senior Amateur Championship in 1955. A number of senior women's associations had come into being on various levels, but no one event was open to members of all USGA Regular Member Clubs. The request to begin such a competition was approved by the Executive Committee in January, 1962. The addition of the Championship to the USGA schedule meant that every man, woman and child golfer has an opportunity to compete in a USGA Championship. The format decided upon was a 54-hole stroke play competition over three days. Competitors must be at least 50 years old. In addition to the Championship proper, it was decided to award prizes in three age groups: A—50 through 54 years; B—55 through 59 years; and C—60 years and over. The first Championship was played at the Manufacturers' Golf and Country Club, Oreland, Pa. Miss Maureen Orcutt, Englewood, N.J., a reporter for the New York *Times,* won the Championship with a total of 240 through three consecutive rounds of 80. Miss Orcutt had twice been runnerup in the Women's Amateur Championship and was a member of the first four Curtis Cup Teams. Her first USGA Championship victory was particularly gratifying. The runnerup, seven strokes behind, was Mrs. Glenna Collett Vare, Bryn Mawr, Pa., six times the Women's Amateur Champion. In the age group competitions Mrs. Allison Choate, Rye, N.Y., was the A winner; Miss Orcutt won in the B section; and Mrs. Theodore W. Hawes earned the C award. The entry for the first Championship was 96.

1963—Mrs. Allison Choate, of Rye, N.Y., went four extra holes on top of an 18-hole playoff round to win the second Championship at The Country Club of Florida, Delray Beach. Mrs. Choate trailed Miss Maureen Orcutt of Englewood, N.J., the defending Champion, by four strokes with five holes to play in the regulation 54-hole stroke play Championship and capped her rally with a 25-foot putt for a birdie 4 on the 54th hole. Each scored 239. Next day in a playoff, Miss Orcutt again seemed on the way to a successful defense when she led by three strokes with three holes remaining. A bunker at the 16th proved her undoing and they finished in a tie at 81. On the fourth extra hole of sudden death, Mrs. Choate won with a birdie 2. Mrs. William R. Kirkland, Jr., of New York, was third with 241 for 54 holes. Next, at 247, was the six time former Women's Amateur Champion, Mrs. Glenna Collett Vare, of Philadelphia. In age class competition, Mrs. Choate repeated in Class A (50-54); Miss Orcutt in Class B (55-59) and Mrs. Vare won Class C (60 and over).

1964—Mrs. Hulet P. Smith, of Pebble Beach, Calif., a former national

*Reprinted from USGA Record Book.

badminton champion who took up golf in 1944, won the third Senior Women's Amateur Championship with 81-79-87—247 at the Del Paso Country Club, Sacramento, Calif. Her winning margin was one stroke over Mrs. William R. Kirkland, Jr., of New York. The defending Champion, Mrs. Allison Choate, of Rye, N.Y., shared third place with Mrs. Maurice Glick, of Pikesville, Md., at 249. Mrs. Smith led by four strokes after 36 holes, lost the lead to Mrs. Kirkland midway of the final round, then regained the advantage with three holes to play. Mrs. Kirkland's 15-foot putt for a tie barely slipped past the hole at the final green. In age class competition, Mrs. Smith won Class A (50-54); Mrs. Aubrey E. Babson, of Novato, Calif., won Class B (55-59) with 254; and Mrs. C.D. Lee, of El Paso, Texas, won Class C (60 and over) with 251.

1965—Mrs. Smith became the only 1964 USGA Champion to defend a title successfully when she won her second straight Senior Women's Amateur Championship. Her score was 242 at the Exmoor Country Club, Highland Park, Ill., an improvement of five strokes over her winning total in 1964. In second place was Mrs. John S. Haskell, Titusville, Pa., with 245. The Champion and the runnerup, paired together on the final round, were all even with three holes to play. The critical hole was the 16th, a par-4, where Mrs. Smith made a birdie as Mrs. Haskell went one over par. In age class competition, Mrs. Smith won Class A (50-54); Mrs. Frank D. Mayer, Glencoe, Ill., won Class B (55-59) with 249; and Mrs. E.L. Cooley, Winnetka, Ill., won Class C (60 and over) with 257.

1966—Miss Maureen Orcutt, of Englewood, N.J., became the second repeat winner at the Lakewood Country Club, New Orleans, when she scored 242. Miss Orcutt won the first Senior

Women's Amateur in 1962. Mrs. Frank Goldthwaite of River Crest, Texas, was second with 248 and Mrs. Glenna Collett Vare, of Philadelphia, was third with 250. Mrs. Hulet P. Smith, of Pebble Beach, Calif., did not defend her title. She had won the two previous years. The three leaders represented different age classes. Miss Orcutt represented the 55-59-year-old division, Mrs. Goldthwaite the 50-54-year-old division, and Mrs. Vare the 60-and-over division.

1967—Mrs. Marge Mason, of Englewood, N.J., set a new scoring record of 236 at the Atlantic City Country Club, Northfield, N.J., and won by four strokes over Mrs. Hulet P. Smith, of Pebble Beach, Calif. The old record was 239, set by Mrs. Allison Choate and Miss Maureen Orcutt, of Englewood, N.J., in 1963. Mrs. Mason opened with 77 and was two strokes ahead of Mrs. Smith. She added 80 in the second round to take a three-stroke lead, and had a closing 79. Miss Orcutt, the defending Champion, scored 258 and never was a factor

1968—Mrs. Philip Cudone, of Myrtle Beach, S.C., led every round at the Monterey Peninsula Country Club, Pebble Beach, Calif., and won by a record 10 strokes over Mrs. Hulet P. Smith, of Pebble Beach. Mrs. Cudone scored 236 for the 54 holes and equalled the Championship record set in 1967 by Mrs. Marge Mason, of Ridgewood, N.J. Mrs. Smith, the Champion in 1964 and 1965, scored 246, one stroke better than Mrs. Mason, who had 247. Mrs. Cudone was pursued closely by Mrs. Smith throughout the first two rounds. Mrs. Cudone started with 80 and was two strokes ahead of Mrs. Smith. Both ladies scored 79 in the second round. Mrs. Cudone had 77 in the last round; Mrs. Smith's final round was 85. In age class competition, Mrs. Cudone won

Class A (50-54); Mrs. Smith won Class B (55-59); and Miss Maureen Orcutt, Englewood, N.J., won Class C (60 and over) with 255. For the first time the entry limit of 120 was surpassed and some entries had to be rejected.

1969—Mrs. Cudone won her second consecutive Championship by defeating Mrs. Lowell D. Brown, of Tyler, Texas, in a playoff at the Ridglea Country Club, Fort Worth, Texas. Both ladies equalled the scoring record of 236 for the regulation 54 holes, and then Mrs. Cudone scored 76 in the playoff against 84 by Mrs. Brown. With nine holes to play in the Championship proper, Mrs. Cudone held a five-stroke lead over Mrs. Brown. Mrs. Cudone then lost four strokes to Mrs. Brown in the first three holes of the home nine, and three-putted the 18th to allow Mrs. Brown to pull even. In the playoff Mrs. Cudone birdied three of the first four holes and was never caught.

1970—Mrs. Philip Cudone, of Myrtle Beach, S.C., became the first player since the 1930s to hold a USGA Championship three consecutive years by winning the Senior Women's Amateur Championship at the Coral Ridge Country Club, Fort Lauderdale, Fla., with a record score of 231 for 54 holes. Mrs. Cudone won previously in 1968 and 1969. Miss Virginia Van Wie was the last to win three straight—the 1932-33-34 Women's Amateur Championships. Mrs. Cudone played rounds of 78-75-78 and bettered by five strokes the record 236 she shared with Mrs. Marge Mason. Mrs. Paulette Lee, of Coral Gables, Fla., was second with 239. Mrs. Cudone was four strokes ahead after the first round, and increased the margin to 12 after the second. The entry was 68.

1971—Mrs. Philip Cudone, of Myrtle Beach, S.C. became the first player ever to win a USGA Championship four consecutive years. Mrs. Cudone scored 77-81-78—236 at the Sea Island Golf Club, Sea Island, Ga., to win by one stroke over Mrs. Ann Gregory, of Chicago. Mrs. Gregory's scores were 77-82-78—237. Four other players had won three. They were Willie Anderson (Open, 1903-04-05), Mrs. Glenna Collett Vare (Women's Amateur, 1928-29-30), Miss Virginia Van Wie (Women's Amateur, 1932-33-34), and Miss Hollis Stacy (Girls' Junior, 1969-70-71). Mrs. Cudone led Mrs. Gregory by one stroke going into the final round and birdied the par 3 third. Mrs. Gregory went one over on the hole and Mrs. Cudone's lead was then three strokes. It remained the same until the 17th where Mrs. Cudone bogied and Mrs. Gregory birdied, cutting the lead to one stroke again. Mrs. Cudone got down in two from 35 feet on the home hole to win.

1972—Mrs. Philip Cudone, of Myrtle Beach, S.C., won her fifth consecutive Championship at the Manufacturers' Golf and Country Club, Oreland, Pa., and thus established a record. Only seven other players won as many as three consecutive USGA Championships. They are Willie Anderson (Open); Miss Beatrix Hoyt (Women's Amateur); Mrs. Alexa Stirling Fraser (Women's Amateur); Mrs. Glenna Collett Vare (Women's Amateur); Miss Virginia Van Wie (Women's Amateur); Miss Hollis Stacy (Girls' Junior), and Carl Kauffmann (Amateur Public Links). After a shaky first round of 82 which left her in a tie for seventh place with Mrs. I. Wayne Rutter, of Buffalo, N.Y., Mrs. Cudone scored 76-73 in the last two rounds for a total of 231, six strokes ahead of Mrs. Rutter, the runnerup. Mrs. Rutter kept pace with Mrs. Cudone for the first two days, shooting 82-76, but scored 79 in the last round against Mrs. Cudone's 73. In third place was Mrs. Helen Sigel Wilson, of Philadelphia, runnerup in the Women's

Amateur Championship in 1941 and 1948. Mrs. Cudone's 231 total equalled the Championship record she set in 1970 at Coral Ridge Country Club, Fort Lauderdale, Fla. Her 73 was a competitive course record for the difficult Manufacturers' course, which measured 5,832 yards. The field was composed of 85 players from 24 states. This was the second Senior Women's Championship for the Manufacturers' Golf and Country Club, which was host to the inaugural in 1962.

1973—For the first time in five years the senior women had a new Champion. Mrs. David L. Hibbs, of Long Beach, Calif., concluded Mrs. Philip J. Cudone's custody of the Senior Women's Amateur Championship by scoring 229 for 54 holes at the San Marcos Country Club, Chandler, Ariz., in her second attempt. Mrs. Cudone, of Myrtle Beach, S.C., had won every year since 1968. Mrs. Hibbs' 229 set a record; the previous record of 231 was first set by Mrs. Cudone in 1970. Mrs. I. Wayne Rutter, of Williamsville, N.Y., was runnerup for the second year in succession, with a score of 235, six strokes back of Mrs. Hibbs, Mrs. James Roessler, of San Jose, Calif., was third at 236, while Mrs. Cudone finished fourth, one stroke back with 237. In the first round Mrs. Hibbs scored a four-over-par 8 on the first hole, but recovered and shot 77, tying Mrs. Cudone for second place. At the end of the day, Mrs. Rutter led with 76. Mrs. Hibbs shot even par 74 the second day and took a six-stroke lead with a 36-hole score of 151. Mrs. Rutter was second with 157, Mrs. Cudone third at 158, and Mrs. Roessler fourth with 159. Both Mrs. Rutter and Mrs. Cudone shot 81, while Mrs. Roessler had an 80. Mrs. Hibbs shot 78 in the final round to Mrs. Rutter's 78 and Mrs. Cudone's 79. Mrs. Roessler had a 77 to pull ahead of Mrs. Cudone into third place. There

were 98 entries—the second largest in the 12-year history of the Championship—and 95 starters.

1974—Mrs. Justine B. Cushing, of New York City, won the Championship with a 54-hole score of 231 over the 5,992-yard Lakewood Golf Club course in Point Clear, Ala. Mrs. Philip Cudone was second at 233. Mrs. Cudone had won the Championship five consecutive years and was bidding to join Mrs. Glenna Collett Vare as the only woman to have won six United States Championships, Mrs. Vare won the United States Women's Amateur Championship six times between the years 1922 and 1935. Mrs. Gwen Hibbs, of Long Beach, Calif., the defending Champion, and Mrs. Mark Porter, of Cinnaminson, N.J., tied for third place, seven strokes back. Mrs. Nancy Rutter, runnerup in 1972 and 1973, had a lost ball and scored a 10 on the first hole. She finished with a total score of 245 and placed sixth in the field of 118. Mrs. Cushing and Mrs. Hibbs started with 77 in the first round. Mrs. Cudone shot a 79, while Mrs. Porter, the 1949 United States Women's Amateur Champion, scored a 78. Mrs. Porter took the lead, by one stroke, after the second round with a 76 for 154 while both Mrs. Cushing and Mrs. Cudone shot 78s, giving them 155 for 36 holes. Mrs. Hibbs shot 83 in the second round and never challenged again. Mrs. Porter increased her lead to two strokes after three holes of the final round, but she lost two strokes to par on the fourth as Mrs. Cushing birdied. Mrs. Cushing was then a stroke ahead, and when she finished the first nine in 36, she led Mrs. Cudone by three strokes and Mrs. Porter by five. Mrs. Cushing finished the round with a 76, matched by Mrs. Cudone, while Mrs. Porter shot 84. The 76s by Mrs. Cushing and Mrs. Cudone in the last round and by Mrs. Porter in the second were the low rounds of the

Championship. Mrs. Porter's 76 included an eagle three on the 408-yard, par-5 ninth hole. There were 122 entries and 119 starters from 28 states and Washington, D.C.

1975—Mrs. Albert Bower, of Pelham, N.Y. improved her score in every round and won the 14th United States Senior Women's Championship a day before her 53rd birthday. Mrs. Bower opened with 82, shot 79 in the second round, and finished with 73, for a 54-hole score of 234. Mrs. Bower won by six strokes over Mrs. Philip Cudone, of Myrtle Beach, S.C., with 240. Mrs. Mark Porter, of Cinnaminson, N.J., was third with 243. The Championship was played at the Rhode Island C.C., West Barrington, R.I. The defending champion, Mrs. Justine B. Cushing, from Glen Head, N.Y., finished in a tie for eighth place. Mrs. Cudone, who won this event five times, shared the first day lead with Mrs. Porter at 79, and took sole possession of first place with another 79 the second day. Mrs. Bower was in third place, three strokes behind Mrs. Cudone. Mrs. Porter shot an 81—despite six three-putt greens—and was in second place. In the final round Mrs. Bower picked up nine strokes on Mrs. Cudone, who could manage only an 82. Mrs. Porter added an 83. Mrs. Bower's closing 73, the only sub-par round of the Championship, included four birdies. There were 109 entrants, representing 22 states and the District of Columbia.

1976—Mrs. Cecile Maclaurin, of Savannah, Ga., the only player in the field to break 80 in every round, won the Championship over the Dunes Course of the Monterey Peninsula Country Club, Pebble Beach, Calif. Playing in her first Senior Women's Championship, Mrs. Maclaurin scored 78-75-77—230 for the 54 holes and won by seven strokes over Mrs. Lyle Bowman, from San Rafael, Calif. Mrs. Wayne Rutter, of Williamsville, N.Y., runnerup in 1972 and 1973, finished third at 240. The 1975 Champion, Mrs. Albert Bower, of Pelham, N.Y., was fifth with 246. Mrs. Philip J. Cudone, who won the Championship five times (1968 through 1972) and was runnerup twice, finished in fourth place. Mrs. Bowman led by three strokes after the first round with a par 75, Mrs. Maclaurin was second at 78, and Mrs. Justine Cushing, 1974 Champion, was third at 79. Mrs. Maclaurin took the lead after the second round with 153, while Mrs. Bowman added an 80 to her opening 75 for 155. Mrs. Rutter moved into third place at 157. There was no change in the top three the final day. Mrs. Maclaurin shot a 77, while Mrs. Bowman had 82 and Mrs. Rutter scored 83. One hundred-sixty-one players filed entries—a record; the field, however, is limited to 120, based on handicaps.

Senior Amateur Records*

USGA SENIORS' CHAMPIONSHIP

Year	Winner, Runner-up	Score
1955	J. Wood Platt, Philadelphia	
	d. Geo. Studinger, S.F.	5 and 4
1956	Fred Wright, Watertown, Mass.	
	d. J. C. Espie, Indianapolis	4 and 3
1957	J. C. Espie, Indianapolis	
	d. Fred Wright, Watertown, Mass.	2 and 1
1958	Thomas G. Robbins, Pinehurst, N.C.	
	d. J. Dawson, Palm Springs, Cal.	2 and 1
1959	J. C. Espie, Indianapolis, Ind.	
	d. J. W. Brown, Sea Girt, N.J.	3 and 1
1960	Michael Cestone, Jamesburg, N.J.	1 up (20)
	d. David Rose, Cleveland	
1961	Dexter Daniels, Winter Haven, Fla.	
	d. Col. Wm. K. Lanman, Golf, Ill.	2 and 1
1962	Merrill Carlsmith, Hilo, Hawaii	
	d. Willis Blakely, Portland, Ore.	4 and 2
1963	Merrill Carlsmith, Hilo, Hawaii	
	d. Bill Higgins, San Francisco	3 and 2
1964	Bill Higgins, San Francisco	
	d. Eddie Murphy, Portland, Ore.	2 and 1
1965	Robert Kiersky, Oakmont, Pa.	
	d. Geo. Beechler, Prineville, Ore.	19 holes
1966	Dexter Daniels, Winter Haven, Fla. d. George Beechler, Prineville, Ore.	1 up
1967	Ray Palmer, Lincoln Park, Mich.	
	d. Walter Bronson, Oak Brook, Ill.	3 and 2
1968	Curtis Person, Memphis, Tenn.	
	d. Ben Goodes, Reidsville, N.C.	2 and 1
1969	Curtis Person, Memphis, Tenn.	
	d. David Goldman, Dallas, Tex.	1 up
1970	Gene Andrews, Whittier, Calif.	
	d. Jim Ferrie, Indian Wells, Calif.	1 up
1971	Tom Draper, Troy, Mich.	
	d. Ernie Pieper, S.J. Batista, Cal.	3 and 1
1972	Lew Oehmig, Lookout Mtn., Tenn.	
	d. Ernie Pieper, San Jose, Calif.	1 up (20)
1973	Bill Hyndman, Huntingdon Valley, Pa.	
	d. Harry Welch, Salisbury, N.C.	3 and 2
1974	Dale Morey, High Point, N.C.	
	d. Lew Oehmig, Lookout Mtn., Tenn.	3 and 2
1975	Bill Colm, Pebble Beach, Calif.	
	d. Steve Stimac, Walnut Creek, Calif.	4 and 3
1976	Lew Oehmig, Lookout Mtn., Tenn.	
	d. John Richardson, Laguna Niguel, Calif.	4 and 3

U.S. SENIORS CHAMPIONSHIP

Site: Apawamis Country Club, Rye, N.Y.

Year	Winner, club
1955	John W. Roberts, Columbus, O.
1956	Franklin G. Clement, Fort Sheridan, Ill.

1957 Franklin G. Clement, Fort Sheridan, Ill.

1958 John Dawson, Palm Desert, Calif.

1959 John Dawson, Palm Desert, Calif.

1960 John Dawson, Palm Desert, Calif.

1961 Joseph Morrill, Jr., Great Barrington, Mass.

1962 George Dawson, Glen Ellyn, Ill.

1963 Jack Westland, Seattle, Wash.

1964 J. Wolcott Brown, Sea Girt, N.J.

1965 Fred Brand, Pittsburgh, Pa.

1966 George Haggarty, Detroit, Mich.

1967 Robert Kiersky, Winnetka, Ill.

1968 Curtis Person, Memphis, Tenn.

1969 William Scott, San Francisco, Calif.

1970 David Goldman, Dallas, Tex.

1971 Jim Knowles, Greenwich, Conn.

1972 David Goldman, Dallas, Tex.

1973 Bob Kiersky, Delray Beach, Fla.

1974 James Knowles, So. Londonderry, Vt.

1975 Dale Morey, High Point, N.C.

1976 Dale Morey, High Point, N.C.

WORLD SENIOR AMATEUR

Year	Winner, Runner-up	Score
1960	Harry Strasburger, Coffeyville, Kan.	
	d. John Roberts, Chicago	2 and 1
1961	Howard Creel, Colo. Springs, Colo.	
	d. Adrian McManus, Pasadena	19 holes
1962	Howard Creel, Colo. Springs, Colo.	
	d. Adrian McManus, Pasadena	7 and 5
1963	George Haggarty, Detroit, Mich.	
	d. Fred Siegel, Scottsdale, Ariz.	3 and 2
1964	Dorsey Nevergall, Pompano Bch., Fla.	
	d. Jack Barkel, Australia	5 and 4
1965	Jack Barkel, Australia,	
	d. Adrian French, Los Angeles	1 up
1966	Cecil Dees, Glendale, Calif.	
	d. Walter Dowell, Walnut Ridge, Ark.	4 and 3
1967	Cecil Dees, Glendale, Calif.	
	d. James Quinn, Kansas City	2 and 1
1968	David Goldman, Dallas, Tex.	
	d. Walter Dowell, Walnut Ridge, Ark.	2 up
1969	David Goldman, Dallas, Tex.	
	d. Curtis Person, Memphis	3 and 1
1970	Merrill Carlsmith, Hilo, Hawaii	
	d. Jack Walters, Tacoma, Wash.	2 and 1
1971	Jude Poynter, Beverly Hills, Calif.	
	d. Curtis Person, Memphis, Tenn.	6 and 4
1972	Howard Everitt, Jupiter, Fla.	
	d. Merrill Carlsmith, Hilo, Hawaii	1 up
1973	W.F. Colm, Bakersfield, Calif.	
	d. Truman Connell, Boynton Bch., Fla.	3 and 2
1974	Larry Pendleton, Glendale, Calif.	

d. O. W. Nelson,
Wheat Ridge, Colo.   4 and 3
1975 Truman Connell, Pompano Beach, Fla.
d. Merrill Carlsmith,
Hilo, Hawaii   5 and 3
1976 Robert Willits, Kansas
City, Mo.
d. Richard Stevenson,
Borrego Springs, Calif.   4 and 3

AMERICAN SENIORS' CHAMPIONSHIP

Year   Winner, Runner-up

1955 Edward H. Randall,
Rochester, N.Y.
d. Judd Brumley,
Greeneville, Tenn.
1956 Judd Brumley,
Greeneville, Tenn.
d. Peter Snekser,
Rochester, N.Y.
1957 Leon R. Sikes, W.
Palm Beach, Fla,
d. Frank D. Ross,
Hartford, Ky.
1958 Edward Randall,
Rochester, N.Y.
d. John Roberts, Col-
umbus, O.
1959 Leon R. Sikes, W.
Palm Beach, Fla.
d. Bruce Coffin,
Marblehead, Mass.
1960 Egon Quittner, Rydal,
Pa.
d. J. W. Brown, Sea
Girt, N.J.
1961 Jack Russell,
Clearwater, Fla.
d. George Haggarty,
Detroit, Mich.
1962 John Roberts, Chicago,
Ill.
d. Clyde Haynie,
Largo, Fla.
1963 Bruce Coffin,
Marblehead, Mass.
d. Jack Russell, Clear-
water, Fla.
1964 Robert Kiersky, W.

Palm Beach, Fla.
d. Adrian McManus,
Windermere, Fla.
1965 Dr. John Mercer,
Sarasota, Fla.
d. Jack Russell, Clear-
water, Fla.   1 up
1966 Walter A. Dowell,
Walnut Ridge, Ark.
d. Adrian McManus,
Windermere, Fla.   4 and 3
1967 Joel Shepherd,
Kalamazoo, Mich.
d. J. W. Brown, Sea
Girt, N.J.   2 and 1
1968 Walter Dowell, Walnut
Ridge, Ark.
d. J. W. Brown, Sea
Girt, N.J.   4 and 2
1969 Curtis Person, Mem-
phis, Tenn.
d. J. W. Brown, Sea
Girt, N.J.   1 up
1970 J. Wolcott Brown, Sea
Girt, N.J.
d. Robert Loufek,
Moline, Ill.   2 and 1
1971 Truman Connell,
Boynton Beach, Fla.
d. J.B. Davis, Melvin
Village, N.H.   5 and 4
1972 Howard Everitt,
Tequesta, Fla.
d. Truman Connell,
Boynton Beach, Fla.   4 and 2
1973 Bill Hyndman, Hun-
tingdon Valley, Pa.
d. John D. McCue,
Winter Park, Fla.   2 and 1
1974 Ray Palmer, Largo,
Fla.
d. Dick Giddings,
Fresno, Calif.   1 up
1975 Edward Ervasti,
Canada
d. John Pottle, Lin-
ville, N.C.   2 and 1
1976 Neil Croonquist,
Edina, Minn.
d. Dick Remson, Lo-
cust Valley, N.Y.   3 and 1

NORTH & SOUTH SENIORS

Site: Pinehurst (N.C.) Country Club

Year	Winner, Runner-up	Score
1955	Benjamin K. Kraffert, Titusville, Pa. d. John Roberts, Columbus, O.	1 up
1956	Tom Robbins, Larchmont, N.Y. d. J.W. Platt, Philadelphia	1 up
1957	J.W. Platt, Philadelphia d. J. Ackerman, Princeton, N.J.	4 and 2
1958	J.W. Brown, Sea Girt, N.J. d. Jack Brittain, Hempstead, N.Y.	7 and 6
1959	Walter Pease, Plainfield, N.J. d. Paul Dunkel, Hackensack, N.J.	4 and 3
1960	Tom Robbins, Pinehurst, N.C. d. J.W. Brown, Sea Girt, N.J.	2 and 1
1961	Robert Bell, Worthington, O. d. Dr. J. Mercer, Fitchburg, Mass.	1 up
1962	William K. Lanman, Glenview, Ill. d. Frank Ross, W. Hartford, Conn.	2 and 1
1963	James McAlvin, Lake Forest, Ill. d. Merrill Carlsmith, Hilo, Hawaii	2 and 1
1964	James McAlvin, Lake Forest, Ill. d. J.W. Brown, Sea Girt, N.J.	2 and 1
1965	David Goldman, Dallas, Tex. d. Curtis Person, Memphis, Tenn.	1 up
1966	Curtis Person, Memphis, Tenn. d. David Goldman, Dallas, Tex.	2 and 1
1967	Bob Cochran, St. Louis, Mo. d. Dr. John Mercer, Sarasota, Fla.	1 up (20)
1968	Curtis Person, Memphis, Tenn. d. Mickey Bellande, Biloxi, Miss.	1 up
1969	Curtis Person, Memphis, Tenn. d. David Goldman, Dallas, Tex.	2 and 1
1970	Bob Cochran, St. Louis, Mo. d. John Pottle, Linville, N.C.	1 up
1971	Dave Goldman, Dallas, Tex. d. Byron Jilek, Worthington, O.	4 and 2
1972	Bill Hyndman, Huntingdon Valley, Pa. d. Curtis Person, Memphis, Tenn.	3 and 2
1973	Ray Palmer, Grosse Ile, Mich. d. Tom Draper, Troy, Mich.	1 up
1974	David Goldman, Dallas d. Harry Welch, Salisbury, N.C.	2 and 1
1975	Harry Welch, Salisbury N.C. d. Neil Croonquist, Edina, Minn.	3 and 2
1976	Paul Severin, Richmond, Va. d. Jim Tingley, Glen Cove, N.Y.	3 and 2

USGA NATIONAL WOMEN'S SENIOR*

Year	Site	Winner
1962	Manufacturer's C.C.	Maureen Orcutt
1963	C.C. of Florida	Mrs. Allison Choate
1964	Del Paso C.C.	Mrs. Hulet P. Smith
1965	Exmoor C.C.	Mrs. Hulet P. Smith
1966	Lakewood C.C.	Maureen Orcutt
1967	Atlantic City C.C.	Mrs. Marge Mason
1968	Westchester C.C.	Mrs. Philip Cudone
1969	Ridglea C.C.	Mrs. Philip Cudone
1970	Coral Ridge C.C.	Mrs. Philip Cudone
1971	Sea Island G.C.	Mrs. Philip Cudone
1972	Manufacturer's G.&C.C.	Mrs. Philip Cudone
1973	San Marcos C.C.	Mrs. David Hibbs
1974	Lakewood C.C.	Mrs. Justine Cushing
1975	Rhode Island G.C.	Mrs. Albert Bower
1976	Monterey Penin. C.C.	Mrs. Cecile Maclaurin

*Reprinted from *Golf Digest*, February 1977.

INDEX